I0465783

ISBN: 9781795753340

Engineer 4.0

The Life Science Handbook for Medical Devices

and Biopharmaceuticals

Des O'Brien

Published by EX Enterprises

FORWARD

This book is inspired by the move towards Industry 4.0, the Fourth Industrial Revolution representing the technological advances that are transforming the manufacturing industry, in particular the capabilities within Life sciences. While the benefits of big data, automation and machine learning will be self-evident, I believe there is a place for underscoring the important of Good Engineering Practice (GEP), quality and compliance in both medical device engineering and Pharmaceuticals. It is important that advancement are based on sound science with an understanding of historical precedencies and regulatory requirements of medical devices and treatments.

As powerful new technologies become more mainstream, above all, engineers and scientists must deliver safe and effective products that meet quality and regulatory expectations. This handbook can serve as a physical asset in an age of cloud -based services and information technology. It is based on over 15 years of experience of working in life science industry. It pulls together current practices, regulatory guidance and industry know-how to provide a concise handbook useful for reference or general guidance purposes.

ISBN: 9781795753340

Special thanks to

Priscilla Browne, B. Cooper and the validation and engineering working group at EXP Solutions Ltd.

© Copyright 2019, First Edition

This book is for general guidance purposes only.

The author or contributors shall take no responsibility for the application or interpretation of content contained herein. Legal and regulatory compliance is the responsibility of manufacturers, distributers and suppliers of regulated products.

Recommended Reading

Bioprocessing Piping and Equipment Design: A Companion Guide for the ASME BPE Standard (Wiley-ASME Press Series), William M. (Bill) Huitt

Jurans Quality Handbook, The complete Guide to Performance Excellence, Joseph A, Defeo.

Basic Statistics, Tools for Continuous Improvement, Forth edition, Kiemele, Schmidt, Berdine.

Leading Change, John Kotter.

The Goal, A process of ongoing Improvement, Eliyahu Goldratt.

Engineering Mathematics, K.A. Stroud.

GLOSSARY

Contents

PART 5 VALIDATION

1.1 SI Units and Measurement

Introduction

This chapter covers the international system of units (SI units) and the basic principles of linear measurement. Communicating numbers and values is a basic requirement to allow engineering to occur. SI units provide a system that ensures different companies or sub-contractors across the globe can interrupt drawings and specifications.

SI Units

The Seven SI base units, which are comprised of:

m - Meter - Length
s - Second - Time
mol - Mole - Amount of Substance
A - Ampere - Electric Current
K - Kelvin - Temperature
cd - Candela - Luminous Intensity
kg - Kilogram – Mass

Benefits of SI Units

A benefit of the International System of Units (SI) is that written technical information is effectively communicated, overcoming the variations of language, including spelling and pronunciation. Arabic numerals describe the quantity. A quantity is then paired with a unit symbol, often with a prefix symbol that modifies unit magnitude.

Basic application of units and symbols

- Units: Names of units are made plural only when the numerical value that precedes them is more than one. Examples: 0.25 liter (quantity is less than one) and 250 milliliters (quantity is more than one).
- Symbols also remain the same despite the value. For example, 250 mm = 250 millimeters, is not 250 mms.

- The names of all units should start with a lower case letter except, at the beginning of a sentence. There is one exception: in "degree Celsius" (symbol °C) the unit "degree" is lower case but "Celsius" is capitalized. E.g. 37 degrees Celsius.
- Unit symbols are written in lower case letters except for liter and those units derived from the name of a person (m for meter, but W for watt, Pa for pascal, etc.).
- Symbols of prefixes that mean a million or more are capitalized and those less than a million are lower case (M for mega (millions), m for milli (thousandths).
- The dot or period (full stop) is used as the decimal point within numbers. In numbers less than one, zero should be written before the decimal point. Examples: 7.038 g; 0.038 g.

The prefix "kilo" stands for one thousand of the named unit. It is not a stand-alone term in the metric system. The most common misuse of this is the use of "kilo" for a "kilogram" of something. The word "micron" is obsolete and no longer used with "micrometer" the correct term. Also "degree centigrade" is no longer the correct unit term for temperature in the metric system; it has been replaced by degree Celsius.

The SI unit of time (actually time interval) is the second (s) and should be used in all technical calculations. When time relates to calendar cycles, the minute (min), hour (h), and day (d) might be necessary. For example, the kilometer per hour (km/h) is the usual unit for expressing vehicular speeds.

Linear Measurement

This involves the comparison of the piece under test with a known standard. These days, many manufacturing processes utilise automated measurement or complex vision systems. However, an understanding of the basic principles of measurement along with traditional measurement methods is fundamental to the engineer. Many of these traditional measurement methods are *contact* in nature (opposed to non-contact). While modern vision systems tend to be *non-contact* in nature, which offers multiple benefits.

Working standards are calibrated against master standards meaning that a comparison between the absolute standard and the particular measuring instrument is achieved. The calibration and testing of all measuring tools and instruments usually requires special equipment and should be completed in accordance with a recognised calibration standard.

Non-contact methods of measurement do not require physical contact with the part under measurement. The measurement is achieved by the use of optics, lasers or magnification. This can be an advantage as the part does not get exposed to physical forces during measurement. Non-contact methods also eliminates the "feel factor" and resulting human error of hand gages.

Contact measuring methods include micrometers and plug gages where accuracy and consistency are subject to the amount of pressure applied to the part during measurement.

Care & Maintenance of Instruments

1. Keep all instruments clean, treat them with care and avoid misuse.
2. Place instruments in cases or fit covers when not in use.
3. Keep the inside of the instrument case clean. The case is meant to protect the instrument.
4. Do not attempt to dismantle an instrument. If it is not functioning correctly return it to the appropriate department for servicing.
5. Choose the instrument in keeping with the tolerance of the dimension to be measured.
6. Wherever possible use an instrument that gives a direct reading.
7. Do not use worn or damaged instruments.
8. Remember that the graduations on a measuring instrument (resolution) are not necessarily the accuracy to which it can be used. Quite often the in-built inaccuracies of measuring instruments exceed their resolution.

Manual Measuring Instruments

The ease with which a component can be measured and the accuracy to which it need be or can be measured depends on the correct choice of measuring instrument. Factors such as the shape of the component and the position of the dimension to be measured influence the choice of instrument but nevertheless, certain basic rules should be followed.

Instrument or Measurement	Type of Measurement	Range	Value of Smallest Graduation (Resolution)	Suggested Reliability/ Accuracy
Steel rule	direct	150-500mm.	.5 mm.	±.5 mm.
Depth gauge	direct	150 mm.	.5 mm.	±.5 mm.
Callipers	direct	150 mm.	none	±.5 mm.
Vernier callipers	direct	600 mm.	.01 mm.	±.05 mm.
Vernier depth gauge	direct	300 mm.	.01 mm.	±.05 mm.
Vernier height gauge	direct	600 mm.	.01 mm.	±.05 mm.
Micrometres				
25-50 mm.	direct		.01 mm.	±.01 mm.
150-300 mm. plain	direct		.01 mm.	±.01 mm.
150-300 mm. plain	direct		.01 mm.	±.02 mm.
Inside micrometres	direct		.01 mm.	±.01 mm.
Depth micrometre	direct		.01 mm.	±.01 mm.
Telescopic gauges	transfer	150 mm.	none	±.02 mm.
Slip gauges	end standard	Up to 100 mm.	.001 mm.	±.0005 mm.
Dial test indicator	comparison	5 mm	.01 mm.	±.01 mm.
Dial test indicator	comparison	1.0 mm	.001 mm.	±.001 mm.

The volume of measurements is also a limiting factor. While a properly trained individual may be able to make repeated measurements, automation should be considered for repetitive inspection if possible.

Slip Gauges

Slip gauges or block gauges are used as standards for precision length measurement throughout the engineering industry. The gauges are usually made in sets and consist of a number of hardened steel blocks. Each block has two of the opposite faces lapped flat and parallel to a definite size within an extremely tight tolerance. In building up packs of slip gauges, errors can occur if care is not exercised. There are three main causes of errors:

1. Errors due to deviation from true size.
2. Errors due to grease or dirt between the wringing faces.
3. Errors due to expansion caused by excessive handling or leaving gauges exposed to strong sunlight or electric lamps.

The Sine Bar

The sine bar, which is one of the most effective methods of precision measurement of angles consists of a rectangular section bar to which are attached two hardened rollers of the same diameter, such that the common centre line of the rollers is parallel to the top face of the bar. The principle of the sine bar is based on the fact that in a right angled triangle the function known as the sine of the angles is the relationship of the hypotenuse to the side opposite the angle.

$$\text{Sin } A = \frac{a}{c}$$

To set up a sine bar to a required angle:
(a) Select and clean the appropriate bar.
(b) Determine the sine of the angle required by reference to the sine tables.
(c) Calculate the slip gauges required, i.e. multiply the sine of the angle by the size of the sine bar, i.e. the distance between the centres of the rollers.
(d) Select and clean the necessary slip gauges including protector slips.
(e) Wring the slip gauges together.
(f) Clean the surface plate and set up the bar by inserting the slip gauges under the roller as shown.

Other Applications of the Sine Principle

Sine centres are a very useful adaption of the sine bar principle. Adjustable centres are mounted on the face of the sine bar allowing such items as taper gauges and taper shafts to be inserted between centres and be measured for both angle and concentricity.

Vision Systems

While manual measurement via the previous instrumentation is useful in certain circumstances of low volume manufacturing, the requirements of high volume manufacturing and reliable quality calls for automated vision measuring systems to meet the volumes of inspection required.

Coordinate Measuring Machines

Coordinate measuring machines are available in a wide range of sizes and accuracy and can meet most precision 3D measuring applications for today's needs.

Depending on the system setup and capability, a wide range of contact and non-contact probes can allow numerous kinds of measurements to be performed. CMM software assists in the analysis and interpretation of measurement results, which is particularly useful with increasing quantities of measurement.

Optical Comparators

Often referred to simply as a comparator, they are used to measure parts dimensionally by using optics and projection. Measurement is achieved by overlaying limits or graduations over the image projected. Inspection with comparators is relatively quick and is most useful when looking for a pass/fail result.

However, the rise of vision-based inspection systems makes the manual comparator, even equipped with modern capabilities, an oft overlooked tool. This is particularly the case when the requirement is to inspect large quantities of parts at once, since a vision system allows you to place multiple parts for inspection on the stage at the same time. For many simple measurements on two-dimensional parts with clearly defined edges, the optical comparator is quite suited.

Optical comparators can provide more information than just simple dimensions. Length and width measurements of the part shown above, for example, can be quickly obtained from two separate measurements by using a micrometer. These superficial measurements, however, might not reveal burrs, scratches, indentations or undesirable chamfers. Such imperfections are best detected on a comparator. In addition, a comparator's screen can be simultaneously viewed by more than one person and provide a medium for discussion, just as a white board might facilitate a conference.

Advantages

- Fast length and width measurement
- Length and width measurement can be done simultaneously
- Burrs and chamfers can be detected
- Screen can be viewed by more than one person

Digital Comparators

Digital comparators allow Pass/Fail inspection comparisons, Files are simply uploaded to the equipment in order to compare measurement results against a CAD overlay - the overlay moves with the datum, so you don't even need to use an optical comparator / profile projector.

Geometric Tolerancing

Before an object measured, complete information about both the size and dimensions and tolerances of the object must be available. The shape of an object is communicated through orthographic drawings, which are developed following standard drawing practices. The process of adding size information to a drawing is known as dimensioning.

Geometric Dimensioning and Tolerancing (GD&T) is a way of defining and communicating engineering tolerances. The geometry, tolerances and other information such as surface finish, concentricity or parallelism are expressed using symbols and text. In turn, properly created engineering drawings communicate the degree of accuracy and precision needed on each controlled feature of the part.

Tolerancing specifications define the allowable variation for the form and possibly the size of individual features, and the allowable variation in orientation and location between features. Two examples are linear dimensions and feature control frames using a datum reference (both shown above).

Standards define GD&T rules and drawing conventions include:

- American Society of Mechanical Engineers (ASME) Y14.5-2009
- International Organization for Standardization (ISO) ISO 2768:1989 General tolerances
  -Part 1: Tolerances for linear and angular dimensions without individual tolerance indications
  -Part 2: Geometrical tolerances for features without individual tolerance indications
- International Organization for Standardization (ISO) ISO 286:2010 Geometrical Product Specification

However, it should be noted that ASME Y14.5 standard provides a fairly complete set of standards for geometric dimensioning and tolerancing in one document. Where the ISO standards address a single topic at a time under different standards.

Dimension: a numerical value that defines the size or geometric characteristic of a feature.

Basic dimension: a numerical value defining the theoretically exact size of a feature. ' Reference dimension is the numerical value enclosed in parentheses provided for information only and is not used in the fabrication of the part.

Leader line: a thin solid line used to indicate the feature with which a dimension, note, or symbol is associated.

Datum: a theoretically exact point used as a reference for tabular dimensioning.

Tolerance: the amount a particular dimension is allowed to vary.

Dimension line: a thin solid line which shows the extent and direction of a dimension. Arrows are placed at the ends of dimension lines to show the limits of the dimension.

Extension line: a thin solid line perpendicular to a dimension line indicating which feature is associated with the dimension.

1.2 Basic Statistics

Data is used to inform engineers and make decisions on failure modes, corrective actions or as justifications to make adjustments to process parameters. The type of data (variable or attribute) mostly depends on how the characteristic or feature is measured and recorded. For example, a product dimension or characteristic may be a continuous variable feature (e.g. thickness of 2.200mm), but if a go/no-go gauge is used to disposition the characteristic it is recorded as attribute data.

Variable data is typically measured and recorded on a continuous scale that provides precision greater than the number of digits required by the technical specification.

Examples of variable data include:

> Dimension in mm of an orthopaedic fixing plate
> Tensile force of a spring in Newtons
> Total drug content of a substance

Attribute characteristics tend to be a pass/fail, presence or absence of a characteristic. Examples include:

> Optical lens is scratched (cosmetic defect)
> Part is laser marked or not
> The product defective or not
> The tablet is broken or not

Inspection by Attributes

This is a method of inspection wherein either the unit of product is classified simply as conforming or nonconforming, or the number of nonconformities in the unit of product is counted with respect to a given requirement or set of requirements. Defining a defect is fundamental to sampling by inspection for attributes.

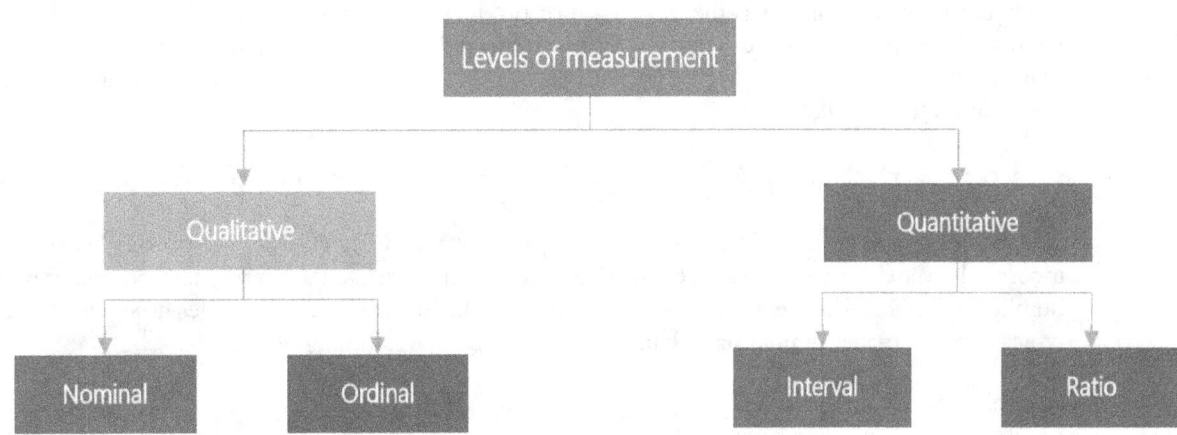

There are two types of qualitative data- nominal and ordinal. Nominal represents categories that cannot be put in any order, while ordinal represents categories that can be ordered. There are also two types of quantitative levels - interval and ratio. They both represent "numbers" however, ratios have a true zero but intervals don't.

Test for Normality

Statistical methods such as t-test, ANOVA, regression analysis and as confidence and reliability are based on sampling from a population with specific parameters including the mean and the standard deviation. They are known as parametric methods. Parametric methods should conform to certain conditions, such as the requirement of normality for them to be suitable for use. Therefore, it is necessary to determine if the data distribution is normal or symmetric.

One such means of testing for normality is the Chi-Square test. This assesses a given data set and determines if the assumption of normality is true or false. Standardized Skewness and kurtosis also provides a test of normality to determine if the distribution is normal to use statistical tolerance limit calculations (confidence and reliability). Values of skewness and kurtosis outside the range of -2 to +2 indicate significant non normal data which would deem statistical tests regarding standard deviation unsuitable.

Hypothesis Testing

For statistical analysis two hypothesis must be considered. The hypotheses are defined as:

Null Hypothesis, Ho
Alternative Hypothesis, Ha

If the null hypothesis is rejected, the alternative hypothesis is concluded to be true. However, if the null hypothesis is not rejected, no conclusion is reached; the null hypothesis is not concluded to be true. The only way that a conclusion is reached is if the null hypothesis is rejected.

Acceptance Sampling

Acceptance sampling involves the inspection of product or raw material before in enters or as it leaves the factory. Its purpose is to ensure that defective products do not reach the patient or consumer. Similar to the type of data, there are two main types of sampling plans (1) attribute and (2) variable. Sampling plans involve either an AQL or RQL.

An acceptable Quality Level (AQL) is the defect rate you are willing to accept for a high percentage of the time.
The probability that a lot will be incorrectly rejected at the AQL. RQL is the probability that a lot will be incorrectly passed. An attribute acceptance sampling plan consists of a sample size and accept number. If the number of defects found exceeds the accept number, the lot is rejected. Variables acceptance sampling plans typically have smaller sample sizes than attribute acceptance sampling plans.

Statistical sampling and inspection is used for new product or service design, verification of those designs, incoming inspection, in-process inspection and final inspection. The level of sampling and inspection depends on the characteristics of the product, capability of the process, and the level of assurance required.

Lot inspection plans are determined based on the quality characteristic defect level and the manufacturing process used when assigning sampling plans and Acceptable Quality Level (AQL) and/or Index Values (I.V.) level(s).

Pp (Process Performance) and Cp (Process Capability): quantify the stability of a process, i.e., the amount of variation in the output

Ppk (Process Performance Index) and Cpk (Process Capability Index) : These represent both the degree of variability AND the degree that the output is centred between the lower and upper specification limits.

Pp and Ppk are calculated using the standard deviation of **the entire population** and represent a long term performance. (Typically measured over a number of batches/runs and reflects both normal and special causes of variation).

Cp and Cpk are calculated using **a sample** standard deviation and represent a potential that could be achieved if sources of variation are eliminated.

Pp and Ppk values are used to describe the Process expected over the long term.

Cp and Cpk are more commonly used during process optimisation studies as these represent the potential capability that could be achieved if the process were made stable by reducing special causes of variation. They may be applied during Operational Qualification. For stable processes (i.e. those where special cause variation is eliminated or minimised) Pp/Ppk and Cp/Cpk values will be similar in value.

Short term capability: Results from collecting data over a short period of time, (typically using only a single operator, a single manufacturing line, a single lot of material, etc.). **Typically expressed as Cp and Cpk.**

Long term capability: Based on reasonably independent observations collected from a process exhibiting normal production variation, for example using multiple operators, lines and lots of material over time. **Typically expressed as Pp and Ppk.**

Confidence Level

- Confidence Level is expressed as a percentage and represents the probability that the conclusion of the test is correct. A 95% confidence level means you can be 95% certain that the conclusion is correct.

- OQ typically should have a confidence level of 90%

- PQ typically should have a confidence level of 95%

AQLS

The AQL of a sampling plan is the Process Performance Level routinely **accepted** by the sampling plan.

- AQL is generally defined as the Process Performance Level that the sampling plan will **accept** 95% of the time - a 95% probability that the process will be accepted, i.e. the validation passes.

- This means validations for processes/equipment producing devices with a Process Performance Level at or better than the AQL are accepted at least 95% of the time and rejected at most 5% of the time.

  It describes the risk (5%) associated with rejecting a good process.

- The **producer** would like to design a sampling plan such that there is a **high probability** of **accepting** a validation

- The RQL of a sampling plan is the Process Performance Level routinely **rejected** by the sampling plan. RQL0.10 is defined as the Process Performance Level that the sampling plan will **reject** 90% of the time -**a 90% probability that the process will be rejected ,i.e. the validation fails.**

- This means that validations for processes/equipment producing devices with a **Process Performance Level at or worse than the RQL are rejected at least 90% of the time and accepted at most 10% of the time.**

- RQL0.05 is defined as the Process Performance Level that the sampling plan will **reject** 95% of the time.

- **It describes the risk associated with releasing/accepting a bad process.**

- The **consumer** would like the sampling plan to have a **high probability** of **rejecting** a validation with a Process Performance Level (% nonconforming) greater than, or equal to, the RQL.

Part 1

-GXP & COMPLIANCE-

2.1 Good Manufacturing Practices

History Of GMP

1906: The Pure Food and Drug Act: Establishment of regulatory agency- FDA. The act makes it illegal to sell "adulterated" or "misbranded" food and drugs.

1938: Federal Food, Drug and Cosmetic (FD&C) Act: Sulfanilamide containing poisonous solvent causes 107 deaths.

1941: Insulin Amendment: requires FDA to test and certify purity and potency of insulin.

1962: Kefauver-Harris Drug Amendments Tragedy: Thalidomide tragedy causes severe birth defects in thousands of European babies.

1978: CGMPs Final Rule for drugs and devices (21 CFR Parts 210–211 and 820) establishes minimum Current Good Manufacturing Practices for manufacturing, processing, packing, or holding drug products and medical devices.

1979: GLPs Final Rule (21 CFR Part 58) -Good Laboratory Practices (GLP) for conducting non-clinical laboratory studies that support applications for research or marketing permits for human and animal drugs, medical devices for human use and biological products.

1982: Tamper-resistant packaging regulations issued for over-the-counter medicinal products: Acetaminophen-capsule poisoning by cyanide causes several deaths.

1987: Guideline on General Principles of Process Validation are issued by the FDA.

1996: Proposed Revision to US CGMPs for Drugs and Biologics (21 CFR Parts 21–211) adds detail for validation, blend uniformity, prevention of cross-contamination and out-of-specification results.

1997: Electronic Records Final Rule (21 CFR Part 11) requires controls that ensure security and integrity of all electronic data. 1998 Draft Guidance on Manufacturing, Processing, or Holding Active Pharmaceutical Ingredients and Investigating Out-of-Specifications and Test Results.

What is GMP?

Good Manufacturing Practices are a set of practices that are required in order to comply with industry standards and regulations. GMP helps to minimise the risks involved during manufacturing and helps to ensure products meet quality and regulatory standards. A GMP quality system ensures that products are consistently produced and controlled according to predefined quality standards. It is designed to minimise the risks involved in any pharmaceutical production that cannot be eliminated through testing the final product.

Often, a broader term is used in industry -GxP-where the "x" is used as an umbrella letter representing different subjects or disciplines in industry. Some prime examples include GLP (Good Laboratory Practice),

GDP (Good Documentation Practice), GEP (Good Engineering Practice) and GMP (Good Manufacturing Practices). Furthermore, the use of a lower case "c" as a prefix indicates "current" or "up-to-date". cGMP stands for "Current Good Manufacturing Practices. This means that some conventions or practices are subject to change within the industry. Therefore, it is important to be up-to-date in the application of cGxP or Cgmp. There are multiple regulators and organisations that provide definitions of "Good Manufacturing Practices". They include Organisations such as the World Health Organisation (WHO) and the International Society of Pharmaceutical Engineering (ISPE). Other definitions are offered by bodies such as the American competent authority for Food and Drug Administration. It is good to have an awareness of how organisations, bodies and competent authorities define GMP, and one should always review the "local" regulatory landscape. Below some definitions are provided to provide a feel for GMP and highlight the common thread between definitions.

W.H.O. World Health Organisation-"Good Manufacturing Practices (GMP, also referred to as 'cGMP' or 'current Good Manufacturing Practice') is the aspect of quality assurance that ensures that medicinal products are consistently produced and controlled to the quality standards appropriate to their intended use and as required by the product specification."

Food and Drug Administration: cGMP refers to the Current Good Manufacturing Practice regulations enforced by the US Food and Drug Administration (FDA). cGMPs ensure systems are properly designed and monitored, safeguarding the control of manufacturing processes and facilities. Adherence to the cGMP regulations ensures the identity, strength, quality, and purity of drug products by requiring that manufacturers of medications adequately control manufacturing operations. This includes establishing strong Quality Management Systems, obtaining appropriate quality raw materials, establishing robust operating procedures, detecting and investigating product quality deviations and maintaining reliable testing laboratories. This formal system of controls at a pharmaceutical company, if adequately put into practice, helps to prevent instances of contamination, mix-ups, deviations, failures and errors. This assures that drug products meet their quality standards.

MHRA (Medicines and Healthcare Products Regulatory Agency) defines GMP as follows:

"Good Manufacturing Practice (GMP) is that part of quality assurance which ensures that medicinal products are consistently produced and controlled to the quality standards appropriate to their intended use and as required by the marketing authorisation (MA) or product specification. GMP is concerned with both production and quality control. Many of the drivers of GMP in effect are also benefits to the manufacturer. Good manufacturing practices are an expected practice in regulated industries and a manufacturer must meet all relevant GMP regulations if they wish to manufacture within a country or sell to a particular market. It is important to maintain accurate, complete, up-to-date and consistent information to ensure patient safety and reduce any potential risks."

GMP Failures

FDA Warning Letters

This is an official message from the United States Food and Drug Administration (FDA) that usually states that it has found that a manufacturer or other organisation has violated some rule of the Quality System regulations.

FDA 483s

An FDA 483 letter typically includes a summary of findings and observations in relation to an audit or inspection where the FDA representatives have reason to believe GMP or other regulations have been violated or are not being met. In response to an FDA 483 letter, the company should address each item and provide a timeline for correction or request clarification of what changes are required.

Below are the top three items of concern:

- Procedures not written or are not fully followed
- Poor investigations of discrepancies or failures (CAPA)
- Absence of written procedures

Consent Decree

A consent decree is a binding order issued by a judge that stipulates the voluntary agreement by the participants in a case of litigation. Decrees are sometimes issued after one party voluntarily agrees to cease a particular action without admitting to any illegality of the action to date.

Product Recall

Product recalls can be initiated in circumstances where a manufacturer becomes aware of a manufacturing defect, packaging or mislabelling defect or otherwise. Product recalls can have a serious impact on the financial stability and future of a company due to bad publicity and loss of sales.

Plant Injunction

An injunction is a judicial process initiated to stop or prevent violation of the law, such as to halt the flow of violative products in interstate commerce and to correct the conditions that caused the violation to occur. (FDA 21 U.S.C. 332; Rule 65, Rules of Civil Procedure).
If a firm has a history of violations and has promised correction in the past but has not made the corrections, the injunction is more likely to succeed. However, the freshness of the evidence is critical.

For an injunction action to be credible in the eyes of the Department of Justice (DOJ), the U.S. Attorney and the court, the evidence must be current. Timeliness is an important factor when considering an injunction action, with or without a Motion for Preliminary Injunction or a temporary restraining order (TRO). However, case quality and credibility must not be sacrificed to meet guideline time frames. The purpose of the guideline time frames is to limit, as much as can reasonably be expected, the need to update evidence. Updating entails extra work at all levels of the case development and review process and more importantly, delays obtaining an injunction which is intended to stop violations that adversely affect the safety or quality of products in commerce. When an injunction is granted, FDA has a continuing duty to monitor the injunction and to advise the court if the defendants fail to obey the terms of the decree. (FDA 6-2-Injuctions)

Debarment
The FDA has the authority to "disqualify," or remove, researchers from conducting clinical testing of new drugs and devices when the agency determines that the researcher has repeatedly or deliberately not followed the rules intended to protect study subjects and ensure data integrity. Further, the FDA can disqualify a clinical investigator who has repeatedly or deliberately submitted false information to the agency or study sponsor in a required report.
Under its statutory debarment authority, the agency may also ban, or "debar" from the drug industry individuals and companies convicted of certain felonies or misdemeanours related to drug products. Once individuals have been subjected to "debarment," they may no longer work for anyone with an approved or pending drug product application at FDA. Debarred companies may no longer submit abbreviated drug applications.

Documentation Creation

The principles of GDP should be applied at the document creation stage. As most people are familiar with softcopy or electronic documents, some of these points are obvious but nonetheless need to be made. All documents should be electronically written and not handwritten except for execution of protocols, test results and adding entries. Documents that are approved controlled should be:

Accurate and free from errors
Have revision or version controlled
Should have an effective date or date of release

Approval of Documents

Document approval must be completed by trained and appropriately experienced personnel. Often companies will use an approval matrix which explains which people are required to approve each document. For example, an EHS (Environment Health and Safety) officer would be required to approve a risk assessment.

Signatures

A signature on any document is legally binding so remember to read and understand what is being signed for. Every signature should also include the date in the correct format. If a signature appears within the same document alongside initials, substituting a full signature with initials and date is generally acceptable. This practice is common when large documents are being completed.

Date and Time Format

A standardised approach to dates and times is important especially within large global organisations. For instance, in the USA, the norm is to place the month before the date, whereas in Ireland and Great Britain it is common to write the day of the month followed by the month. Most companies would define their date and time format in an SOP or procedure.

The date and time format can also be configured in Word documents and Excel worksheets to align with a companies preferred date and time format.

Handwritten Entries

When a handwritten entry is required such as a signature or a test result, indelible ink must be used. Many companies will have an SOP or procedure that states the specific ink colour required. If an entry of a test result or test data isn't completed at the time of execution, this constitutes a late entry. Backdating an entry or signature is forbidden. Always use the correct and current date.

How Are Mistakes Corrected?

This is a critical area of GDP. Failure to follow the requirements of GDP when correcting mistakes is the most common failure when it comes to documentation in industry. The method of correcting mistakes using GDP allows for a person looking at the document for the first time to clearly see the original entry and the corrected entry. This maintains the integrity of the document. In order to identify the changes and corrections, certain rules must be followed. No overwriting is allowed and white-out or Tipp-Ex is not allowed.

Accuracy

Accuracy of information provided in documents is critical in the life science industry. As the end user is a patient, inaccurate records or documents could cause serious injury or death. Controlled documents are also legal documents and could be called upon if recalls, litigation or investigations arise.

Many documents used in the manufacture of medical devices are designed to record information or test results. These test results are then used to disposition (pass or fail) batches of product. Inaccurate information could risk the release and distribution of defective product. This has a potential impact on both the business and the patient or user.

Blank Spaces or Blank Fields

On completion of a document such as a logbook or record, no blanks spaces should be left unfilled. This is to avoid late entries and also to prevent confusion. Blank spaces or blank fields should have a diagonal line drawn neatly across the space, the letters "N/A" written and the entry signed and dated. If the reason for "N/A" is not evident then it is wise to include an explanatory note or sentence.

Data Transcription

Transcribing is the process of transferring data from one source to another. This is often required when raw data is involved. When data is in raw format it may need to be entered into a Microsoft Excel sheet. When transcribing data remember that all original raw data must be stored in case it is needed at a future date. After the data is transcribed it must be verified by a second person to check for any errors or omissions.

Revision Control

Controlled documents should always have a version number or revision number electronically on each page of the document. This is similar to books which always list what edition they are e.g. first edition or second edition. Revision control is a key element of the Quality Management Systems in place in regulated industries. As the need for changes in the document arises, the controlled document can be amended/updated. With each update the version number revises also. Some companies will use alphabetic revision control and to a lesser extent numeric revision control (Version A, Version B or Version 01, Version 02).

Management of Attachments

Attachments to controlled documents can include training records, data sheets, lab results and so on. It is important that attachments are identified for traceability purposes. If the attached becomes detached from the main document, the attachment should be identifiable. It is best practice to include a reference number on the attachment if available. If the attachment consists of several pages, each page should be numbered in Page X of Y format if not electronically done so. And remember, hand written entries must be accompanied by a signature and date. Always use staples to attach documents together. Glue or paper clips are not acceptable.

Management of Documents through Their Lifecycle

GDP applies to all the different stages of a document's lifecycle. These stages include creation, review, approval, issuing, completion of records, revision, updating, retirement and storage. Storage a.k.a. retention is an important stage and often a legal requirement for medical devices and pharmaceutical products. For consumer OTC medicines a 5-year retention of quality records often suffices. For implants such as TKRs or Total Knee Replacements, a 90-year retention period is required. This ensures that traceability and a quality record is available if the need arises.

Test Results

This section provides an overview on the correct handling of test results. Test results can be generated from various types of product testing such as visual inspection, dimensional inspection and chemical analysis. The recording of all test results should be completed on an approved form. This is to ensure that the correct information is being recorded and the same approach is taken by different people who might have to complete testing.

Calculations

There are different ways calculations can be completed. Many simple calculations can be done by an individual using a calculator, alternatively, a software package such as Minitab or an Excel sheet can be used to complete complex calculations. It should be clear to the reader what calculation is required, what the formula is and how the calculation is completed.

If the formula used is not included on the sheet, it should be referenced in a controlled document. Care is also required when recording numbers of several decimal places in length, as rounding error can be introduced.

Units of Measurement

The most important thing to remember is consistency in units of measurement when recording data or making calculations. Consult your company procedure if available to determine the correct units of measurement. Many U.S. companies use imperial units e.g. inches, pounds etc. In Europe the International System of Units or SI is used, e.g. millimetres and kilograms.

What Is Quality?

Quality can be defined as the ability to consistently produce products meeting the same specifications time after time. Products must be safe, pure, uniform and effective. Specifications can be set down internally within a company, however, depending on the product, external specifications from regulators or standards may be required.

Patient safety is the primary focus of any pharmaceutical drug or medical device. This is the expectation of any patient or user. Secondly, the patient or user is interested in receiving an effective product. It is product specifications that ensure these criteria are accounted for.

What Is A Quality Management System?

A Quality Management System, often abbreviated to (QMS) is any system based on a collection of business processes that are primarily focused on providing safe and quality products that consistently meet customer requirements. The core themes of a QMS are outlined below.

Customer and Regulatory Focus

An understanding of the customer needs and requirements should be evident within the organisation and with the future vision of the company. The company should have an understanding of the regulatory landscape as this is subject to change over time. In turn the company should be positioned to respond to that change.

Leadership

To truly lead, one must be accepted in the hearts and minds of those they lead. Authentic leadership pays off. A leader should foster a sense of togetherness and common vision. A leader is anyone who influences or directs people either formally or informally. We are all leaders to some extent.

Involvement

Engagement by everyone across an organisation is now recognised as being key in the successful deployment of any Quality Management System. Everyone should have a voice within the company. As the saying goes "we are only as strong as the weakest link" is very apt.

The Process Approach

ISO 9001 and ISO 13485 are standards that are based on process approaches. A process approach essentially utilises methods or standardised tools to help drive consistency.

Systems Management

This essentially means that systems are defined and described in writing along with the appropriate responses to expected issues that arise. Effective systems management must ensure that the various systems work in support of each other and communicate effectively with one another.

Decision Making

In order to make the right decision, the person empowered to make the decision must be informed. To be correctly informed one must have the correct details and facts available. In a manufacturing environment the facts are essentially data and the analysis of data. During manufacturing or processing, data is generated as a result of monitoring and measurement of products and the related processes.

Supplier Management

Don't ruffle your suppliers' feathers. Security of supply is key in delivering products to customers or patients again and again, Raw materials or sub-components sourced from external suppliers must always be sourced at the right price and time with the emphasis on getting the best quality possible.

Continuous Improvement

For IS0 13485 continuous improvement refers to improving the effectiveness of the Quality Management System. It is harder to drive improvement of the product due to regulatory and practical requirements. This is a key difference in contrast to ISO 19001:2008 as there is a requirement to continually improve both product and processes.

Quality Management Systems

The key elements of a QMS are listed below. The ISO Standard, ISO 9001, is a global Quality Management standard used by thousands of organisations and companies. This standard sets out the requirements of a QMS.

Quality Policy: A company will document their commitment and approach to quality within their organisation. It usually sets out how they plan to achieve a high and consistent standard of quality. It should in some way speak to the customer or end user.

Quality Objectives: Quality objectives can be documented in a Quality Plan at site or organisational level. An effective way of defining quality objectives is use of the SMART method. SMART stands for Specific, Measurable, Achievable, Realistic and Timely.

Quality Manual: An in-house guidance document to provide a framework for achieving the quality objectives.

Organisational Structure and Responsibilities: Organisational charts can be used to map out the company structure. Roles and responsibilities can be documented in site quality plans, job descriptions and Standard Operating Procedures.

Data Management: A coherent approach to the provision, storage and maintenance of data.

Processes: Processes are defined and documented.

Resources: Resources must be properly understood, allocated and linked across the organisation.

Product Quality & Customer Satisfaction: The proper management and investigation of complaints is important to reduce future instances from reoccurring. Continual engagement with the end user or customer is critical.

Continuous improvement including corrective and preventive action- where continuous improvement projects and initiatives are encouraged and supported. The application of a CAPA system to ensure quality is maintained and consistent.

Maintenance: A Preventative Maintenance schedule is in place and managed accordingly.

Sustainability: All work practices are sustainable and consistent throughout the lifecycle of processes and products.

Auditing: Systems are auditable and maintained to allow internal or external review and audit.

Engineering Change Control: Where changes are required to validated processes or equipment, changes are managed and introduced under change control.

A common acronym used to highlight the aims of Good Manufacturing Practices (GMP) is SPUE which stands for Safe-Pure-Uniform-Effective. This definition is particularly suited to pharmaceutical products as the chemicals and drugs used need to be pure and free of contaminants. Furthermore, they need to be uniform, meaning they will have the same constituents from tablet to tablet and batch to batch. A description of each word is shown below:

SAFE- the product has the right ingredients if it is a drug product. It is packaged as intended and correctly labelled in order to provide identification and safe use.

PURE- it is free of contaminants, foreign matter, chemicals and harmful microbes.

UNIFORM- The product is manufactured consistently and will have the same quality between batches manufactured on different days.

EFFECTIVE- Ultimately, the product must be effective in treating the medical condition. To be effective, it requires the correct ingredients, the correct amount of ingredients and correct packaging to maintain the product stability over time.

Ten Rules for GMP

Rule #1: Get the facility design right from the start

Rule #2: Validate processes

Rule #3 : Write good procedures and follow them

Rule #4 : Identify who does what

Rule #5: Keep good records

Rule #6 : Train and develop staff

Rule #7 : Practice good hygiene

Rule #8 : Maintain facilities and equipment

Rule #9: Build quality into the whole product lifecycle

Rule #10: Perform regular audits

Rule #1: Get the facility design right from the start

Every food, drug, and medical device manufacturer aims to operate their business in accordance with the principles of Good Manufacturing Practice (GMP). It's much easier to be GMP compliant if the design and construction of the facilities and equipment are right from the start. It's important to embody GMP principles and use GMP to drive every decision.

Facility layout

Lay out the production area to suit the sequence of operations. The aim is to reduce the chances of cross contamination and to avoid mix-ups and errors. For example, don't have final product passing through or near areas that contain intermediate products or raw materials. Aim to:

- remove unnecessary traffic in the production area
- segregate materials and products
- minimise potential for mix-ups and errors

Facilities Design

The following points should be considered at the facility design stage. The impact of choices and decisions here can must be understood. The scope and type of manufacturing will determine many of the building and facility requirements.
Risk assessments should be considered as a tool in identifying the right materials and design features.

Materials of construction
Windows
HVAC requirements
Utilities to be supplied
Emergency generators/UPS systems
Access to site and area's
Entrances

FDA Requirements

211.42 Design and Construction Features
(a) "Any building or buildings used in the manufacture, processing, packing, or holding of a drug product shall be of suitable size, construction, and location to facilitate cleaning, maintenance, and proper operation."

211.46 Ventilation, Air Filtration, Air Heating and Cooling

(b) "Equipment for adequate control over air pressure, micro-organisms, dust, humidity, and temperature shall be provided when appropriate for the manufacture, processing, packing, or holding of a drug product."

Equipment

Design, locate, and maintain equipment to suit its intended use. The equipment should be:

- easy to repair and maintain
- designed and installed in an area where it can be easily cleaned
- consistent with the intended use
- calibrated at defined intervals (if required)

Environment

It's important to control the air, water, lighting, ventilation, temperature, and humidity within a plant so that it does not impact product quality. You should design facilities to reduce the risk of contamination from the environment.

Make sure that:
- lighting, temperature, humidity and ventilation are appropriate
- walls, floors and ceilings are smooth, free from cracks and do not shed particulate matter
- interior surfaces are easy to clean
- pipe work, light fittings, and ventilation points are easy to clean
- drains are sized adequately and have trapped gullies.

Rule #2: Validate processes

Validation is defined as "Establishing documented evidence that provides a high degree of assurance that a specific process will consistently produce a product meeting its pre-determined specifications and quality attributes." It's a GMP requirement to prove control of the critical aspects of certain operations. New facilities and equipment, as well as significant changes to existing systems, require validation.

All validation activities should be well planned and clearly defined. This is usually by means of a Validation Master Plan, or VMP. Before you get to this stage consider all the critical parameters that may be affected and impact product quality; what happens if the stirring speed is changed? How does this affect temperature or pH? Once this is complete, define the testing and documentation required.

Validation usually is made up of three components:

- Installation Qualification, or IQ, which is testing to verify that the equipment is installed correctly as per manufacturers recommendations.

- Operational Qualification, or OQ, which is testing to verify that the equipment operates correctly as described in the user requirement specification.

- Performance Qualification, or PQ, which is testing to verify and confirm that product(s) can be consistently be produced to specification under anticipated conditions.

The FDA defines 4 types of validation. (1) Prospective, (2) Concurrent, (3) Retrospective and (4) Revalidation. These various types of validation form what approach the validation takes. E.g. is it a new process that will be validated in advance of commercial manufacturing, or, will the process be validated in a staged basis – concurrently. Etc. The 4 types of process validation are explained in the definitions section.

Rule #3 : Write good procedures and follow them

Within a regulated company, many documents are used to instruct, track, test and record information on the manufacturing process. Any document that can impact the quality of the product or product safety is treated as a controlled document. A controlled document is classified as a legal document. These controlled documents must incorporate certain requirements such as the date of approval, revision control and appropriate levels of review and approval. The accuracy and content of these documents can be subject to review by regulatory bodies Including the FDA in the US and the MHRA in the UK It is important that there are no errors or "questions marks" over the content. Examples of controlled documents include:

- Policies
- SOPs
- Specifications
- MFR (Master Formula Record)
- BMR
- Validation protocols and reports
- Forms
- Logbooks
- Records
- Bills of Materials (BOMs)
- Test Methods

Written procedures are controlled documents that provide detailed step-by-step instructions for the user. Written procedures promote consistency as they allow the same task to be performed in the same way, even by different people. They also act as a reference. If changes or improvements are identified, having a procedure in place creates a clear starting point which can be improved or modified in a controlled manner.

-Procedures should be written using clear and concise language
-Steps should be numbered clearly and individually to make them easy to follow
-Remember, written procedures are only effective if they are followed correctly, consistently and at all times by everyone

Never deviate from written procedures, these controlled documents ensure consistency and accuracy is maintained over time.

Documenting Work

As the saying goes "Good Science starts with good documentation", meaning that data should be recorded as soon as available and accurately, following principles of GDP (Good Documentation Practices). The rule of thumb "If it's not documented, it didn't happen" is an important statement. To provide evidence or proof that a process is producing quality product, it must be verifiable in a document. Documentation requires that you record, sign and date every step of the operation you perform. It is a regulatory requirement for companies manufacturing medical products to keep accurate documents and records relating to products and their manufacture. To ensure documents meet a high standard of compliance both prompt recording and accurate recording is paramount. Prompt recording- recording of test results and data should take place as soon as possible and as soon as available. This is to minimise risk or errors with late entries which could lead to quality issues or audit findings.

Rule #4: Identify who does what

All employees understand what tasks and activities need to be done each day. Documenting these responsibilities is key to ensuring people fulfil their roles and responsibilities. Job descriptions for each role and employee should be created and should detail the following:

- job title
- job objective
- duties and responsibilities
- skill requirements

An organisational chart helps to document and display roles and functions.

Rule #5: Keep good records (Documentation)

Good science starts with good documentation. As part of GMP it is essential to keep accurate records, and during an audit, it helps convey that you are following procedures. It also demonstrates that processes are known and under control. Guidelines on Good Documentation and record keeping:

- Record all information immediately upon completion of a task

- Write legibly with indelible ink. By signing records you are certifying that the record is correct and that you have performed the task as per the defined procedure

- Correct mistakes as per GDP. Draw a single line through any error, and initial and date the correction. Include a reason for the correction at the bottom of the page if the reason is not obvious

- Record details if you deviate from a procedure. Ask your supervisor or the Quality Department for advice should a deviation occur

- If it's not documented then it didn't happen!

Rule #6: Train staff

Training should be provided for all employees who work within manufacturing, production or laboratories or where they may have an impact the quality of the product. Basic training on GMP is a fundamental requirement, followed by any job specific training as required. All training should be documented with a record of when the training occurred, dates, attendees and results of any assessments. Training records are often drawn upon by external auditors and are an essential record.

Personnel are central to the application of CGMP and compliance to regulations. At every level throughout an organisation, people interact with materials, equipment and processes in order to deliver products to the market and patient. Personnel must therefore be suitably qualified and equipped to carry out their responsibilities effectively.

Table: GMP Personnel Requirements Overview

Overview of Personnel GMP Requirements					
Item	**PICS/s**	**EudraLex**	**FDA**	**WHO**	**ICH**
Reference	GMP Guide Intro/ Part 1	V4 Part 1, Chapter 2	21 CFR Part 210 Subpart B-- Organisation and Personnel	Annex 2	Q7 ICH
Key Headings	Principle General Key Personnel Training Personnel Hygiene Consultants	Key Personnel Personnel Hygiene Consultants	Responsibilities of quality control unit. Personnel qualifications. Personnel responsibilities. Consultants	Personnel Personnel Hygiene	Personnel Qualifications Personnel Hygiene Consultants
Key Words/ Themes	Training Authorised person Head of Quality Control Detailed hygiene programs Consultant records (Name, address, qualifications, type of service)	Key Personnel Qualified Person Training Duties of Quality Control Detailed hygiene programmes Protective Garments	Training in GMP. Adequate number of personnel. Good health habits. Consultant records (Name, address, qualifications, type of service)	Key Personnel with heads of departments Authorised Persons Direct contact avoided Personnel Hygiene Procedures	Appropriate Education Training Responsibilities defined Good health habits. Suitable Clothing Smoking eating etc. restricted to designated areas.

Provisions in guidance and regulations are therefore made for personnel in a Quality Management System. Despite advances in automation and computerised systems, people are centrally involved in day-to-day decisions. For this reason, there must be sufficient and suitably qualified personnel to carry out all the tasks. Individual responsibilities should be clearly defined and understood by the persons concerned and recorded formally in procedures and job descriptions. It may be an obvious point; however, manufacturers must ensure an adequate number of personnel with the necessary qualifications and practical experience are resourced to manufacturing. Having a broad base of people with the experience, knowledge and skills reduces the risk of quality issues. Responsibilities placed on any one individual should not be so extensive as to present any risk to quality.

Personnel should have specific duties recorded in written descriptions and adequate authority to carry out their responsibilities. Their duties may be delegated to designated deputies with a satisfactory level of qualifications.

Personal Hygiene

All personnel should be trained in the practices of personal hygiene. A high level of personal hygiene should be observed by all those concerned with manufacturing processes. Personnel should be instructed to wash their hands before entering production areas. Signage should be in place along with hand washing facilities. Hand washing demonstrations and training should be provided by a suitably qualified QC analyst or microbiologist. Any person experiencing an illness or exhibiting open lesions or wounds that may adversely affect the quality of products should not be allowed to handle starting materials, packaging materials, in-process materials or medicines until the condition is no longer a risk to quality or patient safety. Direct contact should be avoided between the operator's hands and starting materials, primary packaging materials and intermediate or bulk product.

Rule #7: Practice good hygiene

In order to reduce the risk of product contamination, good hygiene by every person is required. A culture of hygiene awareness should be evident in a GMP compliant facility. The practice of good hygiene should be supported by procedures and monitoring programs.
Depending on the classification of medical device been manufactured, the level of cleanliness required increases as the risk to patient increases. Therefore, sterile product manufacturing will require more strict controls and levels of hygiene.

Points to note:

- Practice good personal hygiene by washing your hands at regular intervals
- Wear the required PPE and protective garments and follow gowning procedures
- If you are ill, inform your manager
- Minimise any contact with product or product contact surfaces and manufacturing equipment.
- Do not eat or drink in GMP areas or where indicated

Rule #8: Maintain facilities and equipment

Preventative maintenance plays an important role in ensuring facilities and equipment remain fit for purpose. Regular maintenance prevents equipment breakdowns and unplanned interventions which disrupt production and cause backlogs within the process. Proper maintenance also reduces the risk of product contamination and helps to maintain the 'validated state' of the equipment and facility. GMP requires accurate records relating to maintenance activities are kept for audit and quality purposes.

Rule #9: Build quality into the whole product lifecycle

Quality must be paramount in any medical device. From its very design through development and into commercial manufacturing, quality must meet acceptable levels. Quality is the responsibility of everyone, not just the quality department.

Quality by Design

The traditional old school of thought focused on 100% inspection where defects are identified by operators. This approach to product quality leads to undetected defects and risk patient safety. This is not to say 100% inspection is not valuable, however, it is more effective is suitable automated systems conduct the inspection.

Online Control or in process control often uses statistical and process controls measures during the manufacturing stage to monitor and respond to drift in process settings and react if defects are produced or detected. The alternative to traditional inspection is to ensure quality is "built in" to the design of a product – Quality by design. Quality by design starts early on within the design and development stage. Potential defects or failure modes are identified and can be designed out of the product or controls put in place to reduce risks or communicate issues.

Control of Components

All materials and components when accepted onsite must meet predefined acceptance criteria. Sampling and accepting testing may be required. Typically, suppliers provide certificates of conformance to ensure the materials or components meet specifications. Suitable storage conditions need also be accounted for. All materials and components must be approved prior to release for manufacturing. Rejected materials or materials that fail inspection should be identified and stored in a secure area to prevent unauthorized use.

Control of the Manufacturing Process

Establish records and procedures to ensure that employees perform the same job every time. Each product must have:

- A master record that outlines the specifications and manufacturing procedures.
- Individual batch records to document conformance to the master record.
- SOPs and procedures for cleaning and maintaining the equipment and areas.

Packaging and Labelling Controls

Proper Packaging and labelling is necessary to identify how materials are packed, stored and labelled. Distinctive labels and accurate descriptions help prevent mix-ups and errors. Labelling also supports traceability, to different batches or different products.

Rule #10: Perform regular audits

Audits are conducted to assess if GMP rules and regulations are followed. External bodies such as the Food and Drug Administration (FDA) conduct formal audits also known as external audits. A Corrective Action Preventative Action (CAPA) system is required to manage and fix issues identified during an audit.

Simply put, an audit is a review activity that examines if a company or organizations processes are been followed. It also allows the identification and improvement of any concerns or non-compliances. Audits can be internal (conducted by internal staff) or external, (external-regulatory audits or certification bodies)

Audits are a key element of a quality management system. The process of establishing an internal audit process can be aided with reference to ISO 19001. This standard provides guidance and lots of examples on implementing and maintaining audit systems.

Benefits of Auditing

Audits provide a means of assessing a company's quality management system, and how well it is in compliance with the processes and procedures within the company.

Some key benefits of audits:

- Audits help verify compliance and conformity to requirements laid out in regulations and industry legislation. (e.g. ISO, FDA, Eudralex etc.)
- They measure the effectiveness of the QMS and the engagement of Top Management
- Audit help identify opportunities for improvements
- Audits promotes awareness of the Quality Management System

Quality Management

Quality: Degree to which a set of inherent properties of a product, system or process fulfils requirements. Risk: defined as the combination of the probability of occurrence of harm and the severity of that harm.

Management: Systematic process for the assessment, control, communication and review of risks to the quality of the drug (medicinal) product across the product lifecycle. Achieving an effective product design, requires in depth knowledge of the customer requirements, clinical or medical need, regulatory requirements, and the manufacturing technology to be used. This collective knowledge or "knowledge space" ensures a robust and quality product is more likely to be designed that will meet the market requirements. Literature, engineering studies and the qualification and experience of employee's all contribute to the knowledge space. From this knowledge space, the most stable and effective design should be selected with product quality and safety as key factors. Furthermore, the quality of the product is then controlled and maintained within what can be described as the control space. During the design stage of product development, specifications are created to describe the attributes and features of the product. The output of the design stage is to have the required product specification documents available as inputs to equipment selection and process selection. Examples of some specifications include Raw material specifications, intermediate product specifications and finished product specifications. Specifications contain information on various features and product attributes such as dimensions, formulation, purity, cleanliness, surface finish and so on. The critical requirements stated in specification are often referred to as Critical Quality Attributes (CQAs) and Critical Process Parameters (CPPs)

Critical Quality Attributes (CQA): a particular property of a material, product or output of a process that is key to the product performance and safety.

Critical Process Parameters (CPP): a process parameter such as temperature or time that when varied it impact the quality or CQA of a product.

Quality Control

Quality Control is that part of Good Manufacturing Practice which is concerned with sampling, specifications and testing, and with the organisation, documentation and release procedures which ensure that the necessary and relevant tests are actually carried out and that materials are not released for use, nor products released for sale or supply, until their quality has been judged to be satisfactory. The basic requirements of Quality

Control are that:

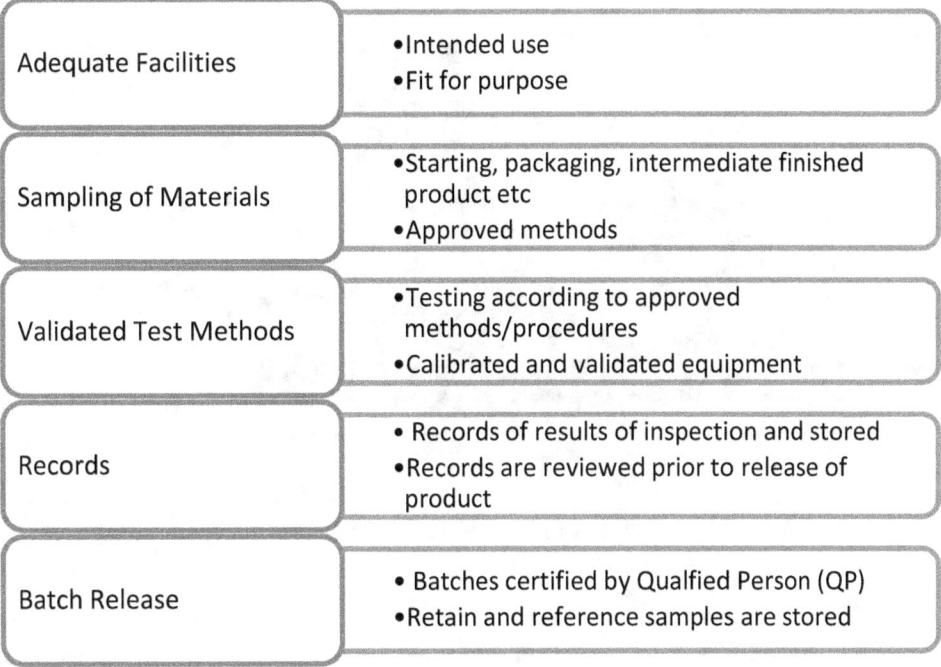

Adequate Facilities	• Intended use • Fit for purpose
Sampling of Materials	• Starting, packaging, intermediate finished product etc • Approved methods
Validated Test Methods	• Testing according to approved methods/procedures • Calibrated and validated equipment
Records	• Records of results of inspection and stored • Records are reviewed prior to release of product
Batch Release	• Batches certified by Qualfied Person (QP) • Retain and reference samples are stored

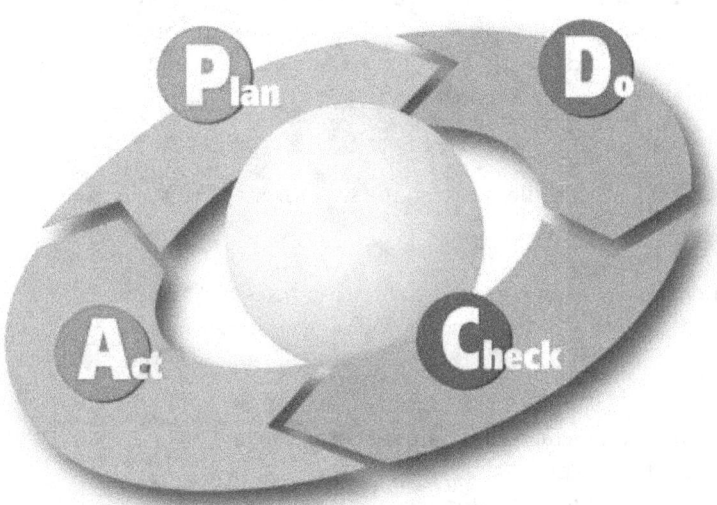

Figure: Plan, Do, Check Act, PDCA- continuous improvement methodology.

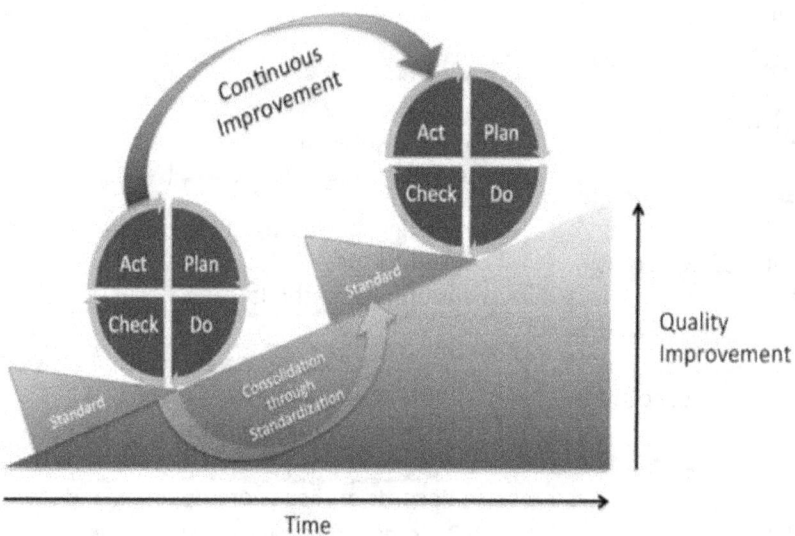

Figure: The PDCA implemented continuously over time

Ongoing Stability

After marketing, the stability of the medicinal product should be monitored according to a continuous appropriate programme that will permit the detection of any stability issue (e.g. changes in levels of impurities or dissolution profile) associated with the formulation in the marketed package.

The purpose of the ongoing stability programme is to monitor the product over its shelf life and to determine that the product remains, and can be expected to remain, within specifications under the labelled storage conditions.

The protocol for an ongoing stability programme should extend to the end of the shelf life period and should include at a minimum, the following:

(i) Number of batch(es) per strength and different batch sizes, if applicable
(ii) Physical, chemical, microbiological and biological test methods

(iii) Acceptance criteria

(v) Description of the container closure system(s) or packaging
(vi) Testing intervals (time points)
(vii) Description of the conditions of storage

A summary of all the data generated, including any interim conclusions on the programme, should be written and maintained. This summary should be subjected to periodic review.

What Is RFT?

Right First Time strives to create a culture of excellence. People are challenged with performing their tasks always in the correct manner to achieve the correct results always - *right the first time*. RFT is the enabler to providing customers worldwide with accessible, high quality and advanced healthcare solutions which comply with cGMP requirements.

RFT in Practice
Achieve excellence *rather than* "that's good enough"
Prevent defects *rather than* "detect defects"
Right first time *rather than* rework

5S

What is 5S?

5S is a Japanese methodology of organising and storing items in a work or lab environment. It has been adopted by many Western companies as a tool to help maintain standards and reduce errors and mix-ups. The "5s" represents each stage of the method.

Sort

Sorting out any items that are not in use and removing to a more appropriate area or to storage or the bin.

Set-in-Order

The idea of "Set-in-Order" is to be always organised. "A place for everything and everything in its place. "If we "set-in-order" we can help to make live processing and testing more efficient and reduce the risk of errors, omissions and accidents.

Shine

Regular cleaning is an important practice and it is always helpful to "Clean as you go."

Standardise

Implement standard practices through SOPs and training. Standardisation can also be applied to work station layout.

Sustain

Make it a habit! After implementing a 5s methodology, it is only effective if continuous efforts are made to "sustain" the changes.

Sort- Set-in-Order- Shine – Standardise – Sustain

Counterfeit Protections

The definition according to the WHO of a medical counterfeit product is *"when there is a false representation in relation to its identity (e.g. any misleading statement with respect to name, composition, strength, or other elements), its history or source (e.g. any misleading statement with respect to manufacturer, country of manufacturing, country of origin or the marketing authorisation holder)."*

This definition applies to the product, the container and other packaging or labelling information. Counterfeit products may include:

products with the correct active ingredient(s), in the correct proportions;
products with the correct active ingredient(s), but
with the wrong dosage;
products without active ingredient(s);
products with impurities or toxic ingredients.

Packaging Technology Designed to Reduce Counterfeiting

DataMatrix: In January 2011, traceability technology based on bar codes (DataMatrix) was put in place in Europe. The ultimate objective and purpose of the DataMatrix is to help ensure traceability of each box in its supply chain to the pharmacist until the end-user, the patient.

Unique Device Identification (UDI) System for Medical Devices: The purpose of the (UDI) is intended to assign a unique identifier to medical devices sold within the United States. It was signed into law on September 27, 2007, as part of the Food and Drug Administration Amendments Act of 2007. The unique identifier enables identification of the device through distribution and use. A secondary benefit of the UDI is it helps to identify counterfeit products by improving the ability to distinguish between authentic and counterfeit devices.

PICS/s Manufacturing Principles for Medicinal Products:

Pharmaceutical Inspection Convention and Pharmaceutical Inspection Co-Operation Scheme (PIC/S): The Pharmaceutical Inspection Convention and Pharmaceutical Inspection Co-Operation Scheme (jointly known as PIC/S) develop international standards between countries and pharmaceutical inspection authorities, to provide a harmonised and constructive co-operation in the field of Good Manufacturing Practices. PIC/S provides an active and constructive cooperation in the field of GMP and related areas. The purpose of PIC/S is to facilitate:

➢ networking between participating authorities
➢ maintenance of mutual confidence
➢ exchange of information and experience
➢ mutual training of GMP inspectors.

The guide consists of an introduction section along with two parts and a number of annexes.

- **Guide to Good Manufacturing Practice for Medicinal Products –** Introduction

  o Introduction
  o Adoption and entry into force
  o Revision history

- **Guide to Good Manufacturing Practice for Medicinal Products -** Part I
  Part I covers GMP principles for the manufacture of medicinal products

  1. Quality management
  2. Personnel
  3. Premises and equipment
  4. Documentation
  5. Production
  6. Quality control
  7. Contract manufacture and analysis
  8. Complaints and product recall
  9. Self-inspection

- **Guide to Good Manufacturing Practice for Medicinal Products** - Part II
  Part II covers GMP for active substances used as starting materials

  1. Introduction
  2. Quality management
  3. Personnel
  4. Buildings and facilities
  5. Process equipment
  6. Documentation and records
  7. Materials management
  8. Production and in-process controls
  9. Packaging and identification labelling of APIs and intermediates
  10. Storage and distribution
  11. Laboratory controls
  12. Validation
  13. Change control
  14. Rejection and re-use of materials
  15. Complaints and recalls
  16. Contract manufacturers (including laboratories)
  17. Agents, brokers, traders, distributors, repackers and relabellers
  18. Specific guidance for APIs manufactured by cell culture / fermentation
  19. APIs for use in clinical trials
  20. Glossary

The annexes provide detail on specific areas of activity and are listed below:

- **Technical interpretation of PIC/S GMP guide Annex 1 -** Manufacture of sterile medicinal products

  PIC/S has published a recommendation for the technical interpretation of Annex 1 on the manufacture of sterile medicinal products.

  This recommendation summarises the interpretations an inspector adopts during an inspection of the manufacture of sterile medicinal products. It reflects the most important changes introduced in the revised Annex 1, but is not intended to address all changes in the revision.

  - Document history
  - Purpose and scope
  - Basics
  - Definitions and abbreviations
  - New texts and their interpretation
  - Revision history

- **Guide to Good Manufacturing Practice for Medicinal Products – Annexes**
  - Annex 1 - Manufacture of sterile medicinal products

- o   Annex 2 - Manufacture of biological medicinal products for human use
- o   Annex 3 - Manufacture of radiopharmaceuticals
- o   Annex 4 - Manufacture of veterinary medicinal products other than immunologicals
- o   Annex 5 - Manufacture of immunological veterinary medical products
- o   Annex 6 - Manufacture of medicinal gases
- o   Annex 7 - Manufacture of herbal medicinal products
- o   Annex 8 - Sampling of starting and packaging materials
- o   Annex 9 - Manufacture of liquids, creams and ointments
- o   Annex 10 - Manufacture of pressurised metered dose aerosol preparations for inhalation
- o   Annex 11 - Computerised systems
- o   Annex 12 - Use of ionising radiation in the manufacture of medicinal products
- o   Annex 13 - Manufacture of investigational medicinal products
- o   Annex 14 - Manufacture of products derived from human blood or human plasma
- o   Annex 15 - Qualification and validation
- o   Annex 16 - Qualified person and batch release
- o   Annex 17 - Parametric release
- o   Annex 18 - GMP guide for active pharmaceutical ingredients (This annex no longer exists)
- o   Annex 19 - Reference and retention samples
- o   Annex 20 - Quality risk management
- o   Glossary

EudraLex - Volume 4 - Good Manufacturing Practice (GMP) Guidelines

Volume 4 of the rules governing medicinal products in the European Union contains guidance for the interpretation of the principles and guidelines of good manufacturing practices for medicinal products for human and veterinary use laid down in Commission Directives 91/356/EEC, as amended by Directive 2003/94/EC, and 91/412/EEC respectively.

EudraLex V4 is made up of the following parts:

- ➢   Introduction
- ➢   Part I - Basic requirements for medicinal products
- ➢   Part II - Basic requirements for active substances used as starting materials
- ➢   Part III - GMP related documents

The Commission Directive 2003/94/EC, of 8 October 2003, sets out the principles and guidelines of good manufacturing practice in respect of medicinal products for human use and investigational medicinal products for human use.

Part I - Basic Requirements for Medicinal Products

- o   Chapter 1 - Pharmaceutical Quality System
- o   Chapter 2 - Personnel
- o   Chapter 3 - Premise and Equipment
- o   Chapter 4 - Documentation
- o   Chapter 5 - Production
- o   Chapter 6 - Quality Control
- o   Chapter 7 - Outsourced Activities

o Chapter 8 - Complaints and Product Recall
o Chapter 9 - Self-Inspection

Part II - Basic Requirements for Active Substances Used as Starting Materials

Basic requirements for active substances used as starting materials.

Part III - GMP Related Documents

Site Master File
Q9 Quality Risk Management
Q10 Note for Guidance on Pharmaceutical Quality System
MRA Batch Certificate

Annexes

Annex 1- Manufacture of Sterile Medicinal Products
Annex 2- Manufacture of Biological Active Substances and Medicinal Products for Human
Annex 3- Manufacture of Radiopharmaceuticals
Annex 4- Manufacture of Veterinary Medicinal Products Other than Immunological Veterinary Medicinal Products
Annex 5- Manufacture of Immunological Veterinary Medicinal Products
Anne 6- Manufacture of Medicinal Gases
Annex 7- Manufacture of Herbal Medicinal Products
Annex 8- Sampling of Starting and Packaging Materials
Annex 9- Manufacture of Liquids, Creams and Ointments
Annex 10- Manufacture of Pressurised Metered Dose Aerosol Preparations for Inhalation
Annex 11- Computerised Systems
Annex 12- Use of Ionising Radiation in the Manufacture of Medicinal Products
Annex 13- Manufacture of Investigational Medicinal Products
Annex 14- Manufacture of Products Derived from Human Blood or Human Plasma
Annex 15-Qualification and Validation (in operation since 1 October 2015)

Annex 16- Certification by a Qualified Person and Batch Release
Annex 17- Parametric Release

Annex 19- Reference and Retention Samples

FDA

The FDA publishes regulations and guidance documents for industry in the Federal Register. The FDA's website also contains links to the cGMP regulations and guidance documents as well as various resources to help drug companies comply with the law. The FDA also conducts onsite audits and public outreach through presentations at national and international meetings and conferences on the subject of cGMP requirements.

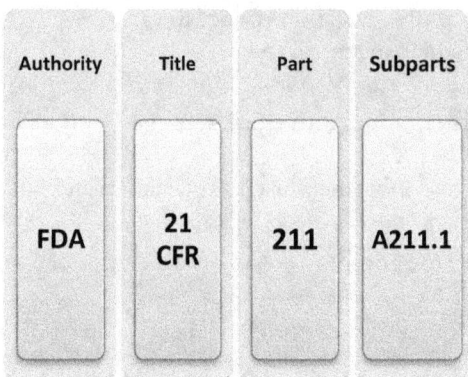

Figure: The FDA organises its regulations under titles. Within titles there are parts and subparts.

Pharmaceutical quality affects every American. The FDA regulates the quality of pharmaceuticals very carefully. The main regulatory standard for ensuring pharmaceutical quality is the Current Good Manufacturing Practice (CGMPs) regulation for human pharmaceuticals. Consumers expect that each batch of medicines they take will meet quality standards so that they will be safe and effective. Most people, however, are not aware of CGMPs, or how the FDA ensures that drug manufacturing processes meet these basic objectives. Recently, the FDA has announced a number of regulatory actions taken against drug manufacturers based on the lack of CGMPs. This paper discusses some facts that may be helpful in understanding how CGMPs establish the foundation for drug product quality.

PART 211   Current Good Manufacturing Practice for Finished Pharmaceuticals
Subpart A--General Provisions
§ 211.1 - Scope
§ 211.3 - Definitions
Subpart B--Organisation and Personnel
§ 211.22 - Responsibilities of quality control unit
§ 211.25 - Personnel qualifications
§ 211.28 - Personnel responsibilities
§ 211.34 - Consultants
Subpart C--Buildings and Facilities
§ 211.42 - Design and construction features
§ 211.44 - Lighting
§ 211.46 - Ventilation, air filtration, air heating and cooling
§ 211.48 - Plumbing
§ 211.50 - Sewage and refuse
§ 211.52 - Washing and toilet facilities
§ 211.56 - Sanitation
§ 211.58 - Maintenance
Subpart D--Equipment
§ 211.63 - Equipment design, size, and location
§ 211.65 - Equipment construction
§ 211.67 - Equipment cleaning and maintenance

§ 211.68 - Automatic, mechanical, and electronic equipment

§ 211.72 - Filters

Subpart E--Control of Components and Drug Product Containers and Closures

§ 211.80 - General requirements

§ 211.82 - Receipt and storage of untested components, drug product containers, and closures

§ 211.84 - Testing and approval or rejection of components, drug product containers, and closures

§ 211.86 - Use of approved components, drug product containers, and closures

§ 211.87 - Retesting of approved components, drug product containers, and closures

§ 211.89 - Rejected components, drug product containers, and closures

§ 211.94 - Drug product containers and closures

Subpart F--Production and Process Controls

§ 211.100 - Written procedures; deviations

§ 211.101 - Charge-in of components

§ 211.103 - Calculation of yield

§ 211.105 - Equipment identification

§ 211.110 - Sampling and testing of in-process materials and drug products

§ 211.111 - Time limitations on production

§ 211.113 - Control of microbiological contamination

§ 211.115 - Reprocessing

Subpart G--Packaging and Labelling Control

§ 211.122 - Materials examination and usage criteria

§ 211.125 - Labelling issuance

§ 211.130 - Packaging and labelling operations

§ 211.132 - Tamper-evident packaging requirements for over-the-counter (OTC) human drug products

§ 211.134 - Drug product inspection

§ 211.137 - Expiration dating

Subpart H--Holding and Distribution

§ 211.142 - Warehousing procedures

§ 211.150 - Distribution procedures

Subpart I--Laboratory Controls

§ 211.160 - General requirements

§ 211.165 - Testing and release for distribution

§ 211.166 - Stability testing

§ 211.167 - Special testing requirements

§ 211.170 - Reserve samples

§ 211.173 - Laboratory animals

§ 211.176 - Penicillin contamination

Subpart J--Records and Reports

§ 211.180 - General requirements

§ 211.182 - Equipment cleaning and use log

§ 211.184 - Component, drug product container, closure, and labelling records

§ 211.186 - Master production and control records

§ 211.188 - Batch production and control records

§ 211.192 - Production record review

§ 211.194 - Laboratory records

§ 211.196 - Distribution records

§ 211.198 - Complaint files

Subpart K--Returned and Salvaged Drug Products

§ 211.204 - Returned drug products

§ 211.208 - Drug product salvaging

World Health Organisation GMP Guideline Annexes

The WHO Essential Medicines and Health Products (EMP) Department works with countries to promote affordable access to quality, safe and effective medicines, vaccines, diagnostics and other medical devices. As part of this effort, the WHO publishes a number of guidance annexes that describe best practice quality requirements for specific areas within the life science industry.

List of WHO GMP Annexes:

- WHO Good Manufacturing Practices for Pharmaceutical Products: Main Principles
  Annex 2, WHO Technical Report Series 986, 2014
- Active Pharmaceutical Ingredients (Bulk Drug Substances)
  Annex 2, WHO Technical Report Series 957, 2010
- Active Pharmaceutical Ingredients - Bulk Drug Substances: Additional Clarifications and Explanations
- Pharmaceutical Excipients
  Annex 5, WHO Technical Report Series 885, 1999
- WHO Good Manufacturing Practices for Sterile Pharmaceutical Products
  Annex 6, WHO Technical Report Series 961, 2011
- WHO Good Manufacturing Practices for Biological Products
  Annex 3, WHO Technical Report Series 996, 2016
- WHO Good Manufacturing Practices for Blood Establishments (jointly with the Expert Committee on Biological Standardisation)
  Annex 4, WHO Technical Report Series 961, 2011
- Pharmaceutical Products Containing Hazardous Substances
  Annex 3 WHO Technical Report Series 957, 2010
- Investigational Pharmaceutical Products for Clinical Trials in Humans
  Annex 7, WHO Technical Report Series 863, 1996
- Herbal Medicinal Products
  Annex 3, WHO Technical Report Series 937, 2006
- Radiopharmaceutical Products
  Annex 3, WHO Technical Report Series 908, 2003
- Water for Pharmaceutical Use
  Annex 2, WHO Technical Report Series 970, 2012
- WHO Guidelines on Good Manufacturing Practices for Heating, Ventilation and Air-Conditioning Systems for Non-Sterile Pharmaceutical Dosage Forms
  Annex 5, WHO Technical Report Series 961, 2011
- Validation
  Annex 4, WHO Technical Report Series 937, 2006

- Guidelines on Good Manufacturing Practices: Validation, Appendix 7: Non-Sterile Process
  Validation
  Annex 3, WHO Technical Report Series 992, 2015

International Council for Harmonisation, ICH, GMP Guide

The International Council for Harmonisation of (Technical Requirements) for Pharmaceuticals for Human Use (ICH) brings together the regulatory authorities and pharmaceutical industry to discuss scientific and technical aspects of drug registration. Since its inception in 1990, ICH has gradually evolved, to respond to the increasingly global face of drug development.

ICH Q7 Good Manufacturing Practice Guide for Active Pharmaceutical Ingredients

2.2 Data Integrity and Principles of Compliance

Introduction

Data generated by or used in GxP impacting activities must be handled and protected in accordance with international and national regulatory requirements. The application of data integrity applies to many industries and products that touch the lives of patients and end users across the globe. Some examples of products that must meet data integrity regulations include (1) active pharmaceutical ingredients, (2) medical devices, (3) medicinal products, (4) vaccines and (5) cosmetics.

The below agencies and regulatory authorities provide specific requirements on data integrity:

> ➢ EU GMP – EudraLex – Rules Governing Medicinal Products in the European Union Volume 4 – Guidelines to Good Manufacturing Practice for Medicinal Products for Human Use – Products for Human and Veterinary Use, Annex 11: Computerised Systems – (1, 7.2, 17)

> ➢ FDA – 21 CFR Part 11 – Food and Drug Administration – Electronic Records; Electronic Signatures – Scope and Application (C)

> ➢ FDA- 21 CFR Part 211 – Food and Drug Administration – Code of Federal Regulations - Good Manufacturing Practices - 211.188a, 211.194.2, 211.194.8

> ➢ ICH E6 – International Conference on Harmonisation - Guideline for Good Clinical Practice (5.2.1, 8.1, 8.3)

> ➢ MHRA – United Kingdom - Medicines and Healthcare Products Regulatory Agency - GMP Data Integrity Definitions and Guidance for Industry (2015)

> ➢ PIC/S Guidance PI 011-] – Pharmaceutical Inspection Convention Scheme - Good Practices for Computerised Systems in Regulated "GXP" Environments

Within the life science industry the saying goes "if it's not written down, it didn't happen". This is a powerful message that is a suitable starting point for data integrity. In the current and present day, the mere mention of data integrity quickly conjures an image of Excel sheets, big data, databases and computers in our minds. However, it has a broader impact with its roots in the basics of good science – good documentation.
Data integrity indeed does apply to "soft" or electronic data but also applies to paper-based systems and records. GxP is the umbrella acronym that stands of "good practices" in all our tasks and activities, be it laboratory testing, process engineering and so on. A core element in meeting GxP is abiding by "Good Documentation Practices" (GDP). Having good written records is fundamental to patient and product safety within the pharmaceutical, biopharmaceutical and medical devices industries. So, data integrity begins with the small stuff — real-time data collection, real-time review, honest and accurate recording of data and events.

The integrity of data relies on several factors. It can be influenced by a company's culture or approach to doing business. It can also be affected by the level of experience or knowledge within a company. Many traditional engineering companies outside the regulated life science community simply do not have the need to be so thorough in their handling of data and information. Within a GxP environment, controls, training and the design and operation of systems and processes influence data integrity on a day-to-day basis. Most of the time, those affected by the controls or systems do not think of them, but they can either support or inhibit data integrity and the reliability of data. Obviously, equipment, systems and processes should play a key role in making data reliable and accurate.

Key Terms

Configuration Identification

Software and hardware packages should be identified by a unique product identifier and a version number. For the software end-user, the parts of an automated system that are subject to configuration management should be clearly identified. The system should therefore be broken down into configuration items. These should be identified at an early phase of development so that a complete list of configuration items is defined and maintained. The application-specific items should have a unique name or version ID. The depth of detail when specifying the elements is decided by the needs of the system, and the organisation developing that system.

Requirements for the User ID and Password

User ID: The user ID of a system should have a minimum length agreed with the customer and should be unique within the system.

Password: A password should always consist of a combination of numeric and alphanumeric characters. When setting up passwords, the number of characters and a period after which a password expires should be stipulated. The structure of the password is normally selected to suit the specific customer. The configuration is described in the security settings section of password policy. Criteria for the structure of a password are as follows:

> ➢ Minimum length of the password
> ➢ Use of numeric and alphanumeric characters
> ➢ Case sensitivity

Audit Trail

The audit trail is a control mechanism of a system that allows all data entered or modified to be traced back to the original data. A reliable and secure audit trail is particularly important in conjunction with the creation, change or deletion of GMP-relevant electronic records. In this case, the audit trail must archive and document all the changes or actions made along with the date and time. Typical contents of an audit trail must be recorded and describe the procedures "who changed what and when" (old value/new value).

Data: any data (numerical or otherwise) which is collected or processed as part of GxP activities in order to generate GxP documents and records using a paper-based or electronic process.

Data Handling: Any GxP task that involves creation, entry, review, approval, analysis, reporting, storage, archival, retrieval, or disposal of GxP data.

Data Integrity: The degree to which a collection of GxP data is managed through effective organisational, operational, and technical mechanisms to ensure GxP data reliability.

Data Life Cycle: Starts from the time of data creation to the point of use and during its retention, archival, retrieval and eventual disposal

GxP Impacting: Any action that can impact the quality or safety of a product or critical process.

Application: Software installed on a defined platform/hardware providing specific functionality.

Bespoke/Customised Computerised System: A computerised system individually designed to suit a specific business process.

Commercial Off-the-Shelf Software: Software commercially available, whose fitness for use is demonstrated by a broad spectrum of users.

IT Infrastructure: The hardware and software such as networking software and operation systems, which makes it possible for the application to function.

Life Cycle: All phases in the life of the system from initial requirements until retirement including design, specification, programming, testing, installation, operation and maintenance.

Process Owner: The person responsible for the business process.

System Owner: The person responsible for the availability, and maintenance of a computerised system and for the security of the data residing on that system.

Third Party: Parties not directly managed by the holder of the manufacturing and/or import authorisation.

The Life Cycle of Data

Regulations that speak to GxP and data integrity can apply to many different streams within the life science sector as previously mentioned. From medical devices to pharmaceuticals, all act in different manners, with long and short term applications. Take the example of a total knee replacement. Many designs now ensure their effectiveness in excess of ten years, even up to twenty years depending on individual circumstances. This requires many key records within manufacturing to be kept for several decades. Thus, data retention requirements specify the retention periods of such documents. The integrity of GxP data must be protected during the entire data life cycle, from creation of the data and records to the eventual destruction of data after the retention period is fulfilled. Data integrity equally applies to:

➢ Equipment
➢ Computerised systems
➢ Test records
➢ Inspection records
➢ Material certificates

Data integrity ensures that patient safety, product quality, and product supplies are generated by the product life cycle processes.

<u>Process Design</u>

Failure to maintain data integrity can occur throughout the life cycle of data; however, a thoughtful design of systems can prevent breaches in data and restrict the severity of any attempts to alter data. Therefore, design should aim to include controls and preventative measures. At a high level, this can be achieved by:

➢ Limiting access to GxP events and data
➢ Standard Operating Procedures (SOPs)
➢ Training
➢ System owners

Data Reliability

Data reliability is the foundation to achieving cGxP data integrity. The FDA's ALOCA model can be used to enforce data reliability.

Accuracy: the GxP data is recorded, calculated, analysed, and reported as found and correctly.

Attributable: any actions or calculations performed on GxP data can be attributed to or traceable to the person that performed the actions and the date and time at which they were performed.

Legible: the GxP data is recorded in a clear and human-readable form.

Contemporaneity: the GxP data is recorded at the same time as the observation/measurement is made or as soon as possible after the event.

Original: the initial data recorded is available and not altered.

An additional point to make it that of trustworthiness. It is assumed that engineers and scientists etc. working across the life science industries are ethical and do not falsify data or information. Typically companies can implement a code of practice or ethical behaviour programme to desist people from intentional unethical behaviour or the falsification of records.

Data Creation: The point at which the values or data is created. The data and information is original (raw).

Data Authentication: Within a GxP environment, authentication refers to the approval of data (electronic signatures). E-signatures are key controls within software that prompt the user to enter a unique username and password to acknowledge a recording or action. The e-signature should create a permanent link with the electronic record that cannot be removed and can be viewed through an audit trial.

Data Protection: Once the data is created, the handling of the data must ensure data integrity. For electronic data, this includes access control to computer systems. Other practical restrictions can also be made such as limiting room and site access to authorised personnel.

Data Retention: This refers to the controlled storage, backup and arching of data. Retention of records may be required for several decades depending on the type of data and the regulatory requirements relating to the particular product or industry.

Technical Controls

The benefits of modern software and computerised systems allow robust and complex data handling and calculations to be completed. With this modern capability that is becoming more powerful comes more responsibility with regard to the use of data.

The computerised systems used to generate, gather or interpret GxP data must fulfil several criteria. First and foremost, they must be fit for the intended use. The software and hardware must be validated and proven to be consistent and reliable. Some general considerations for the use of computerised systems include:

> ➤ Systems designed to foster integrity of GxP data

➢ User requirements specification detailing the intended use and required functionality
➢ An approved vendor with certification to ISO 9001 or other quality management standards
➢ Software should meet the requirements of regulations such as FDA 21 CFR Part 11
➢ Written procedures on how automated processes function

It should not be an easy process for personnel to alter or corrupt data when using computerised systems. GxP-impacting computer systems should have controls that prevent unauthorised access along with audit trail history.

Audit trail design and configuration capture key critical processes, events, settings and information. This enables any investigations of quality events impacting data integrity to be reviewed and analysed.

<u>Computer System Design and Development</u>

For computer systems, software requirements are typically stated in functional terms and are defined, refined and updated during the development phase. Success in accurately and completely documenting software requirements is a crucial factor in successful validation of the resulting software. A specification* is defined as "a document that states requirements." It may refer to or include engineering drawings or other relevant documents *21 CFR 820.3(y).

There are different kinds of written specifications:

➢ User requirements specifications
➢ System requirements specification
➢ Software requirements specification
➢ Software design specification
➢ Software test specification
➢ Functional design specification

All of these documents establish "specified requirements" and are design outputs for which various forms of verification or validation are required. The URS must also define non-software requirements and hardware. Non-functional requirements such as maintainability and usability can also be included. There should be a clear distinction between mandatory regulatory requirements and optional features. Proper definition at this stage ensures the system meets data integrity requirements and prevents costly updates down the line.

<u>Practical Elements to Data Integrity</u>

Facilities and systems must be configured in a way that encourages compliance with principles of data integrity. Examples include:

➢ Availability of clocks for recording times.
➢ Access points to allow swift reference to GxP records at locations where tasks are completed.
➢ Control of raw data.
➢ Control of approved documents.

<u>Organisational Controls</u>

Regulated companies such as medical device, pharmaceutical and biotechnology companies are required to operate under a quality management system. For medical devices, ISO 13485 serves as a quality management system. Likewise, the FDA Code of Federal Regulations 21 CFR Part 211 functions as a QMS for finished pharmaceuticals.

Organisational controls for Data Integrity can address:

➢ Assessment of GxP computerised systems
➢ Management of GxP computerised systems
➢ Electronic Records Implementation and handling
➢ Use of Electronic signatures
➢ Quality Risk Management

Operational Factors

Operational factors refer to process or manufacturing errors, deviations or non-compliance to established procedures that may impact data integrity.

GxP data handling activities should be designed to limit human intervention. As with human intervention there can be errors or omissions. Furthermore, it may call into question the reliability of the data. Mistake-proofing methodologies should be developed to avoid human error related breaches in data integrity. As with any system or technology, training is a fundamental step. Building upon training, exposure to GxP data systems and on-the-job training all play a part in delivering a system that is robust and meets regulatory requirements. It is important to remind ourselves that while regulations are the driving force to comply with data integrity, the ultimate goal is always the protection and safety of the patient or end user of the product, medicine or treatment.

Software Validation

Where there is the potential to affect product conformance to requirements or where software or IT systems provide support to aspects of quality management, validation is required. Most companies categorise software validations to account for the different applications of software and IT systems. For example, enterprise systems, such as the drawing package SolidWorks would be validated in a different manner to manufacturing systems that contain software (a.k.a. embedded software).
"Embedded" software is where the software is integrated into the manufacturing equipment. Embedded software is typically validated during the equipment qualification stage, process validation stage or test method validation. Enterprise software falls outside of equipment or process validation but does require validation if it impacts product quality or is used to make quality decisions. Standalone systems such as ERP (Enterprise Resource Planning) systems also require validation.

Software Validation and GAMP

Good Automated Manufacturing Practice (GAMP) is a set of guidelines for manufacturers and users of automated systems in regulated industries. GAMP specifically impacts the medical device, pharmaceutical and biopharmaceutical industries. The application of GAMP and validation of automated systems in manufacturing helps ensure that regulated medical devices and medicinal products have the required quality and are manufactured according to good practices, meet regulatory and legal requirements and ensure patient safety. GAMP ensures quality is in-built into each stage of the manufacturing process. Therefore, GAMP has a place in all aspects of automation and production, including the handling of raw materials, control of facilities and equipment etc.

Key Terms

Automated System: Term used to cover a broad range of systems, including automated manufacturing equipment, control systems, automated laboratory systems, manufacturing execution systems and computers running laboratory or manufacturing database systems. The automated system consists of the hardware, software and network components, together with the controlled functions and associated documentation. Automated systems are sometimes referred to as computerised systems; in this guide the two terms are synonymous.

Commercial Off-the-Shelf (COTS): Configurable programs and stock programs that can be adapted to specific user applications by "filling in the blanks" without (COTS) altering the basic program.

Computer System Validation: A process that confirms by examination and provision of objective evidence that the computer system conforms to user needs and intended uses. System validation is a process for achieving and maintaining compliance with GxP regulations and fitness for intended use by adoption of life cycle activities, deliverables, and controls.

GAMP 5: A set of guidelines that offers a risk-based approach to ensuring the compliance of GxP-impacting computerised systems.

V- Model: A development process which sets out a roadmap of stages and deliverables during a project. 21 CFR Part 820: FDA requirements pertaining to medical devices.

User Requirement Specification, URS: The URS is a critical document that defines the requirements of the computerised system and agreement to the requirements.

Software Requirement Specification, SRS: An SRS can be written to interpret the requirements of a URS and how they relate to the requirement or how the requirement is met in practical terms regarding software.

Functional Design Specification, FDS: A functional design specification is a document that specifies how particular requirements are met — this can be a combination of how the equipment/process operates mechanically/automatically etc. An FDS is typically written in response to a URS.

Computer System Validation Life Cycle

The computer system validation life cycle refers to all activities from initial concept to retirement of a computer system. The life cycle of the system includes the defining of, and performance of activities in a systematic way from conception, requirements, development or configuration, testing, release and operational use. The four GAMP life cycle phases include:

> ➢ Concept
> ➢ Planning and project stage
> ➢ Operation
> ➢ Retirement

The concept stage is concerned with understanding the need or the problem to be addressed. We will see that the user requirement specification (along with other specifications) and the initial risk assessment help to drive a project forward in a systematic manner. The most common life cycle approach for computerised and automated systems is the V-Model. The GAMP-based V-model lays out a roadmap which facilitates the validation of equipment and automated systems.

The planning and project stage involves the planning of the validation effort required to implement the system into the business area(s) based on identification and approval of system concept. This phase includes assessments of the regulatory and system risks, supplier assessment, development of validation strategies, identification of deliverables that will be generated, definition of the business process the system will support as well as the user requirements which the system will fulfil.

Design, development and configuration of the hardware and software is also required to meet the system requirements as per specifications. In the case of custom software components, this effort could also include detailed software design and developmental testing to ensure readiness for verification testing.
The verification stage confirms that specifications have been met and releases the system for use. This phase will involve multiple stages of reviews and testing depending on the system type, the development method applied and its use. Once verification activities have begun, any changes to the system must be captured through change control.

On successful completion of the verification activities, the system is then released for effective use. The test strategy and other verification activities will vary widely between simple equipment and more complex customised/configurable systems. The verification and validation approach is typically agreed and detailed at the validation planning stage. The VP can be updated accordingly as the project develops with more detail being added. Alternatively, a test strategy document or matrix could be written to provide more specific test plans.

Verification deliverables vary based on the complexity and level of customisation of the system in question. Corporate or company specific procedures also shape the required activities to be completed and reported. Some generic deliverables are listed below.

> Approval, execution and review of test protocols
> Writing and approving SOPs for operation and maintenance of the system
> Traceability matrix
> Completion of any risk mitigations (e.g. updates to FMEA etc.)
> Validation summary report(s)

Validation reporting requirements vary depending upon the scope of the system and should also be driven by a procedure and template. The validation plan can also outline the deliverables and what needs to be addressed in the report. A Validation Summary Report (VSR) should be written to summarise the results of executing the VP, the documents created for the validation activities and the testing performed. Finally, the VSR indicates the acceptance of the system/equipment by the user and the project team and states that the equipment is released for commercial operation/production.

The operation phase supports the need to maintain compliance and fitness for intended use after the system is released for normal use. It is important to ensure the system remains within a continued validated state. All proposed or necessary changes to the system must be assessed and controlled as part of a change control process. Once the system has been accepted and released for use, the operation phase begins. This phase consists of maintaining the system's compliant state and fitness for intended use through the control of the procedures supporting the system's operational use.

During the operation phase, the below activities are typically completed:

> Ongoing training
> Preventative maintenance
> Service management and performance monitoring.
> Change control
> Periodic review
> Maintaining system security
> Records management

➢ Calibration

The retirement phase involves the planning and proper management of activities relating to the removal of systems from service (shutdown). The retirement should take into account the storage of any data and any data migration that needs to occur prior to retirement. The retirement plan, if needed, will outline the retirement strategy from the roles and activities that will be conducted to the removal of the system for use. A retirement summary report is produced that documents the results of the activities defined in the retirement plan including:

➢ Retirement plan and timelines.
➢ Summaries of any data migration activities.
➢ Identification of the storage location of documentation relating to the system.
➢ Obsoleting of SOPs.

It must be stressed that GAMP is a set of principles, a set of guidelines that aim to achieve compliant computerised systems that are fit for intended use. GAMP guidelines differ to 21 CFR QSR regulations as they are not legal or statutory requirements. However, they represent industry best practice and complement the validation efforts that are legal requirements and statutory requirements.

Regulatory Review

Software validation is a requirement of the quality system regulation, 21 Code of Federal Regulations (CFR) Part 820. Validation requirements apply to:

(1) software used as components in medical devices,
(2) software that is itself a medical device, and
(3) software used in production of the device or in implementation of the device manufacturer's quality system.

Note: EU GMP Annex 11, provides information on the inspection of 'Computerised Systems'.

In addition, computer systems used to create, modify, and maintain electronic records and to manage electronic signatures are also subject to the validation requirements. Such computer systems must be validated to ensure accuracy, reliability, consistent intended performance, and the ability to discern invalid or altered records. The regulated user should be able to demonstrate through the validation evidence that they have a high level of confidence in the integrity of both the processes executed within the controlling computer system and in those processes controlled by the computer system within the prescribed operating environment.

System Categorisation

GAMP 5 makes provision for four categories of software in order to distinguish the level of tcustomisation/configurability that exists across software serving different functions:

GAMP Software Category 1, Operating Systems
GAMP Software Category 2, Non-configured software
GAMP Software Category 4, Configurable software packages
GAMP Software Category 5, Custom Software

GAMP Software Category 1, Operating Systems

Category 1, operating systems, covers established commercially available operating systems. These systems are not subject to validation themselves. The name and version of the operating system must, however, be documented and verified during Installation Qualification (IQ). Application software hosted on operating systems needs to be validated.

GAMP Software Category 3, Non-Configured Software

Category 3 covers commercially available, standard software packages and "off the-shelf" solutions for certain processes. The configuration of the software packages should be limited to adaptation to the runtime environment (for example network and printer connections) and the configuration of the process parameters. The name and version of the standard software package should be documented and verified in an installation qualification (IQ). Special user requirements, such as security, alarms, messages, or algorithms must be documented and verified in an operational qualification (OQ).

GAMP Software Category 4, Configurable Software Packages

GAMP Software Category 4, Configurable Software Packages Category 4 covers configurable software packages that allow special business and manufacturing processes. This involves configuring predefined software modules. These software packages should only be considered as belonging to Category 4 if they are well-known and mature. Normally, a supplier audit is necessary. If this is not available, the software packages should be handled as Category 5. The name, version, and configuration should be documented and verified in an installation qualification (IQ). The functions of the software packages should be verified in terms of the user requirements in an operational qualification (OQ). The validation plan should take into account the life cycle model and an assessment of suppliers and software packages.

GAMP Software Category 5, Custom Software

GAMP Software Category 5, Custom Software Custom/Bespoke Software (GAMP Software Cat 5) is software that contains custom code designed or modified specifically for a particular customer. As the code is custom, it presents a greater risk. This risk must be mitigated with the right approach to the validation.

GAMP Considerations

Correctly assigning a GAMP software category to equipment, systems or processes is an important activity that should be completed early on in the planning stage of a project. There must of some degree of familiarity with the equipment or system. The manufacturer or vendor can be a source of information that may help the designation. In many cases, companies create tools or processes that help determine what GAMP software category applies. These have different names such as questionnaires, screening tools, planning tools etc.

Risk Assessments

A risk assessment process should be applied to cGxP computerised systems in order to identify and mitigate potential risks to (1) patient safety, (2) product quality and (3) data integrity. Results identified through a risk assessment help to determine the validation strategy, the effort and time required, and allow better targeting of the validation activities to the highest risks.

The risk assessment should be revised during the software development lifecycle (SDLC) if the functionality, requirements or intended use of the system changes. The risk assessment activity should also be evaluated during system build-up as well as when implementing changes. Risk assessment tools for cGxP computerised systems are typically completed during the planning stage, specification stage and post-qualification if a change or update is required.

Planning Stage

Initial Impact/Risk Assessment – takes place during the planning phase to identify the level of impact and GxP relevance of the system/equipment. (Tools used: High Level Risk Assessment).

Specification Stage

Functional or Quality Risk Assessment – takes place during the specification phase and identifies potential risks and possible mitigations to be to be introduced to the process. (Tools used: Quality Risk Matrix, (p)FMEA).

Changes to the System

Impact Assessment of Changes – takes place as part of the change control process in the system operational phase.

Quality Risk Matrix

A QRM is a risk assessment that identifies and manages the risk to patient safety, product quality and data integrity that relate to system processes. Risk scenarios or potential causes should be developed for each identified function or process step and then assessed for the impact on patient safety, product quality or data integrity. Risk mitigations and controls should then be introduced to address both medium and high levels of risk. The QRM requires three "assessments" in order to produce an estimation or overall risk (low, medium, high),

> Assess likelihood
> Assess detectability
> Assess severity

Traceability Matrix

A traceability matrix should be prepared as required in accordance with company and internal policy. It is also recommended by GAMP guidelines, ASTM E2500 and ISPE risk-based approaches to validation. The matrix links the user requirements and specifications to testing and validation activities. A traceability matrix illustrates that all user requirements are traceable to the verification/validation activity or vendor documents as relevant (FDS if applicable, design specifications etc.) Generally, individual organisations will have an approved template to work from. However, the URS structure can form the basis of the template, with additional columns added to document the test/verification method or reference documents (such as FDS and vendor specifications and design documents)

21 CFR Part 11

This section specifically covers the regulatory requirements of part 11 of Title 21 of the Code of Federal Regulations; Electronic Records; Electronic Signatures (21 CFR Part 11). Part 11 of the FDA CFR is relevant to "records in electronic form that are created, modified, maintained, archived, retrieved, or transmitted under any records requirements set forth in agency regulations."
As of 2007, several sections of the regulation have been identified as excessive and the FDA announced in guidance that it will exercise enforcement discretion on some parts of 21 CFR part 11. This has been welcomed by some manufacturers but it has also caused a degree of confusion. The requirements relating to

access controls are the most fundamental requirements and are routinely enforced. The "predicate rules" that required organisations to keep records in the first place are still in effect. If electronic records are illegible, inaccessible, or corrupted, manufacturers are still subject to those requirements.

If a regulated firm keeps "hard copies" of all required records, those paper documents can be considered the authoritative document for regulatory purposes. This then means that the computer system is not in scope for electronic records requirements, although subject to predicate rules which still require validation. If the "hard copy" is to be identified as the authoritative document, the "hard copy" must be a complete and accurate copy of the electronic source. The manufacturer must use the hard copy (rather than electronic versions stored in the system) of the records for regulated activities.

Definition of Records

The FDA has deemed the following records or signatures in electronic format subject to 21 CFR part 11:

Records that are required to be maintained under predicate rule requirements and that are maintained in electronic format in place of paper format. On the other hand, records (and any associated signatures) that are not required to be retained under predicate rules, but that are nonetheless maintained in electronic format, are not part 11 records. Records that are required to be maintained under predicate rules, that are maintained in electronic format in addition to paper format, and that are relied on to perform regulated activities.

Records submitted to FDA, under predicate rules (even if such records are not specifically identified in agency regulations) in electronic format (assuming the records have been identified in docket number 92S-0251 as the types of submissions the agency accepts in electronic format). However, a record that is not itself submitted, but is used containing nonbinding recommendations in generating a submission, is not a part 11 record unless it is otherwise required to be 205 maintained under a predicate rule and it is maintained in electronic format.

Electronic signatures that are intended to be the equivalent of handwritten signatures, initials, and other general signings required by predicate rules. Part 11 signatures include electronic signatures that are used, for example, to document the fact that certain events or actions occurred in accordance with the predicate rule (e.g. approved, reviewed, and verified).

The above definitions are taken from the FDA guidance document entitled "FDA Guidance for Industry: 21 CFR Part 11 - Electronic Records and Electronic Signatures: Scope and Application, August 2003." This document also provides recommendations on documenting key decisions that may be taken in relation to 21 CFR Part 11 applicability and compliance.

Requirements and Specifications

The need for compliance to 21 CFR depends on the type of technology and level of automation and computerisation involved in the manufacturing process or other actives that are GxP-impacting. Does the system store electronic records? Does the system require a login? Is there an audit trial? If a complex system is to be procured, the requirements need to be communicated to the manufacturer as part of a user requirement specification and/or software requirement specification.

General Guidance on Requirement Specifications

While the quality system regulation states that design input requirements must be documented, and that specified requirements must be verified, the regulation does not further clarify the distinction between the terms "requirement" and "specification." A requirement can be any need or expectation for a system or for its software. Requirements reflect the stated or implied needs of the customer, and may be market-based, contractual, or statutory, as well as an organisation's internal requirements.

There can be many different kinds of requirements (e.g., design, functional, implementation, interface, performance, or physical requirements). Software requirements are typically derived from the system requirements for those aspects of system functionality that have been allocated to software. Software requirements are typically stated in functional terms and are defined, refined, and updated as a development project progresses. Success in accurately and completely documenting software requirements is a crucial factor in successful validation of the resulting software. *Page 6 Guidance for Industry and FDA Staff General Principles of Software Validation A Specification* is defined as "a document that states requirements." (21 CFR 820.3(y)). It may refer to or include drawings, patterns, or other relevant documents and usually indicates the means and the criteria whereby conformity with the requirement can be checked.

There are many different kinds of written specifications, e.g., system requirements specification, software requirements specification, software design specification, software test specification, software integration specification, etc. All of these documents establish "specified requirements" and are design outputs for which various forms of verification are necessary.

Validation of Computerised Systems

The requirement for computerised systems to be compliant to 21 CFR part 11 needs to be identified early on in the project to ensure that the vendor or supplier of the systems or equipment can develop and build a system that meets the requirements of 21 CFR part 11. Computer system validation can be divided into three distinct phases: (1) planning, (2) design and development, (3) verification and (4) retirement.

Planning: This phase involves the planning of the validation effort required to implement the system and identification of key milestones and requirements. It requires supplier assessments, assessments of the regulatory and system risks, supplier development of a validation approach and the identification of deliverables that will be generated to support the implementation and operation of the system.

Design and Development: This phase consists of the design, development and configuration of the hardware and software required to meet the system requirements. In the case of custom software, design and developmental testing is important to ensure proper functionality prior to verification testing.

Verification: This phase confirms that requirements and specifications have been met. Testing is required to ensure the system operates as intended. Upon successful testing and verification, the system can be released for use. Once verification activities have begun, any changes to the system must be managed through change control. In case of successful completion of the verification activities (i.e. any deviation has been evaluated and addressed), the system is released for effective use. The operation phase supports the need to maintain compliance and fitness for intended use after the system is accepted and released for use.

Retirement: This phase consists of the planning, executing and summarising of the events required for system shutdown. It includes the appropriate handling of the supporting documents and the data contained within the system. While described here as a separate phase, a system's retirement can be handled as part of a new system implementation or as a separate project.

Best practice when it comes to computer system validation is to adopt a life cycle approach which requires the completion of activities in a systematic way from system conception to retirement. Life cycle activities could be scaled according to system impact on product quality, patient safety and data integrity, system complexity and novelty, supplier assessment and business risk.

Definitions

Computer System: A computer/automated system consisting of the hardware, software, and network components, together with the controlled functions (personnel, procedures, and equipment) and associated documentation.

Computer System Validation: A process that confirms by examination and provision of objective evidence that the computer system conforms to user needs and intended uses. Computer system validation is a process for achieving and maintaining compliance with GxP regulations and fitness for intended use by adoption of life cycle activities, deliverables, and controls.

GxP-Regulated Computer Systems: Computer systems determined to have a potential impact on product quality, patient safety and data integrity; these systems are required to comply with the relevant GxP regulations.

Data Integrity: The degree to which data is reliable and without error. Data must be accurate, attributable, contemporaneous, original, legible and available. A breach of data integrity occurs when any person manipulates or distorts data and submits the results of that data as valid.

Predicate Rules: A predicate rule is any FDA regulation that requires companies to maintain certain records and submit information to the agency as part of compliance.

To gain a better understanding of the validation of computerised systems, consult the following publication: "FDA's Guidance for Industry and FDA Staff General Principles of Software Validation." See also industry guidance such as the GAMP 5 guide issued by ISPE for a useful reference.

Electronic Records

When it comes to the regulated industries such as the medical device industry, every process and procedure must be documented. Documentation ensures that everyone is working in the same manner with the same procedures. However, documentation is more than just writing down procedures and processes. It is also concerned with how documents are controlled, how they are updated and how they are stored.

Electronic Document management systems

Electronic document management systems aka EDMS are now the norm and gold standard for most medium to large organisations. Many companies that provide medical device manufacturers with an EDMS that can be customised to match the business processes particular to an organisation. With configurable or customisable software, validation and proper verification is important to ensure the system operates as intended. There are also regulatory requirements that stipulate the expectations and requirements of such systems. For example, the application of electronic signatures and the presence of audit trials. FDA 21 CFR Part 11 details the requirements with regard to electronic records and electronic signatures. For medicinal products in Europe, GMP V4 Annex 11 specifies similar requirements.

Record Retention

With regard to the part 11 requirements for the protection of records to enable their accurate and ready retrieval throughout the records retention period (11.10 (c)), persons must also comply with all applicable predicate rule requirements for record retention and availability such as (211.180(c) general requirements. The decision to follow 21 CFR part 11 should be justified and documented as part of a risk assessment and based on the value of the records over time. The FDA does not object to archiving of required records in electronic format to non-electronic media such as paper, or to a standard electronic file format (examples of such formats include, but are not limited to, PDF, XML, or SGML). Persons must still comply with all predicate rule requirements, and the records themselves and any copies of the required records should preserve their content and meaning. As long as predicate rule requirements are fully satisfied and the content and meaning of the records are preserved and archived, you can delete the electronic version of the records. In addition, paper and electronic record and signature components can coexist as long as predicate rule requirements are met and the content and meaning of those records are preserved.

Electronic Signatures

Electronic signatures are computer-generated character strings that count as the legal equivalent of a handwritten signature. The regulations for the use of electronic signatures are set out in 21 CFR Part 11 of the FDA. Each electronic signature must be assigned uniquely to one person and must not be used by any other person. It must be possible to confirm to the authorities that an electronic signature represents the legal equivalent of a handwritten signature. Electronic signatures can be biometrically based or the system can be set up without biometric features.

Conventional Electronic Signatures

If electronic signatures are used that are not based on biometrics, they must be created so that persons executing signatures must identify themselves using at least two identifying components. This also applies in all cases in which a chip card replaces one of the two identification components. These identifying components, can, for example consist of a user identifier and a password. The identification components must be assigned uniquely and must only be used by the actual owner of the signature.
When owners of signatures want to use their electronic signatures, they must identify themselves by means of at least two identification components. The exception to this rule is when the owner executes several electronic signatures during one uninterrupted session. In this case, persons executing signatures need to identify themselves with both identification components only when applying the first signature. For the second and subsequent signatures, one unique identification component (password) is then adequate identification.

Audit Trail

Title 21 CFR details predicate rule requirements relating to documentation of, for example, date time, or the sequencing of events, as well as any requirements for ensuring that changes to records do not obscure previous entries. Making the decision on whether to apply audit trails, or other appropriate measures, or on the need to comply with predicate rule requirements should involve a justified and documented risk assessment. Any risk assessment should determine the potential effect on product quality and safety and the integrity of the record.

Change Management

Validation programmes are subject to change control. Each company or organisation should have a procedure detailing the change management process.
Any system, facility, document or process that has the potential to impact product quality and the validated state is generally subject to following a change control process. Another term used in industry is enterprise change control or engineering change control. Essentially, these terms are the same. The intent is to control and manage change consistently.

A change control can take the form of a document which drives the agenda and the specific requirement. Change control is also created with enterprise software such as Kintana, Documentum and SAP. While each company will have varying processes, some basics are common. These include the three stages of change control; pre-implementation, implementation and post implementation (if required).

Validation Deliverables

The deliverables of validation activities should be in accordance with a project validation plan of validation master plan. For small projects or changes to computerised systems, a change control may serve as the validation plan. However, some typical deliverables include the following:

- ➤ GxP assessment (note, some systems may be non GxP applicable)
- ➤ User requirements specification
- ➤ Third party audit
- ➤ Validation plan
- ➤ Design specification such as functional, software, hardware and technical specifications
- ➤ GxP risk assessment
- ➤ Validation protocols
- ➤ Traceability matrix
- ➤ Validation report

Part 3

-MEDICAL DEVICES-

3.1 Device Classification

The manufacturer, in preparing for CE marking, should first determine if their product falls within the scope of the directive or national regulation, either as a medical device or as an accessory to a medical device, as defined in Article 1 of directive 93/42/EEC and Article 2 of the regulation. In order to be classified as a medical device, the product should have a medical purpose and its primary mode of action will typically be physical.

Level of Risk

General medical devices and related accessories must be classified into one of four classes, which are based on the perceived risk of the device to the patient or user. The classification of a device determines the conformity assessment options that are applicable to the device, with higher risk devices undergoing higher levels of assessment.

Class	Risk level
I	Low Risk
IIa	Medium Risk
IIb	Higher Risk
III	Highest Risk

Classification Rules

There are eighteen rules outlined in Annex IX of the directive and related regulation that lay down the basic principles of classification. In MEDDEV 2.4/1 Rev. 8, these rules are further explained and descriptive examples are provided. The eighteen rules are subdivided into four groups as follows:

Rules	Device Type
Rules 1 – 4	Non-invasive Devices
Rules 5 – 8	Invasive Devices
Rules 9 – 12	Active Devices
Rules 13-18	Special Rule e.g. devices containing tissue of animal origin, drug-device combinations

Annex IX and related guidance documents outline a number of key characteristics, listed below, that must be considered to correctly classify a device using the eighteen classification rules:

General Principles of Device Classification

- Medical devices are defined as articles which are intended to be used for a medical purpose. It is the intended purpose that determines the class of device and not the particular technical characteristics of the device. The intended purpose of the device should be substantiated (if required) and be representative of the technical characteristics of the device.
- It is the intended and not the accidental use of the device that determines its class.

o   It is the intended purpose assigned by the manufacturer to the device that determines the class of device and not the class assigned to other similar products.

o   Accessories are classified separately from their parent device.

o   The mode of action of a medical device should be clear and evidenced with appropriate data to confirm this mode of action.

o   If the device can be classified according to several rules then the highest possible class applies.

o   Multipurpose equipment which may be used in combination with medical devices are not themselves classed as medical devices unless the manufacturer places them on the market with the specific intended purpose as a medical device.

o   If the device is not intended to be used solely or principally in a specific part of the body, it must be considered and classified on the basis of the most critical specified use.

Summary Of Rules

(Source: Guidelines Relating To The Application Of

The Council Directive 93/42/EEC On Medical Devices, MEDDEC 2.4/Rev.9 June 2010)

Rule 1

Rule 1: All non-invasive devices are in Class I, unless one of the other 17 rules apply. This is a fallback rule applying to all devices that are not covered by a more specific rule.

This is a rule that applies in general to devices that come into contact only with intact skin or that do not touch the patient.

Some non-invasive devices are indirectly in contact with the body and can influence internal physiological processes by storing, channeling or treating blood, other body liquids or liquids which are returned or infused into the body or by generating energy that is delivered to the body. These must be excluded from the application of this rule and be handled by another rule because of the hazards inherent in such indirect influence on the body.

Rule 2

Rule 2: All non-invasive devices are in Class I, unless one of the other 17 rules apply.

These types of devices must be considered separately from the non-contact devices of Rule 1 because they may be indirectly invasive. They channel or store substances that will eventually be administered to the body. Typically these devices are used in transfusion, infusion, extracorporeal circulation and delivery of anaesthetic gases and oxygen.

In some cases devices covered under this rule are very simple gravity activated delivery devices.

Rule 2: All non-invasive devices intended for channelling or storing blood, body liquids or tissues, liquids or gases for the purpose of eventual infusion, administration or introduction into the body are in Class IIa:

- if they may be connected to an active medical device in Class IIa or a higher class,

-if they are intended for use for storing or channelling blood or other body liquids or for storing organs, parts of organs or body tissues.

- in all other cases they are in Class I.

Rule 3

Rule 3: Non-invasive devices that modify biological or chemical composition of blood, body liquids or other liquids intended for infusion into the body.

These types of devices must be considered separately from the non-contact devices of Rule 1 because they are indirectly invasive. They modify substances that will eventually be infused into the body. This rule covers mostly the more sophisticated elements of extracorporeal circulation sets, dialysis systems and autotransfusion systems as well as devices for extracorporeal treatment of body fluids which may or may not be immediately reintroduced into the body, including, where the patient is not in a closed loop with the device.

Rule 3: All non-invasive devices intended for modifying the biological or chemical composition of blood, other body liquids or other liquids intended for infusion into the body are in Class IIb,

unless the treatment consists of filtration, centrifugation or exchange of gas or heat, in which case they are in Class IIa.

These devices (Rule 3) are normally used in conjunction with an active medical device covered under Rule 9 or Rule 11.

Filtration and centrifugation should be understood in the context of this rule as exclusively mechanical methods.

Rule 4

Rule 4: Non-invasive devices which come into contact with injured skin. This rule is intended to primarily cover wound dressings independently of the depth of the wound. The traditional types of products, such as those used as a mechanical barrier, are well understood and do not result in any great hazard. There have also been rapid technological developments in this area, with the emergence of new types of wound dressings for which non-traditional claims are made, e.g. management of the micro-environment of a wound to enhance its natural healing mechanism.

More ambitious claims relate to the mechanism of healing by secondary intent, such as influencing the underlying mechanisms of granulation or epithelial formation or preventing contraction of the wound. Some devices used on breached dermis may even have a life-sustaining or lifesaving purpose, e.g. when there is full thickness destruction of the skin over a large area and/or systemic effect. Dressings containing medicinal products which act ancillary to the dressing fall within Class III under Rule 13.

Rule 4: All non-invasive devices which come into contact with injured skin:

- are in Class I if they are intended to be used as a mechanical barrier, for compression or for absorption of exudates,

- are in Class IIb if they are intended to be used principally with wounds which have breached the dermis and can only heal by secondary intent.

Products covered under this rule are extremely claim sensitive, e.g. a polymeric film dressing would be in Class IIa if the intended use is to manage the micro-environment of the wound or in Class I if its intended use is limited to retaining an invasive cannula at the wound site. Consequently it is impossible to say a priori that a particular type of dressing is in a given class without knowing its intended use as defined by the manufacturer. However, a claim that the device is interactive or active with respect to the wound healing process usually implies that the device is in Class IIb.

Most dressings that are intended for a use that is in Class IIa or IIb, also perform functions that are in Class

I, e.g. that of a mechanical barrier. Such devices are nevertheless classed according to the intended use in the higher class.

For such devices incorporating a medicinal product or a human blood derivative see Rule 13 or animal tissues or derivatives rendered non-viable see Rule 17.

Rule 5

Rule 5: Devices invasive with respect to body orifices.

Invasiveness with respect to the body orifices (ear, mouth, nose, eye, anus, urethra and vagina) must be considered separately from invasiveness that penetrates through a cut in the body surfaces (surgical invasiveness). For short term use, a further distinction must be made between invasiveness with respect to the less vulnerable anterior parts of the ear, mouth and nose and the other anatomical sites that can be accessed through natural body orifices.

Surgically created stoma, which for example allows the evacuation of urine or faeces, should also be considered as a body orifice.

Devices covered by this rule tend to be diagnostic and therapeutic instruments used in particular specialities (ENT, ophthalmology, dentistry, proctology, urology and gynaecology).

Rule 5: All invasive devices with respect to body orifices, other than surgically invasive devices and which are not intended for connection to an active medical device or which are intended for connection to an active medical device in Class I:

- are in Class I if they are intended for transient use,

- are in Class IIa if they are intended for short term use

except if they are used in the oral cavity as far as the pharynx, in an ear canal up to the ear drum or in a nasal cavity , in which case they are in Class I,

- are in Class IIb if they are intended for long term use,

except if they are used in the oral cavity as far as the pharynx, in an ear canal up to the ear drum or in a nasal cavity and are not liable to be absorbed by the mucous membrane, in which case they are in Class IIa.

All invasive devices with respect to body orifices, other than surgically invasive devices, intended for connection to an active medical device in Class IIa or a higher class, are in Class IIa.

Rule 6

Rule 6: Surgically invasive devices intended for transient use (< 60 minutes)

This rule primarily covers three major groups of devices: devices that are used to create a conduit through the skin (needles, cannulae, etc.), surgical instruments (scalpels, saws, etc.) and various types of catheters, suckers, etc.

This rule primarily covers three major groups of devices: devices that are used to create a conduit through the skin (needles, cannulae, etc.), surgical instruments (scalpels, saws, etc.) and various types of catheters, suckers, etc.

Rule 6: All surgically invasive devices intended for transient use are in Class IIa unless they are:

-intended specifically to control, diagnose, monitor or correct a defect of the heart or of the central circulatory system through direct contact with these parts of the body, in which case they are in Class III

-reusable surgical instruments, in which case they are in Class I

-intended specifically for use in direct contact with the central nervous system, in which case they are in Class III,

- intended to supply energy in the form of ionising radiation in which case they are in Class IIb,

- intended to have a biological effect or to be wholly or mainly absorbed in which case they are in Class IIb,

- intended to administer medicines by means of a delivery system, if this is done in a manner that is potentially hazardous taking account of the mode of application, in which case they are Class IIb.

Rule 7

Rule 7: Surgically invasive devices intended for short-term use (>60 minutes, <30 days).

These are mostly devices used in the context of surgery or post-operative care (e.g. clamps, drains), infusion devices (cannulae, needles) and catheters of various types.

Rule 7: All surgically invasive devices intended for short term use are in Class IIa unless they are intended:

- either specifically to control, diagnose, monitor or correct a defect of the heart or of the central circulatory system through direct contact with these parts of the body, in which case they are in Class III,

- or specifically for use in direct contact with the central nervous system, in which case they are in Class III,

- or to supply energy in the form of ionising radiation in which case they are in Class IIb,

- intended to have a biological effect or to be wholly or mainly absorbed in which case they are in Class III,
- or to undergo chemical change in the body, except if the devices are placed in the teeth, or to administer medicines, in which case they are Class IIb.

Rule 8

Rule 8: Implantable devices and long-term surgically invasive devices (> 30 days). These are mostly implants in the orthopaedic, dental, ophthalmic and cardiovascular fields as well as soft tissue implants such as implants used in plastic surgery.

Rule 8: All implantable devices and long-term surgically invasive devices are in Class IIb unless they are intended:

- to be placed in the teeth, in which case they are in Class IIa,

- to be used in direct contact with the heart, the central circulatory system or the central nervous system, in which case they are Class III,

- to have a biological effect or to be wholly or mainly absorbed, in which case they are in Class III,

- or to undergo chemical change in the body, except if the devices are placed in the teeth, or to administer medicines, in which case they are in Class III.

- Directive 2003/12/EC introduced a derogation from this rule, reclassifying breast implants in Class III Directive 2005/50/EC introduced a derogation from this rule, reclassifying hip, knee and shoulder joint replacements in Class III.

Rule 9

Rule 9: Active therapeutic devices intended to administer or exchange energy.

Devices classified by this rule are mostly electrical equipment used in surgery such as lasers and surgical generators. In addition there are devices for specialised treatment such as radiation treatment. Another category consists of stimulation devices, although not all of them can be considered as delivering dangerous levels of energy considering the tissue involved.

Rule 9: All active therapeutic devices intended to administer or exchange energy are in Class IIa unless their characteristics are such that they may administer or exchange energy to and from the human body in a potentially hazardous way, taking account of the nature, the density and the site of application of the energy, in which case they are in Class IIb. All active devices intended to control or monitor the performance of active therapeutic devices in Class IIb or intended to influence directly the performance of such devices are in Class IIb.

Rule 10

Rule 10: Active devices for diagnosis. This primarily covers a whole range of widely used equipment in various fields, e.g. ultrasound diagnosis, capture of physiological signals and therapeutic and diagnostic radiology.

Rule 10: Active devices intended for diagnosis are in Class IIa:

- if they are intended to supply energy which will be absorbed by the human body, except for devices used to illuminate the patient's body, in the visible spectrum,

- if they are intended to image in vivo distribution of radiopharmaceuticals,

- if they are intended to allow direct diagnosis or monitoring of vital physiological processes,

unless they are specifically intended for monitoring of vital physiological parameters, where the nature of variations is such that it could result in immediate danger to the patient, for instance variations in cardiac performance, respiration, activity of CNS in which case they are in Class IIb.

Active devices intended to emit ionising radiation and intended for diagnostic and therapeutic interventional radiology including devices which control or monitor such devices, or which directly influence their performance, are in Class IIb.

Rule 11

Rule 11: Active devices intended to administer and/or remove medicines, body liquids or other substances to or from the body. This rule is intended to primarily cover drug delivery systems and anaesthesia equipment.

Rule 11: All active devices intended to administer and/or remove medicines, body liquids or other substances to or from the body are in Class IIa, unless this is done in a manner:

- that is potentially hazardous, taking account of the nature of the substances involved, of the part of the body concerned and of the mode of application, in which case they are in Class IIb.

Rule 12

Rule 12: All other active devices. This is a fall-back rule to cover all active devices not covered by the previous rules.

Rule 12: All other active devices are in Class I

Special Rules 12-18

Rule 13: Devices incorporating, as an integral part, a medicinal product or a human blood derivative (See MEDDEV. 2.1/3 for further guidance).

Rule 14: Devices used for contraception or prevention of sexually transmitted diseases.

Rule 15: Specific disinfecting, cleaning and rinsing devices.

Rule 16: Devices to record X-ray diagnostic images.

Rule 17: Devices utilising animal tissues or derivatives.

Rule 18: Blood bags.

3.2 Design Controls

Introduction

Design controls are an important component of the FDAs Quality System Regulation, 21 CFR Part 820. Design controls apply to a wide variety of devices with varying levels of complexity. the regulation does not prescribe the practices that must be used, rather it establishes a framework that manufacturers must use when developing and implementing design controls. Such requirements much be appropriate to ensure that regulation allows design controls to be flexible enough to meet individual manufacturers own design and development processes.

Design controls are a collection of practices and procedures that are incorporated into the design and development process for a product such as a medical device. Based upon quality assurance and engineering principles, they provides a structure and clear path from user needs assessment to product delivery through a step-by-step process. Design controls ensure proper assessment of the design is completed during the design and development phase. It highlights technical issues, conflicts or deficiencies in design input requirements and allows them to be addressed early on in the process. Fixing a design issue early on reduces the cost of doing so at a later point and ensures the resultant design is appropriate for its intended use. Bringing a formal review process (design control) to the table assists engineers and managers in engaging with decisions and understanding them better. It also ensures that when future changes are made, they are documented and reviewed adequately with proper consideration to the design inputs.

Design controls are a requirement of quality systems such as 21 CFR Part 820 (medical devices), and for certain classes of devices and per ISO 13485 - Quality Management Systems.

Benefits of Design Control:

- The intended use of the device is documented and approved
- It ensures inputs align with outputs
- Problems with designs or manufacturability are recognised earlier
- It creates a design "standard" and a "process" to allow benchmarking and consistency within an organisation

Design controls increase the likelihood that the design transferred to production will translate into a device that is appropriate for its intended use.

Design controls allow advanced visibility of the design process. Allowing engineers and managers to respond to problems earlier.

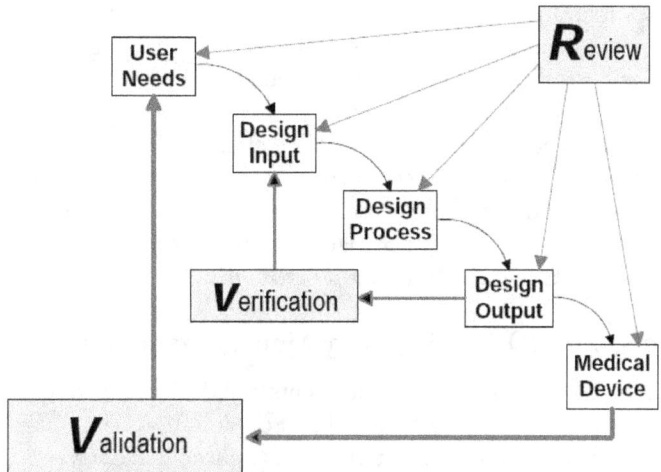

Figure -Design Controls represented via Waterfall Design Process (Ref. Design Control Guidance For medical Device Manufacturers, Mar. 1997 / Medical Devices Bureau, Health Canada)

The development process depicted in the example is a traditional waterfall model. The design proceeds in a logical sequence of phases or stages. Basically, requirements are developed, and a device is designed to meet those requirements. The design is then evaluated, transferred to production, and the device is manufactured. In practice, feedback paths would be required between each phase of the process and previous phases, representing the iterative nature of product development. However, this detail has been omitted from the figure to make the influence of the design controls on the design process more distinct.

The importance of the design input and verification of design outputs is illustrated by this example. When the design input has been reviewed and the design input requirements are determined to be acceptable, an iterative process of translating those requirements into a device design begins. The first step is conversion of the requirements into system or high-level specifications. Thus, these specifications are a design output. Upon verification that the high-level specifications conform to the design input requirements, they become the design input for the next step in the design process, and so on.

This basic technique is used repeatedly throughout the design process. Each design input is converted into a new design output; each output is verified as conforming to its input; and it then becomes the design input for another step in the design process. In this manner, the design input requirements are translated into a device design conforming to those requirements.

Concurrent Engineering

Although the waterfall model is a useful tool for introducing design controls, its usefulness in practice is limited. The model does apply to the development of some simpler devices. However, for more complex devices, a concurrent engineering model is more representative of the design processes in use in the industry.

In a traditional waterfall development scenario, the engineering department completes the product design and formally transfers the design to production. Subsequently, other departments or organizations develop processes to manufacture and service the product. Historically, there has frequently been a divergence between the intent of the designer and the reality of the factory floor, resulting in such undesirable outcomes as low manufacturing yields, rework or redesign of the product, or unexpectedly high cost to service the product.

One benefit of concurrent engineering is the involvement of production and service personnel throughout the design process, assuring the mutual optimization of the characteristics of a device and its related processes. While the primary motivations of concurrent engineering are shorter development time and reduced production cost, the practical result is often improved product quality.

Concurrent engineering encompasses a range of practices and techniques. From a design control standpoint, it is sufficient to note that concurrent engineering may blur the line between development and production. On the one hand, the concurrent engineering model properly emphasizes that the development of production processes is a design rather than a manufacturing activity. On the other hand, various components of a design may enter production before the design as a whole has been approved.

Design Controls and ISO 13485 Quality Management System

Clause 7 of ISO 13485 specifies the requirements for design and development of devices as part of the product realisation process. It should be noted that organisations can opt to exclude specific requirements of ISO 13485, in cases where product realisation is not applicable. However, any such exclusion should be based on sound rationale with the technical case clearly documented. An example of this may be where design and development are not conducted by the manufacturer e.g. contract manufacturers.

Clause 7 (product realisation) of ISO 13485 details requirements for design and development controls. Clause 7 includes the following subparts:

Clause 7.1 Planning of product realisation
Clause 7.2 Customer-related processes
Clause 7.3 Design and development
Clause 7.4 Purchasing
Clause 7.5 Production and service provision
Clause 7.6 Control of measuring devices

Section 7.3 (Design and development) comprises:

Clause 7.3.1 Design and Development Planning
Clause 7.3.2 Design and Development Inputs
Clause 7.3.3 Design and Development Outputs
Clause 7.3.4 Design and Development Review
Clause 7.3.5 Design and Development Verification
Clause 7.3.6 Design and Development Validation
Clause 7.3.7 Control of Design and Development Changes

Definitions

Change Management: a management process where changes to the product, process, facilities or utilities are assessed, planned and reviewed as part of a formal systematic process.

Corrective and Preventative Action (CAPA): when an unplanned or adverse event happens, a corrective and preventative action can be implemented.

Design Phase Review: a process of evaluating the design requirements against the ability of it to deliver the intended device.

Design History File (DHF): an approved list of records that describe the design history of a medical device.

Design Input: the physical and performance requirements of a device that are the basis for the device design.

Design Output: the results of a design effort at each design phase and at the end of the total design effort. The finished design output is the basis for the device master record. The total finished design output consists of the device, its packaging and labelling, and the device master record.

Design Verification: confirmation by examination and provision of objective evidence that specified requirements have been fulfilled.

Design Validation: establishing by objective evidence that device or product specifications conform to user needs and intended use(s) defined in design documentation.

Device Master Record (DMR): a compilation of records containing the procedures and specification for a device. The contents of a DMR can contain local procedures such as SOPs and work instructions along with global or divisional specifications used to detail manufacturing processes, intermediate product or final product.

Design Phase Review: a documented, comprehensive, systematic examination of a design to evaluate the adequacy of the design requirements, the capability of the design to meet those requirements and to identify problems.

Specification: specification means any requirement to which a product, process, service, or other activity must conform.

Validation: validation means confirmation by examination and provision of objective, documented evidence that the particular requirements for a specific intended use can be consistently fulfilled.

Application of Design Controls

Design controls can be applied to any product development process. When the design input has been reviewed and the design input requirements are determined to be acceptable, the process of creating the device design begins. Product specifications are drafted and compared to the design input requirements. They then become the input for the next step in the design process. In the development and drafting of product specifications (e.g. critical quality attributes etc.) due regard must be given to product standards and industry best practices such as ISO and ASTM bodies. For example a catheter manufacturer should develop products with reference to ISO 10555 - intravascular catheters - sterile and single.

The Phase Approach to Design Control

The term "phase approach" is often used when describing the design control process. It simply means that a sequence of tasks needs to be completed, reviewed and approved during the development cycle of a product or medical device. Tasks are grouped into phases or stages. At the beginning, technical issues relating to design input requirements may need to be addressed with solutions identified. Often a range of solutions can be available, utilising different technologies. These different solutions then go on to be reviewed at the design selection process. At design selection, the project team must choose and justify a particular solution. The next phase (such as design verification and validation) ensures that the design is transferred to product launch and commercial supply - no oversights or deviations in the design intent occur. It also ensures that the device meets the user needs and intended uses (design inputs).

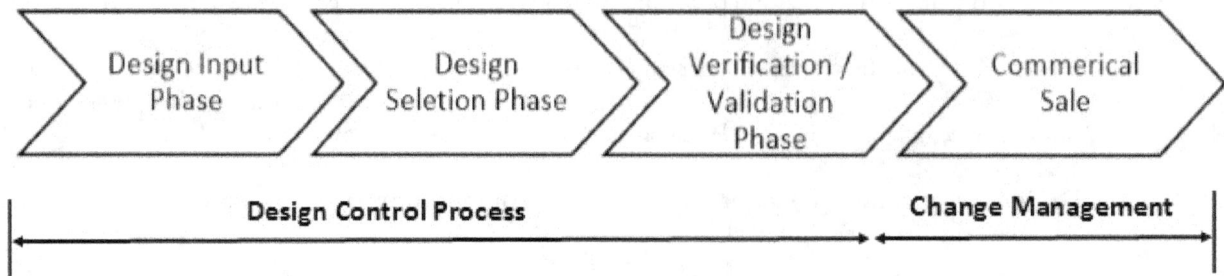

Figure: The above diagram depicts a typical product development process.

Phase Reviews Explained

A phase review is a process of evaluating the progress against the goals and activities of a particular phase. The phase review is typically completed at the end of each phase, but there may be a need to complete interim reviews for long or complex projects. For example, a design phase review is completed to ensure that the design input requirements make sense before they are interpreted into design specifications (design inputs phase review).

Managing Change

Changes made during design control are managed via document control procedures. For products built for commercial sale, the change management process is used to document and manage changes to the validated state of the process or the design of the product itself. While there may be more "flexibility" to make changes during the design phase of a project, diligence must be applied to any proposed change. Changes should be assessed by a multidisciplinary team with a management review.

Risk Management

Risk management involves the systematic application of management policies, practices and procedures that identify, analyse, control and monitor risk.

It is important to recognise that risk management should begin at the outset of the design and development phase of a project. The first step is to identify the user needs and intended use and application of the device. At the design input phase and design selection phase, risk assessments should be in a mature state. This allows the review of potential risks relating to the design of the product. Unacceptable risks can be dealt with by means of revisiting the design or introducing controls or mitigations in order to reduce the risks to acceptable levels. Following on from the design and development phase, the design verification, validation and transfer phases, or the clinical readiness phase, risk management activities and acceptability of the residual risk become the focus and must be approved indicating acceptability. This is often referred to as communicated risk.

In order to apply a risk management strategy, a procedure or SOP on risk management is typically available within manufacturing companies. This should clearly describe the risk management process and the various risk assessment tools, their application and guidance on how to complete them. The content of any risk management procedure or SOP should align with ISO 14971:2007 Medical Devices - Application of Risk Management to Medical Devices. Controlled templates for PFMEAs etc. also bring consistency and continuity to the process.

Risk management begins with the development of the design input requirements. As the design evolves, new risks may become evident. To systematically identify and, when necessary, reduce these risks, the risk management process is integrated into the design process. In this way, unacceptable risks can be identified and managed earlier in the design process when changes are easier to make and less costly.

An example of this is an exposure control system for a general-purpose x-ray system. The control function was allocated to software. Late in the development process, risk analysis of the system uncovered several failure modes that could result in overexposure to the patient. Because the problem was not identified until the design was near completion, an expensive, independent, back-up timer had to be added to monitor exposure times.

The Quality System and Design Controls

In addition to procedures and work instructions necessary for the implementation of design controls, policies and procedures may also be needed for other determinants of device quality that should be considered during the design process. The need for policies and procedures for these factors is dependent upon the types of devices manufactured by a company and the risks associated with their use. Management with executive responsibility has the responsibility for determining what is needed.

Example of topics for which policies and procedures may be appropriate are:

risk management
device reliability
device durability
device maintainability
device serviceability
human factors engineering
software engineering
use of standards
configuration management
compliance with regulatory requirements
device evaluation (which may include third party product certification or approval)
clinical evaluations
document controls
use of consultants
use of subcontractors
use of company historical data

Design and Development Planning

It is the manufacturer's responsibility to establish and maintain plans that describe or reference the design and development activities and define responsibilities for implementation. The plans should identify and describe the interaction with different groups or activities that are part of the design and development process. The maintenance of plans to reflect an accurate state as the design and development progresses is also a key factor. The design and development planning is intended to be prospective in nature. It allows risks to be identified earlier and promotes timely delivery of projects.

Process Inputs and Outputs

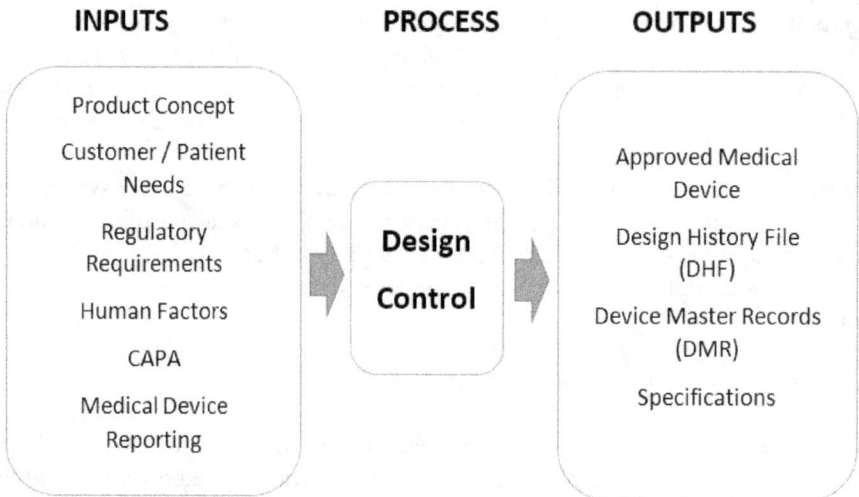

Design Input Phase

The aims of the Design Input Phase are to (1) define and document the user needs and the intended use of the medical device and (2) translate user needs and the intended use of the packaged device into design input requirements. (E.g. engineering specifications and the product requirements specifications.)

The typical documents required when establishing design inputs include:

- The creation of a formal design description detailing the intended use, user requirements and design inputs. (Note: the design description must align with the design input requirements.)
- A design and development plan which provides an estimation of timelines, resources required, responsibilities, project risks and scope of the project.
- Initial risk assessment which contains the user, design and component risks to be mitigated.
- Design concepts and technology overview.
- Business case report addressing the market size and market opportunity.

FDA 21 CFR Requirements – Design Input

21 CFR Part 820.30(C) Design Input

- *Each manufacturer shall establish and maintain procedures to ensure that the design requirements relating to a device are appropriate and address the intended use of the device, including the needs of the user and patient.*

- *The procedures shall include a mechanism for addressing incomplete, ambiguous, or conflicting requirements.*

- *The design input requirements shall be documented and shall be reviewed and approved by designated individuals.*

Incomplete requirements can have a serious and costly effect on the design and ultimate success of a product. If essential design requirements are omitted in error or otherwise, the impact on quality or functionality may not be detected until validation. This presents an expensive problem that may not be easily rectified. If design requirements are missed, a redesign may be necessary before a design can be released to production, thus causing delays to the project. Furthermore, if modifications are required to tooling, or process equipment, timelines can be impacted greatly. However, the safety and quality of the product must be paramount. Keeping one eye on the user requirements and intended use of the product is an important factor in avoiding gross design requirement failings.

What Is Design Input?

An artist's impression or concept documents do not meet the true intent of design input requirements. The purpose of design input is to create a *complete* set of requirements that are written in a technical manner with an engineering and scientific level of detail. The use of qualitative terms in a concept document is both appropriate and practical. This is often not the case for a document to be used as a basis for design. The language used in the creation of Design inputs also has a profound impact on the direction and scope of a product. If a concept document describes the product to be suitable for "outside use", then there will be requirements with regards to insulation, water ingress and operating temperatures and so on.

Scope

Design input requirements must be comprehensive. This may be quite difficult for manufacturers who are implementing a system of design controls for the first time. Design input requirements fall into three categories with most products having requirements within all three categories including:

(1) Functional requirements detailing the operation of the device.
(2) Performance requirements detailing the performance requirements or expectations of the device in relation to accuracy, speed of response times, battery life, device safety and reliability etc.
(3) Interface requirements specifying features of the device which are critical to compatibility with external systems such as the patient interface.

The scope of design input work depends on the complexity of a device and the risk associated with its use.

Tips for Reviewing Design Input Requirements

The ultimate goal of the design input phase is to gain agreement and approve the requirements formally. At this point, the document is a controlled document and subject to change control. Any updates required at a later date will need to be done through the change control process.

Design Input Requirements Should Be Crystal Clear: For example, a medical device may require use of a built-in battery. It would be important to specify the life expectancy of the battery. To say it has an approximate operating life of 2-3 years is too vague. A better description would be to say it has 2000 hours of operation with a software requirement that logs the number of hours the device is powered on. This mitigates the likelihood of failure during use.

Use of Tolerances: For example, a contact lens may have an outer diameter of 14.00mm. While this is the target/nominal value it cannot be ever accurately achieved. There will always be a degree of variation in the diameter measurement. Applying a tolerance, allows an acceptable range in which the measurement is within specification and accepted. If the diameter is specified as 14.00±0.2mm, designers have a basis for determining how accurate the manufacturing processes have to be. In addition, the specification will allow designers determine if the design meets the intended use.

Industry Standards: Design input requirements should meet or exceed industry standards. Compliance to product specific standards should be considered.

Environment: The operating environment of the device should be specified. Take the example of a cardiac defibrillator. If the device is intended for use on a frontline ambulance it may be used outdoors in cold and damp conditions. On the other hand, use within a hospital setting would require greater control of the temperature range and environmental conditions.

Design input is the starting point for product design. The requirements which form the design input establish a basis for performing subsequent design tasks and validating the design. Therefore, development of a solid foundation of requirements is the single most important design control activity.Many medical device manufacturers have experience with the adverse effects that incomplete requirements can have on the design process. A frequent complaint of developers is that "there's never time to do it right, but there's always time to do it over." If essential requirements are not identified until validation, expensive redesign and rework may be necessary before a design can be released to production.

Research And Development

Some manufacturers have difficulty in determining where research ends and development begins. Research activities may be undertaken in an effort to determine new business opportunities or basic characteristics for a new product. It may be reasonable to develop a rapid prototype to explore the feasibility of an idea or design approach, for example, prior to developing design input requirements. But manufacturers should avoid falling into the trap of equating the prototype design with a finished product design. Prototypes at this stage lack safety features and ancillary functions necessary for a finished product and are developed under conditions which preclude adequate consideration of product variability due to manufacturing.

What is the scope of the design input requirements development process and how much detail must be provided? The scope is dependent upon the complexity of a device and the risk associated with its use. For most medical devices, numerous requirements encompassing functions, performance, safety, and regulatory concerns are implied by the application. These implied requirements should be explicitly stated, in engineering terms, in the design input requirements.

There are many cases when it is impractical to establish every functional and performance characteristic at the design input stage. But in most cases, the form of the requirement can be determined, and the requirement can be stated with a to-be-determined (TBD) numerical value or a range of possible values. This makes it possible for reviewers to assess whether the requirements completely characterize the intended use of the device, judge the impact of omissions, and track incomplete requirements to ensure resolution.

For complex designs, it is not uncommon for the design input stage to consume as much as thirty percent of the total project time. Unfortunately, some managers and developers have been trained to measure design progress in terms of hardware built, or lines of software code written. They fail to realize that building a solid foundation saves time during the implementation. Part of the solution is to structure the requirements documents and reviews such that tangible measures of progress are provided.

Assessing Design Input Requirements For Adequacy

Eventually, the design input must be reviewed for adequacy. After review and approval, the design input becomes a controlled document. All future changes will be subject to the change control procedures. Any assessment of design input requirements boils down to a matter of judgment.

Design input requirements should be unambiguous. That is, each requirement should be able to be verified by an objective method of analysis, inspection, or testing. For example, it is insufficient to state that a catheter must be able to withstand repeated flexing. A better requirement would state that the catheter should be formed into a 50 mm diameter coil and straightened out for a total of fifty times with no evidence of cracking or deformity. A qualified
reviewer could then make a judgment whether this specified test method is representative of the conditions of use.

Quantitative limits should be expressed with a measurement tolerance. For example, a diameter of 3.5 mm is an incomplete specification. If the diameter is specified as 3.500±0.005 mm, designers have a basis for determining how accurate the manufacturing processes have to be to produce compliant parts, and reviewers have a basis for determining whether the parts will be suitable for the intended use.

The set of design input requirements for a product should be self-consistent. It is not unusual for requirements to conflict with one another or with a referenced industry standard due to a simple oversight. Such conflicts should be resolved early in the development process. The environment in which the product is intended to be used should be properly characterized. For example, manufacturers frequently make the mistake of specifying "laboratory" conditions for devices which are intended for use in the home. Yet, even within a single country, relative humidity in a home may range from 20 percent to 100 percent (condensing) due to climactic and seasonal variations. Household temperatures in many climates routinely exceed 40 °C during the hot season. Altitudes may exceed 3,000 m, and the resultant low atmospheric pressure may

Design Output Phase

The purpose of the design selection(output) phase is to provide a range of design options and solutions with the relevant evidence to show the effectiveness of the same. Often proof of concept (POC) or proof of principle (POP) trials may be used to verify effectiveness of solutions. POC/POP testing can involve making some limited prototypes. Any documents created in the previous phase, design input, should be reviewed and updated if required. There should be no contradictions or gaps between the documented inputs and outputs.

FDA 21 CFR Requirements – Design Output

21 CFR Part 820.30(D) Design Output

- *Each manufacturer shall establish and maintain procedures for defining and documenting design output in terms that allow an adequate evaluation of conformance to design input requirements.*

- *Design output procedures shall contain or make reference to acceptance criteria and shall ensure that those design outputs that are essential for the proper functioning of the device are identified.*

- *Design output shall be documented, reviewed, and approved before release.*

- *The approval, including the date and signature of the individual(s) approving the output, shall be documented.*

During this phase, product specifications (PS) and the device master record (DMR) are generated to define the design output. Planning for process validation and manufacturing begins during this phase often with the creation of a validation master plan (VMP). In any design office or factory setting, a lot of data and paperwork are generated. Therefore, it is important to be able to make the distinction between what is a design output and what is not. The first way of identifying a design output is to verify if it is listed as a task, a deliverable or listed in the design and development plan. If this is the case, then it is classified as a design output. Furthermore, if it describes or defines a design feature, it can also be classed as a design output.

The quality system requirements for design output can be separated into two elements: Design output should be expressed in terms that allow adequate assessment of conformance to design input requirements and should identify the characteristics of the design that are crucial to the safety and proper functioning of the device. This raises two fundamental issues for developers:

What constitutes design output?
Are the form and content of the design output suitable?

The first issue is important because the typical development project produces voluminous records, some of which may not be categorized as design output. On the other hand, design output must be reasonably comprehensive to be effective. As a general rule, an item is design output if it is a work product, or deliverable item, of a design task listed in the design and development plan, and the item defines, describes, or elaborates an element of the design implementation. Examples include block diagrams, flow charts, software high-level code, and system or subsystem design specifications. The design output in one stage is often part of the design input in subsequent stages. Design output includes production specifications as well as descriptive materials which define and characterize the design.

Production Specifications

Production specifications draw upon many documents that are used to manufacture, test, inspect, install, maintain and service a device. They include: (1) component and material specifications, (2) production and process specifications, (3) work instructions and SOPs, (4) quality plans, specifications and procedures, (4) labelling specifications, and (5) packaging specifications.

Design Review

Formal design reviews are critical to the efficacy of design control, and ultimately, the market success of the device. They should be planned for up front in the design development plan. Changes late in the design cycle are much more expensive than those made early on. Design reviews can play an important role in identifying changes in a timely manner and thus prevent costly redesigns close to the launch date. The FDA QSR clearly specifies the need for independent reviewers. Independent reviewers must be far enough removed from the design in order to provide an objective review.

FDA 21 CFR Requirements- Design Review

FDA CFR Part 820.30(E) Design review

- *Each manufacturer shall establish and maintain procedures to ensure that formal documented reviews of the design results are planned and conducted at appropriate phases of the device's design development.*

- *The procedures shall ensure that participants at each design review include representatives of all functions concerned with the design phase being reviewed and an individual(s) who does not have direct responsibility for the design phase being reviewed, as well as any specialists needed.*

- *The results of a design review, including identification of the design, the date, and the individual(s) performing the review, shall be documented in the design history file (the DHF).*

Key goals of design review:

- provide feedback to designers on existing or emerging problems
- assess project progress
- provide confirmation that the project is ready to move on to the next phase of development

Many types of reviews occur during the course of developing a product. Reviews may have both an internal and external focus.

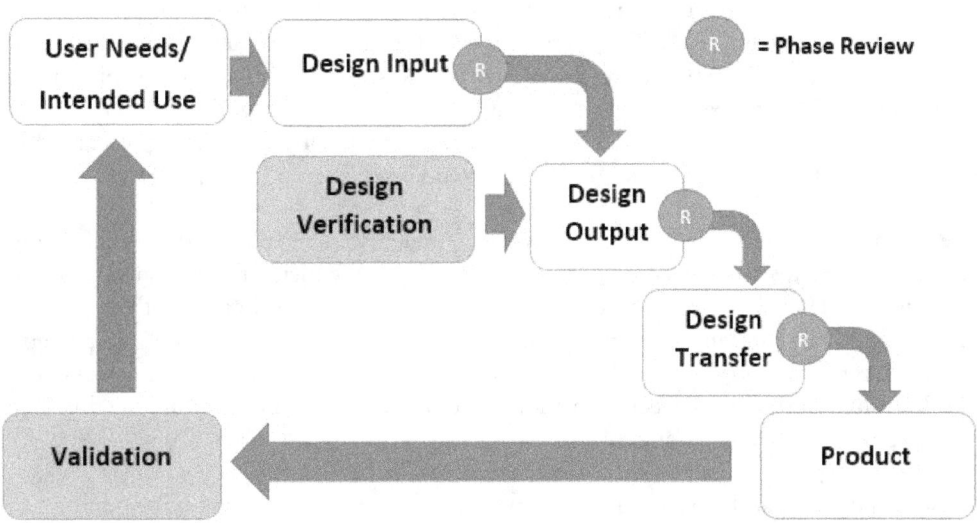

Reviews are important in ensuring that the input requirements are not forgotten as the project progresses. Secondly, there must be "agreement" between the user requirements and design inputs versus the design outputs.

A formal review of the design input requirements early in the development process is normally completed. The number of reviews depends upon the complexity of the device.

☐☐☐ For a simple design, or a minor upgrade to an existing product, it might be appropriate to conduct a single review at the conclusion of the design process.

☐☐☐ For a product involving multiple subsystems, an early design task is to allocate the design input requirements among the various subsystems. For example, in a microprocessor-based system, designers must decide which functions will be performed by hardware and which by software. In another case, tolerance buildup from several components may combine to create a clearance problem. System designers must establish tolerance specifications for each component to meet the overall dimensional specification. In cases like these, a formal design review is a prudent step to ensure that all such system-level requirements have been allocated satisfactorily prior to engaging in detailed design of each subsystem.

Reviewers

In determining who should participate in a formal design review, planners should consider the qualifications of reviewers and the types of expertise required. Often a matrix listing reviewers, functions and responsibilities is documented in a design control procedure or at the outset of a project.

Evaluation of the design

Many formal design reviews take the form of a meeting. At this meeting, the designer(s) may make presentations to explain the design implementation, and persons responsible for verification activities may present their findings to the reviewers. Reviewers may ask for clarification or additional information on any topic, and add their concerns to any raised by the presenters. This portion of the review is focused on finding problems, not resolving them. There are many approaches to conducting design review meetings. In simple cases, the technical assessor and reviewer may be the same person, often a project manager or engineering supervisor, and the review meeting is a simple affair in the manager's office. For more elaborate reviews, detailed written procedures are desirable to ensure that all pertinent topics are discussed, conclusions accurately recorded, and action items documented and tracked.

There is a dangerous tendency for design review meetings to become adversarial affairs. The reputation of the designers tends to be linked to the number of discrepancies found, causing the designers to become defensive, while the reviewers score points by finding weaknesses in the design.
The resulting contest can be counterproductive. An added complication is the presence of invited guests, often clinicians, who are expected to provide the user perspective. These reviewers are often very reluctant to ask probing questions, especially if they sense that they may become involved in a conflict where all the rules and relationships are not evident.

These difficulties can be avoided by stating the goals and ground rules for conducting the formal design review clearly at the outset. While the designers are in the best position to explain the best features of the design, they are also most likely to be aware of the design's weaknesses. If the designers and reviewers are encouraged to work together to systematically explore problems and find solutions, the resultant design will be improved and all parties will benefit from the process. Participants must be encouraged to ask questions, avoid making assumptions, and think critically. The focus must be on the design, not the participants. Not all formal design reviews involve meetings. For extremely simple designs or design changes, it may be appropriate to specify a procedure in which review materials are distributed or circulated among the reviewers for independent assessment and approval. However, such a procedure negates the benefits of synergy and teamwork and should be considered only in cases where the design issues are limited in scope and well defined.

Resolution of concerns

The reviewers consider concerns raised during the evaluation portion of the formal design review and decide on an appropriate disposition for each one. There is wide variation in the way companies implement decision-making processes. In some cases, the reviewers play an advisory role to the engineering manager or other company official, who directs the formal design review and ultimately selects a course of action. In other cases, the reviewers are given limited or broad authority to make decisions and commit resources to resolve problems. The approach used should be documented.

Implementation of corrective actions

Not all identified concerns result in corrective actions. The reviewers may decide that the issue is erroneous or immaterial. In most cases, however, resolution involves a design change, a requirements change, or a combination of the two. If the solution is evident, the reviewers may specify the appropriate corrective action; otherwise, an action item will be assigned to study the problem further. In any case, action items and corrective actions are normally tracked under the manufacturer's change control procedures.

Relationship Of Design Review To Verification And Validation

In practice, design review, verification, and validation overlap one another, and the relationship among them may be confusing. As a general rule, the sequence is: verification, review, validation, review. In most cases, verification activities are completed prior to the design review, and the verification results are submitted to the reviewers along with the other design output to be reviewed. Alternatively, some verification activities may be treated as components of the design review, particularly if the verification activity is complex and requires multidisciplinary review. Similarly, validation typically involves a variety of activities, including a determination that the appropriate verifications and reviews have been completed. Thus, at the conclusion of the validation effort, a review is usually warranted to assure that the validation is complete and adequate.

Design Verification, Validation and Transfer Phase

To illustrate the concepts, consider a building design. In a typical scenario, the senior architect establishes the design input requirements and sketches the general appearance and construction of the building, but contractors typically elaborate and interpret the details into practical terms. Verification refers to the checking at each phase to ensure the output meets the design requirements. For example, if a device is designed to take both AC electrical power and a battery (DC power), the design engineer must verify that these are accounted for in the plans and production specifications.

FDA 21 CFR Requirements - Design Verification

FDA CFR Part 820.30(f) Design Verification

- *Each manufacturer shall establish and maintain procedures for verifying the device design.*

- *Design verification shall confirm that the design output meets the design input requirements.*

- *The results of the design verification, including identification of the design, method(s), the date, and the individual(s) performing the verification, shall be documented in the Design History File.*

The ultimate aim of design verification is to finalise design specification. Examples of verification activities include:

- Design failure modes and effects analysis (DFMEA)
- Fault tree analysis
- Package integrity tests
- Biocompatibility testing
- Bioburden testing of packed products
- Worst case analysis – tolerance stacking of components

Design Validation

Design validation is required for the product to ensure the device meets the user requirements and intended use. Above all, it ensures the device operates reliably and safely. Process validation is required in order to confirm manufacturing specifications and the Device Master Record (DMR). However, process validation is separate to design control and is covered in *Chapter 6 - Process Validation.*

FDA 21 CFR Requirements - Design Validation

FDA CFR 820.30(G) Design Validation

- *Each manufacturer shall establish and maintain procedures for validating the device design.*

- *Design validation shall be performed under defined operating conditions on initial production units, lots, batches, or their equivalents.*

- *Design validation shall ensure that devices conform to defined user needs, intended uses and shall include testing of production units under actual or simulated use conditions.*

- *Design validation shall include software validation and risk analysis, where appropriate.*

- *The results of the design validation, including identification of the design, method(s), the date, and the individual(s) performing the validation, shall be documented in the design history file.*

Verification examines design outputs at the different phases of the process while design validation confirms that all user needs are achieved even when subject to anticipated sources of variation such as materials, processing equipment, suppliers and so on.

Validation Review

Validation may expose deficiencies in the original assumptions concerning user needs and intended uses. A formal review process should be used to resolve any such deficiencies. As with verification, the perception of a deficiency might be judged insignificant or erroneous, or a corrective action may be required.

Validation Methods

Many medical devices do not require clinical trials. However, all devices require clinical evaluation and should be tested in the actual or simulated use environment as a part of validation. This testing should involve devices which are manufactured using the same methods and procedures expected to be used for ongoing production. While testing is always a part of validation,
additional validation methods are often used in conjunction with testing, including analysis and inspection methods, compilation of relevant scientific literature, provision of historical evidence that similar designs And/or materials are clinically safe, and full clinical investigations or clinical trials.

Some manufacturers have historically used their best assembly workers or skilled lab technicians to fabricate test articles, but this practice can obscure problems in the manufacturing process. It may be beneficial to ask the best workers to evaluate and critique the manufacturing process by trying it out, but pilot production should simulate as closely as possible the actual manufacturing conditions.

Validation should also address product packaging and labeling. These components of the design may have significant human factors implications, and may affect product performance in unexpected ways. For example, packaging materials have been known to cause electrostatic discharge (ESD) failures in electronic devices. If the unit under test is delivered to the test site in the test engineer's briefcase, the packaging problem may not become evident until after release to market.

Validation should include simulation of the expected environmental conditions, such as temperature, humidity, shock and vibration, corrosive atmospheres, etc. For some classes of device, the environmental stresses encountered during shipment and installation far exceed those encountered during actual use, and should be addressed during validation.

Design Transfer

The purpose of design transfer is to finalise all deliverables for filing with regulatory agencies.

FDA 21 CFR Requirements - Design Transfer

FDA CFR Part 820.30(H) Design Transfer

- *Each manufacturer shall establish and maintain procedures to ensure that the device design is correctly translated into production specifications.*

As the design output is finalised, the design is transferred into production specifications (drawings, manufacturing, test, and inspection procedures). Production specifications must ensure that manufactured devices are consistently and reliably produced within product and process capabilities, meeting all quality requirements.

No design team can anticipate all factors bearing on the success of the design, but procedures for design transfer should address at least the following basic elements.

First, the design and development procedures should include a qualitative assessment of the completeness and adequacy of the production specifications.

Second, the procedures should ensure that all documents and articles which constitute the production specifications are reviewed and approved.

Third, the procedures should ensure that only approved specifications are used to manufacture production devices.

The first item in the preceding list may be addressed during design transfer. The second and third elements are among the basic principles of document control and configuration management. As long as the production specifications are traditional paper documents, there is ample information available to guide manufacturers in implementing suitable procedures. When the production specifications include non-traditional means, flexibility and creativity may be needed to achieve comparable rigor.

Post-Launch Reviews

A post-launch review is required for each product within one year of initial launch. The purpose of the post-launch review is to confirm that no design or manufacturing changes are required and to document future product development activity. It also considers performance and patient safety.

Some post-launch activities include:

#	Action	Description
1	Post Market Surveillance	Review of data from any studies required due to conditional approval by FDA or other deliverables per Post Market Surveillance Plan
2	Product Performance Analysis	Review product complaints. Compare to equivalent devices if feasible
3	CAPA / Field	Review CAPA data. Review the severity of CAPAs and any re-occurring causes
4	Design Review	Review design or process changes since approval
5	Production Review	Yields, scrap, NCMR, causes of downtime, and any other manufacturing issues
6	Customer Feedback	review patient and user feedback

Change control

Manufacturing change control is usually implemented using a set of standardized procedures similar to the following:
A change request might be originated by a developer, manager, reviewer, marketing representative, user, customer, quality assurance representative, or production personnel, and identifies a design problem which the requester believes should be corrected. Change requests are typically reviewed following the manufacturer's prescribed review process, and the request
might be rejected, deferred, or accepted.

If a change request is accepted and corrective action is straightforward, a change order might be issued on the spot to implement the change. The change order pertains to an explicitly identified document or group of documents and specifies the detailed revision of the document content which will fix the identified problem. Often, the change request results in an assignment to developers to further study the problem and develop a suitable corrective action. If the change is extensive, wholesale revision of affected documents may be warranted in lieu of issuing change orders.

Change requests and change orders should be communicated to all persons whose work might be impacted by the change.

It may not be practical to immediately revise documents affected by a change order. Instead, the common practice is to distribute and attach a copy of the change order to each controlled copy of the original document.

Change control procedures should incorporate review and assessment of the impact of the design change on the design input requirements and intended uses.

A mechanism should be established to track all change requests and change orders to ensure proper disposition.

Design Control Deliverables

This section provides a non-exhaustive list of design documentation deliverables. A brief description of each is provided. This list can be used as a checklist for the design control process or as supplementary information of key activities outlined previously.

Validation Master Plan (VMP): A validation master plan should be written as soon as the project begins. It should describe the product to be manufactured and the process technology. A VMP will also contain generic material such as an outline of the validation approach and the types of validation e.g. prospective, concurrent and so on.

External Requirements: External requirements refer to regulations and industry standards that are relevant to a new product. At the design input phase a list of documents should be created in order to capture essential requirements as early as possible.

Design Development Plan: A design and development plan is an overarching document that describes the design and development, responsibilities, timelines and project scope, list and schedule of major tasks and the phase review details such as the timing and approval requirements.

Product Specification: The product specification is a design output document that is built over the course of the project. Not all information will be final in the early phases, however, having an early draft will help focus minds and generate the right activity in order to define target dimensions, physical attributes and tolerances.

Stability Testing: A document containing a summary of results, testing and analysis should be created and filed as part of the DHF.

Device Master Record: A DMR is an output document and should be available at the design transfer phase. It is a comprehensive list referencing all work instructions, test procedures, test specifications, manufacturing specifications and finished product specifications required to manufacture the product.

Test Method Validations: A list of all validated test methods (functional, analytical, physical etc.) should be available to file in the DHF.

Design History File: The DHF is a repository for all of the documentation generated as a result of the design control process. The DHF serves as a complete record of the design.

Design Control Process via Web-Based Systems

In recent years some companies have entered the market offering web-based design control processes. As mentioned earlier, there are a large amount of documents created during the design control process. Most of the documentation generated is subject to change control and therefore requires review and approval. As with traditional hardcopy approval, this can be time-consuming and complex if approvers are based across different departments or drawn internationally.

All documents also form part of the design history file. Therefore, the proper filing and availability of documents is an important source of concern. The use of an electronic system may mitigate some of these concerns.

Furthermore, some web-based solutions offer integration with existing electronic documentation systems or integration

3.3 ISO 13485

Introduction

ISO 13485 is the quality management standard of choice for manufactures of medical devices. Revised in 2016, ISO 13485:2016 "specifies requirements for a quality management system where an organisation needs to demonstrate its ability to provide medical devices and related services that consistently meet customer and applicable regulatory requirements."[1] The scope of the standard can apply to any organisation or company involved throughout the life-cycle of a product, including design and/or development, production, storage and distribution, installation, or servicing of a medical device and design and development or provision of technical or professional services. [1] International Standards Organisation, www.iso.org

The recent revision is designed to address recent developments in quality management and other updated regulations that relate to the industry. Improvements in the new version of the standard include broadening its applicability to include all organisations involved in the life cycle of the product, from the concept stage to end of life along with greater alignment with regulatory requirements and post-market surveillance including complaint handling.

ISO 13485:2016 is also used by suppliers or external vendors that provide QMS related management system-services. Requirements of ISO 13485:2016 are applicable to organisations regardless of their size and regardless of their type except where explicitly stated. Wherever requirements are specified as applying to medical devices, the requirements apply equally to associated services as supplied by the organisation. If any requirement in Clauses 6, 7 or 8 of ISO 13485:2016 is not applicable due to the activities undertaken by the organisation or the nature of the medical device for which the quality management system is applied, the organisation does not need to include such a requirement in its quality management system. For any clause that is determined to be not applicable, the organisation records the justification as part of their certification and quality management system.

The Process Approach

ISO 13485 is based on a process approach to quality management. A process is any activity that receives inputs and converts them to outputs. For an organisation to function effectively, it has to identify and manage numerous linked processes. Furthermore, many processes impact other processes or downstream processes. The application of a system of processes within an organisation, together with the identification and interactions of these processes, and their management, can be referred to as the "process approach".

Directives Versus Standards

When it comes to regulated industries such as medical devices, it is first important to be familiar with some common terms and definitions and what they really mean. This chapter examines some key terms that are applied widely and relate to regulated industries. They include:

- Directives

- Standards

- Notified Body

• Competent Authority

Directives

Directives are legal requirements which must be met by manufacturers or other bodies within the industry. Directives are based on legislation and are issued at governmental level. It is important to note that standards such as ISO 13485 help companies meet the requirements set up in directives. (See harmonised standards below)

Standards

Standards are not always mandatory. However, they help manufacturers be compliant with directives/legislation.

They also represent the current and best practice in the field of study/industry. Harmonised standards are European standards prepared under a mandate from the European Commission, referenced in the official journal, and drafted so that compliance with their requirements relates to one or more essential requirements of the directive. These standards have special status because, when a manufacturer can show that their products meet the requirements of the standard, there is a presumption that the product conforms to the essential requirements of the directive that is covered by the standard.

What is a Competent Authority?

When it comes to medical devices, a competent authority is the legally delegated authority mandated to monitor compliance to directives and legal requirements within the industry. The competent authority has the power to grant and revoke licenses.

Example of Competent Authorities:

• FDA (Food and Drug Administration) CFR Code of Federal Regulations – U.S.
• MHRA (Medicines and Healthcare Regulatory Agency - UK
• HPRA (Health Products Regulatory Agency) - Ireland
• JPAL (Japanese Regulations for Medical Devices) – Japan

What Is a Notified Body?

A notified body is a certification organisation which the national authority (the competent authority) of a member state designates to carry out one or more of the conformity assessment procedures described in the annexes of the medical devices directives. The Medicines and Healthcare Products Regulatory Agency is the UK competent authority under the three directives.

Organisations and Institutions

Many of the common acronyms that are referenced in literature relate to various standard setting organisations and industry representatives. Some of the more common bodies are listed below:

ISO: Internal Organisation for Standardisation
IMDR (F): International Medical Device Regulators Forum
ASTM: American Society for Testing and Materials
GHTF: Global Harmonisation Task Force

Basic Definitions (Source: Annex IX of Directive 93/42/EEC)

Intended Purpose: Intended purpose means the use for which the device is intended according to the data supplied by the manufacturer on the labelling, in the instructions and/or in promotional materials. (Chapter I section 1 of Annex IX of Directive 93/42/EEC)

Transient: Normally intended for continuous use for less than 60 minutes.

Short Term: Normally intended for continuous use for not more than 30 days.

Long Term : Normally intended for continuous use for more than 30 days.

Invasive Devices: A device which, in whole or in part, penetrates inside the body, either through a body orifice or through the surface of the body.

Body Orifice: Any natural opening in the body, as well as the external surface of the eyeball, or any permanent artificial opening, such as a stoma.

Surgically Invasive Device: An invasive device which penetrates inside the body through the surface of the body, with the aid of or in the context of a surgical operation.

Implantable Device: Any device which is intended:

- to be totally introduced into the human body or,
- to replace an epithelial surface or the surface of the eye, by surgical intervention which is intended to remain in place after the procedure. Any device intended to be partially introduced into the human body through surgical intervention and intended to remain in place after the procedure for at least 30 days is also considered an implantable device.

Medical Device: means any instrument, apparatus, appliance, material or other article, whether used alone or in combination, together with any accessories or software for its proper functioning, intended by the manufacturer to be used for human beings in the:

- diagnosis, prevention, monitoring, treatment or alleviation of disease or injury.

- investigation, replacement or modification of the anatomy or of a physiological process.

- control of conception which does not achieve its principal intended action by pharmacological, chemical, immunological or metabolic means.

A medical device may be assisted in its function by the following means:

Active Medical Device: any medical device relying for its functioning on a source of electrical energy or any source of power other than that directly generated by the human body or gravity.

Active Implantable Medical Device: any active medical device which is intended to be totally or partially introduced, surgically or medically, into the human body or by medical intervention into a natural orifice, and which is intended to remain after the procedure.

Custom-Made Device: means any active implantable medical device specifically made in accordance with a medical specialist's written prescription which gives, under his responsibility, specific design characteristics and is intended to be used only for an individual named patient.

Device Intended for Clinical Investigation: any active implantable medical device intended for use by a specialist doctor when conducting investigations in an adequate human clinical environment.

Intended Purpose: means the use for which the medical device is intended and for which it is suited according to the data supplied by the manufacturer in the instructions.

Putting into Service: means making available to the medical profession for implantation.

Where an active implantable medical device is intended to administer a substance defined as a medicinal product within the meaning of Council Directive 65/65/EEC of 26 January 1965 on the approximation of provisions laid down by law, regulation or administrative action relating to proprietary medicinal products (6), as last amended by Directive 87/21/EEC (7), that substance shall be subject to the system of marketing authorisation provided for in that directive.

Where an active implantable medical device incorporates, as an integral part, a substance which, if used separately, may be considered to be a medicinal product within the meaning of Article 1 of Directive 65/65/EEC, that device must be evaluated and authorised in accordance with the provisions of this directive.

ISO 13485 & Regulations

In chapter 2 the special status of harmonised standards was described which allows companies meet the essential requirements of Directives. In Europe, EN ISO 13485:2013 helps companies meet the requirements of: Directive 93/42/EEC on medical devices. This harmonised standard gives companies the "presumption of conformity" to complying with directives.

EN ISO 13485 was published in February 2013 and harmonised in August 2013 to cover the three directives:

- 90/385/ECC– The Active Implantable Medical Devices Directive (AIMDI)
- 93/42/ECC – The Medical Devices Directive (MDD)
- 98/79/EEC – In Vitro Diagnostic MDD (IVDMDD)

In the United States, medical device manufacturers need to meet the requirements of 21 CFR Part 820 of FDA regulations. While ISO 13485 is not an actual requirement, many companies will seek certification to the standard to support the exporting of products. In Australia, it is a regulatory requirement for manufacturers of medical devices to meet the requirements of ISO 13485. In Canada, certification to ISO13485 is part of the regulatory requirements. The content of ISO 13485 is interpretive (not prescriptive) which gives a degree of scope in how the requirements are applied and met within a company. ISO 13485 provides both a sound and widely recognised basis in meeting regulatory compliance for medical devices. Based off ISO 19001 however, ISO 13485 is a standalone standard for medical devices.

ISO 9001 has requirements and themes relating to continual improvement and customer satisfaction. These have been modified for ISO 13485.

Main differences between ISO 9001 & ISO 13485:

• Customer satisfaction is changed to customer feedback
• Extra requirements regarding procedures for ISO 13485
• Extra requirements for records ISO 13485 (e.g. retention)
• Continual improvement is restricted to continual improvement of the quality management system

ISO 13485 has extra requirements required for regulatory bodies such as post production review and management of advisory events.

ISO 13485 and ISO/TR 14969

ISO/TR 14969 is a technical report that is used for guidance on the application and implantation of ISO 13485. It is recommended for those responsible for the role out of ISO 13485 within their organisation. The content of ISO/TR 14969 is based on several established organisations such as the GHTF, ISO and input from regulatory bodies.

Standard Overview

ISO 13485 has 8 Clauses or Sections which make up the structure of the standard.

Section 0 Normative References, Definitions and Terms
Section 1 Requirements of the Quality Management System (QMS)
Section 2 Normative References
Section 3 Terms and Definitions
Section 4 Requirements of the Quality Management System (QMS)
Section 5 Management Responsibility
Section 6 Resource Management
Section 7 Product Realisation
Section 8 Measurement, Analysis and Improvement

CLAUSE 1: SCOPE

This section refers to the scope and application of the standard.

- The organisation must be able to show its ability to provide medical devices to meet customer requirements and regulatory requirements
- A key aim of the standard is to allow harmonisation to regulatory requirements
- The scope of the QMS must relate to medical devices for a company to be able to use ISO 13485.

Some examples of what's in scope of the standard include (1) the manufacture of hip implants, (2) the design and manufacturing of in-vitro blood testing devices, (3) contact analytical testing (4) consultancy services. The terms "where appropriate" and "if appropriate" are used throughout the standard, therefore, it should be met by the organisation unless a justification is documented.

CLAUSE 2: NORMATIVE REFERENCES

This clause states that when working with ISO 13485, refer to ISO 9000:2000 for fundamentals and vocabulary.

CLAUSE 3: TERMS AND DEFINITIONS

This clause provides terms and definitions. It is very useful in the early days of establishing and implementing ISO 13485 to ensure that terms and definitions are clearly understood.

CLAUSE 4: QUALITY MANAGEMENT SYSTEM

Clause 4 details the general requirements that relate to the quality management system, the documentation requirements and record requirements.

Clause 4 includes:

4.1 General requirements clause

4.2 Documentation requirements clause

CLAUSE 4.1 GENERA REQUIREMENTS

The organisation must implement a Quality Management System, or QMS in order to provide the framework and structure to achieve ISO 13485 roll-out and implementation. However, the role of the QMS does not stop there. After initial roll-out, the requirements of the standard must be maintained and determined to be effective on an on-going basis. The following processes should be documented:

-   List of all processes
-   Process interactions
-   Monitoring of processes
-   Resources to facilitate rollout of processes
-   Measure and monitor effectiveness
-   System of identifying improvements

CLAUSE 4.2: DOCUMENTATION REQUIREMENTS

When it comes to the regulated industries such as the medical device industry, every process and procedure must be documented. Documentation ensures that everyone is working in the same manner with the same procedures. However, documentation is more than just writing down procedures and processes. It is also concerned with how documents are controlled, how they are updated and how they are stored.

Electronic Document Management Systems

Electronic document management systems aka EDMS are now the norm and gold standard for most medium to large organisations. Many companies that provide medical device manufacturers with an EDMS can customise the system to match the business processes particular to an organisation. With configurable or customisable software, validation and proper verification is important to ensure the system operates as intended. There are also regulatory requirements that stipulate the expectations and requirements of such systems. For example, the application of electronic signatures and the presence of audit trials. FDA 21 CRF Part 11 details the requirements with regards to electronic records and electronic signatures. For medicinal products in Europe, GMP V4 Annex 11 specifies similar requirements.

Changes and Updates to Documents

Revision control is a key element of the Quality Management Systems in place in regulated industries. As the need for changes in the document arises, the controlled document can be amended/updated. With each update the version number revises also. Some companies will use alphabetic revision control and to a lesser extent numeric revision control (Version A, Version B or Version 01, Version 02).

Controlled documents should always have a version number or revision number electronically on each page of the document. This is similar to books which always list what edition they are. e.g. first edition or second edition.

Records

Records are generated through the application of processes and procedures. These records can be related in quality inspection and manufacturing. The integrity and quality of records relating to the manufacture of medical devices is important, as it plays a part in safe-guarding the patient or user. Records may also help in the investigation of any quality issues, complaints or adverse events that may arise.

Principles of Good Documentation Practices or GDP, should be applied to records. In particular, handwritten entries should always be accompanied by a signature and date. This is important as traceability must be maintained in the event of an issue or complaint.

CLAUSE 5: MANAGEMENT RESPONSIBILITY

Clause 5 includes:
>5.1 Management Commitment
>5.2 Customer Focus
>5.3 Quality Policy
>5.4 Planning
>5.5 Responsibility, Authority and Communication
>5.6 Management Review

5.1 Management Commitment

It is essential that top management have an authentic and tangible commitment to meeting regulations and the expectations of customers. Quality should be at the forefront of all of activities. Management should encourage discourse and communication on all matters relating to internal processes, quality and the QMS as a whole.

5.2 Customer Focus

Customer –patient/user/doctor/family member

Customer feedback is a requirement of ISO 13485 and as such the manufacturer must engage with the customer. In instances where a defective product is received, the manufacturer must have a complaints process to facilitate proper feedback, communication and investigation.

5.3 Quality Policy

Simple statement /1 pager or more

Often quality policies will be displayed in reception areas etc. Copies should be signed and revision controlled. Quality policy must have a commitment to maintain the effectiveness of the QMS.

5.4 Planning

Top management must plan quality objectives and ensure they are implemented and effective.

Some examples of quality objectives include :

reduce rework by 10%

reduce scrap by 5%

have customer complaints reduced by 2% per year

5.5 Responsibility, Authority and Communication

Roles and responsibilities are defined.

Job descriptions are in place.

Organisational charts are in place and accurate.

5.6 Management Review

The purpose of management review is to ensure the effectiveness of the QMS.

Inputs to management review include:
>(a) Audit results
>(b) Customer feedback
>(c) Process performance and conformity
>(d) Corrective and preventative actions
>(e) Deviations
>(f) Regulatory changes and revisions

CLAUSE 6 : RESOURCE MANAGEMENT

Clause 6 of ISO 13485 is concerned with human resources, infrastructure and work environment.
Clause 6 includes:

    6.2 Human resources

    6.3 Infrastructure

    6.4 Work Environment

People are the key part of any QMS. Therefore, they should have the appropriate level of education, skill and experience. A culture of quality must be lived by everyone.

People must be suitably trained. Training must be documented and consistent throughout an organisation. Training must be seen to be effective. Proper records of education and training must be kept.

Human intelligence, human creativity and human labour are all key inputs to any factory or company manufacturing medical devices. Therefore, an organisation must be properly resourced in order to function correctly, meet the regulatory requirements and customer expectations.

6.2 Human resources

"Change the people or change the people"

With any organisation, it is only as good as the people it has in its make-up. Therefore, the people, operators, engineers, managers etc. all contribute to the quality management system. Clause 6.2.2 also specifies requirements with regards to competence, awareness and training. The person should be matched to the job in terms of their qualifications, experience and training. Typically, job descriptions are used to drive and capture these requirements. Nowadays, most multinational companies will ask for evidence of qualifications, training and experience. These documents are then held on file in the event of an audit. This is recommended practice for medical device companies. While the standard does not specifically call out the need to hold records of degrees and qualifications on file, the company or organisation needs to demonstrate the suitability of the person to their respective roles, and filing the qualification provides the easiest method.

6.3 Infrastructure

Infrastructure has the ability to impact the quality of products and services. Therefore, it must be fit for purpose. It is especially important if the organisation is involved with the manufacture of medical devices. The following element need to be considered with regards to infrastructure:

- Location of equipment and the operating environment
- Equipment installation and validation
- Utilities required for the operation of equipment and systems
- Layout of the factory – flow or raw materials, in-process materials and finished products
- Environmental systems such as HVAC and fire suppression systems

6.4 Work Environment

The work environment is also closely related to infrastructure within a given organisation and they can both affect or impact upon the quality of products manufactured. Risk to product quality and patients is minimised by understanding the work environment and how it can impact the product. When the interactions and risks are understood, work can then be done to eliminate risks or at least control or monitor them. Environmental conditions that can impact upon product quality include:

- Humidity
- Temperature
- Air quality
- Room pressure differentials (negative / positive)
- Air flow/velocity

CLAUSE 7 : PRODUCT REALISATION

Clause 7 includes:

7.1 Planning of product realisation
7.2 Customer-related processes
7.3 Design and development
7.4 Purchasing
7.5 Production and service provision
7.6 Control of measuring devices

7.1 Planning of Product Realisation

Product realisation can be defined as a collection of processes and body of work that delivers a product or service to the customer. Remember, when it comes to medical devices, customers can be patients or users such as doctors and nurses. It should be noted that organisations can opt to exclude specific requirements, in cases where product realisation is not applicable. However, any such exclusion should be based on sound rationale with the case clearly documented. An example of this may be where design and development is not conducted by the manufacturer e.g. contract manufacturers.

7.1 Planning of Product Realisation

Planning is an often underestimated but remains a key element of product realisation. If adequate time and resources are given to planning, it makes all other processes run smoother, and therefore should help to produce improved products and services.

7.2 Customer-Related Processes

There are 3 elements that feed into customer-related processes. They include the following:

Determining the requirements related to the product Clause **7.2.1**
Review of requirements relating to the product-Clause **7.2.2**
Customer communication-clause **7.2.3**

Customer requirements are typically captured in a User Requirements Specification. A requirements specification (URS) documents all of the desired attributes of a product or service. They can be made up by a combination of CQAs, regulatory requirements and design requirements. A URS can then form the basis for review of the product or service requirements.

With regard to customer communication, it is important to remind ourselves that we are concerned with ISO 13485 which as we very well know by now is the standard for medical devices. Therefore, having the right information available to the customer, patient or end user is important. When additional information needs to be transmitted or updates to information need to be communicated, an advisory note can be issued. Another important aspect of customer communication is customer feedback. This communication can be made up of positive feedback from the customer or users, or when there is a query with regard to a product or service. Therefore, processes or systems must be in place to make communication between customer and company both effective and timely.

7.3 Design and Development

Design and Development Verification and Validation ensure that the product is designed, developed and subsequently manufactured meeting all the customer requirements, regulatory requirements and business requirements. These requirements are classed as inputs to the design and development, and verification and validation ensure the inputs have been adequately taken into account.

The design and development testing sometimes replicate the commercial applications of the medical device, hence providing a realistic challenge in order to have confidence in the medical device.

Design Control

Design control is a necessary practice that ensures good engineering principles are maintained throughout the design phase of a product. It also refers to the continual design and development of the product through its very lifecycle. The design and development files and history must be controlled and maintained, with any changes properly assessed, tested and documented.

7.4 Purchasing

Bearing in mind that a quality management system considers all aspects of an organisation's functioning, purchasing and procurement of materials necessitates putting robust controls in place. Simply put, a purchasing process must exist.

7.5 Production and Service Provision

This requirement of ISO 13485 is an extensive section with a great deal of importance associated with it. As we are dealing with the manufacture of medical devices (or other activity associated with medical devices) there are specific requirements for sterile products. If a product is sterile, its use or application is likely to be associated with greater risks to the patient. Therefore, extra safeguards must be in place for sterile medical devices. Key sections of Clause 7.5 include: (1) control of production and service provision – both general and specific requirements, (2) specific requirements for sterile medical devices, (3) validation of equipment and processes for production and service provision, (4) traceability and identification, (5) preservation of product controls with regard to monitoring and measuring medical devices.

7.6 Control of Measuring Devices

This clause requires an organisation to identify what monitoring and measuring is required and to ensure the product or service meets the customer requirements. A calibration procedure must also be maintained to ensure the equipment is accurate and reliable. Calibration must ensure that:

- Equipment used to verify product quality is calibrated to a periodic schedule.
- The calibration is performed to an international standard.
- The calibration status of the equipment is recorded and visible.
- The equipment must be located within a suitable area in order to maintain accurate and reliable results.

If an organisation uses any computer software to monitor or measure outputs, the software must be verified before use via the appropriate validation and qualification activities.

CLAUSE 8 : MEASUREMENT ANALYSIS

Clause 8 includes:

8.1 General requirements
8.2 Monitoring and measurement
8.3 Control of nonconforming products
8.4 Analysis of data
8.5 Improvement

8.1 General Requirements

Measurement, analysis and improvement are the key themes of clause 8. As with all medical devices, inspection and testing both during manufacturing and post manufacturing is necessary to ensure products and services function as intended and without defects. With any type of measurement or inspection analysis, the method used to complete the testing is critical. The method must be fit for purpose, and the equipment must be suitable. This "method validation" typically is done during the design and development phase.

8.2 Monitoring and Measurement

Monitoring and measurement are dependent on the information or feedback provided from various sources. The most important feedback is the post-production feedback that is gathered from customers or the end user. Again, this occurs over the whole lifetime of the product or service in question. There are a number of methods that can be used to obtain feedback. Some examples include:

-Customer surveys
-Customer complaints
-Review of regulatory databases such as MAUDE.
-Repair and servicing information

8.3 Control of Nonconforming Product

Non-conforming product presents a risk to patients or users of medical devices. When a situation arises where non-conforming product is manufactured or detected through inspection processes, the product must be controlled and segregated to prevent unintended use or distribution.

Some examples resulting in non-conformance are:

- When a manufacturing process drifts outside its validation window or operating parameters.
- A certificate of analysis for a raw material is not provided by the supplier or the results are out of specification.
- In-process testing was not completed at the defined intervals.
- Training of personnel completing tests is not current or is inadequate.

8.4 Analysis of Data

In any engineering activity, data and the quality of the data is a key factor in making the right decisions. Provided the data collected is relevant and accurate, analysis of data can provide important insights into process performance, quality control and product functionality. Data should be collated in a consistent way and controlled by a procedure. When it comes to medical device manufacturing, the sources and types of data are multiple. Data can be generated from in-process testing and data can be generated from end of line testing aka finished product testing.

8.5 Improvement

ISO 13485 fosters a culture of continual improvement. As we have seen, each activity can be described as a process. For example, a manufacturing process, a procurement process, a complaints process. The set of processes that make up the quality management system need to be continually reviewed to ensure they are suitable and effective for the task at hand. Typical tools used to keep improvement in mind include:

- Review of the quality policy and quality objectives
- Frequency and category of corrective and preventative actions (CAPA's)
- Customer complaints
- Management review input

CE Marking

In Europe a QMS is required for CE marking of a medical device that is placed on the market in the EU. ISO 13485:2003 is a harmonised standard that can be used by companies to show conformity of their QMS to requirements of directives. EN ISO 13485:2012 was harmonised in August 2012. This allows compliant companies receive an EC Declaration of Conformity.

Summary of the CE Requirements

- Manufacturers of class I devices or their authorised representatives must:
  - review the classification rules to confirm that their products fall within class I (Annex IX of the Directive)
  - check that their products meet the essential requirements (Annex I of the Directive)
- notify the competent authority, in advance, of any proposals to carry out a clinical investigation to demonstrate safety and performance of a device as required by the regulations
- obtain notified body approval for sterility or metrology aspects of their devices and where applicable prepare relevant technical documentation
- Draw up the 'EC Declaration of Conformity' (below) before applying the CE marking to their devices
- Register with the competent authority
- Implement and maintain corrective action and vigilance procedures including a systematic procedure to review experience gained in the post-production phase
- Make available relevant documentation on request for inspection by the competent authority.

In Europe, all medical devices must bear the CE marking of conformity (see Annex XII) of the directive) when they are placed on the market and/or put into service. The CE marking must appear in a visible, legible and indelible form on the device or its sterile pack, where practicable and appropriate, and where applicable on any instructions for use and sales packaging. For 'sterile' and 'measuring' devices, the CE marking must be accompanied by the identification number of the notified body that has acted under the relevant conformity assessment procedure.

EC Declaration of Conformity

In order to affix the CE marking, the manufacturer or their authorised representative must follow the EC declaration of conformity procedure referred to in Annex VII of the directive. This procedure must be completed prior to placing the device on the market. The 'EC declaration of conformity' is the procedure whereby the manufacturer or their authorised representative prepares the required technical documentation, puts into place corrective action and vigilance procedures and declares that the products meet the requirements set out in the directive.

Technical Documentation

The technical documentation should be prepared following review of the essential requirements and must cover all of the following aspects:

Description: A general description of the product, including any variants (for example names, model numbers and sizes).

Raw Materials and Component Documentation: Specifications including, as applicable, details of raw materials, drawings of components and/or master patterns and any quality control procedures.

Intermediate Product and Sub-Assembly Documentation: Specifications including appropriate drawings and/or master patterns, circuits, and formulation specification; relevant manufacturing methods and any quality control procedures.

Packaging and Labelling Documentation: Packaging specifications and copies of all labels and any instructions for use.

Design Verification: The results of qualification tests and design calculations relevant to the intended use of the product, including connections to other devices in order for it to operate as intended.

Risk Analysis: The results of risk analysis to review whether any risks associated with the use of the product are compatible with a high level of protection of health and safety and are acceptable when weighed against the benefits to the patient or user. If biocompatibility is relevant – for example for skin contact and invasive devices – a compilation and review of existing data or test reports based on the relevant standards is required.

Compliance with the Essential Requirements and Harmonised Standards: A list of relevant harmonised standards (for example sterilisation, labelling and information, biocompatibility, electrical safety, risk analysis, product group standards) which have been applied in full or in part of the products. If relevant harmonised standards have not been applied in full, then additional data will be required, detailing the solutions adopted to meet the relevant essential requirements of the directive. The manufacturer may choose to prove conformity with the essential requirements of the directive through the use of their own standards and/or other relevant published standards (ISO, EN, BS). However, the use of such standards does not give similar, immediate presumption of conformity to the essential requirements of the directive. Therefore, using a harmonised standard provides greater protection to the manufacturer.

3.4 Risk Management

Introduction

This chapter covers the risk management requirements of ISO 13485 and the essential requirements of the 93/42/EEC Medical Device Directive, the 98/79/EC IVD Directive and 90/385/ECC – The Active Implantable Medical Devices Directive (AIMDI).

Typically, compliance to the above requirements is met by meeting ISO 14971:2009. This chapter provides an overview of risk management activities for regulated medical devices. It may also be used as a framework for performing risk management for non-regulated products.

Definitions

Risk Management: Systematic application of management policies, procedures and practices to the tasks of analysing,

evaluating, controlling and monitoring risk.

Hazard: A hazard is a potential source of harm (physical injury or damage to the health of people, damage to property or the environment).

Failure Modes: Product or process failures that may lead to a hazard.

Risk: The combination of the probability that harm will occur and the severity of that harm.

Residual Risk: Risk remaining after risk control measures have been taken.

Risk Analysis: The use of available information to identify hazards and estimate risk.

Risk Reduction: Processes, controls or information that will reduce risk.

Risk Evaluation: The process of comparing the estimated risk against given risk criteria to determine the acceptability

of the risk.

Risk Control: The process in which decisions are made and measures implemented by which risks are reduced to, or

maintained within, specified levels.

Severity: A measure of the possible consequences of a hazard.

Probability of Failure: An estimate of the likelihood of failure.

Detection of Failure: An estimate of the likelihood of detection.

Risk Priority Number (RPN): The product of ratings on occurrence, severity and detection.

CAPA: Corrective and Preventative Action.

ALARP: As Low As is Reasonably Practicable.

Verification Plan: A confirmation through the provision of objective evidence, that the specified requirements have

been fulfilled.

Failure Modes and Effects Analysis, FMEA: A formal risk assessment methodology for identifying potential failure

modes, and assigning numerical values to the severity, likelihood of occurrence, and likelihood of escaped detection to

failure modes in order to quantify risk.

Risk Analysis

Risk analysis can be performed using a variety of methodologies such as FMEA/FMECA, HAZOP, HACCP, or other
methods appropriate to the design and function of the product. A common methodology for risk analysis for regulated
products is FMEA. The analysis can be grouped into a product category of similar established device technology.

In estimating risk(s) for each hazard, information may be obtained from the following sources:

- Published standards
- Clinical trial data
- Technical data
- Field data from similar products
- Usability tests
- Results of investigations, e.g. CA/PA investigations, etc.
- Expert opinion
- External assessments or audits

Failure Mode Effects and Analysis (FMEA)

FMEA consists of listing all the potential failure modes associated with the processes followed by a corresponding
list of all the possible causes and effects of each potential failure mode. The impact on the patient, operator,
environment, process, handlers and business should be considered.

Severity (S): the severity score addresses how severe the effect of this failure is on all/or one of the following:
the
patient, operator, environment, process, handlers and business. This is subjectively rated from 1 to 5.

Score	Criteria	Description
5	Very serious	Safety related. Could result in considerable patient harm.
4	Serious	Performance degraded but is operable and safe. Customer dissatisfied.
3	Moderately serious	Minor effect on performance. Customer is slightly annoyed.
2	Not serious	Non-vital fault may be noticed. Customer is not annoyed.
1	Insignificant	No effect.

Occurrence (O): refers to how often the failure is expected to occur. This can be rated either subjectively
(frequent, rare etc.) or via the number of units affected depending on the risk being assessed. This is rated
from 1 to 5.

Score	Criteria	Description

5	Very frequent	$\geq 10^{-3}$
4	Frequent	$< 10^{-3}$ and $\geq 10^{-4}$
3	Rather rare	$< 10^{-4}$ and $\geq 10^{-5}$
2	Rare	$< 10^{-5}$ and $\geq 10^{-6}$
1	Very rare	$< 10^{-6}$

Detection (D): can the problem be detected by the user or patient before it does any damage? This column is subjectively rated from 1 to 4.

Score	Criteria
1	Very high probability of detection.
2	High probability of detection of the defect.
3	Moderate probability of detection of the defect.
4	Low probability of detection of the defect.
5	Very low probability of detection of the defect.

Calculate the Risk Priority Number (RPN) by multiplying the numbers obtained for severity, occurrence and detection together.

$$RPN = O \times S \times D$$

Risk Acceptability

RPN scores of risk acceptability can be divided into three categories:

Acceptable risk – risk is deemed acceptable meaning that there is no need to consider risk reduction measures.

Investigate further risk reduction – investigate if further risk reduction to the "no need to consider level" is practicable.

Unacceptable risk – risk control measures must be implemented to reduce risk.

The table below details the RPN score which determines the risk acceptability category:

Score	Criteria
RPN> 75	Unacceptable Risk
$20 < RPN \leq 75$	Investigate further risk reduction
$RPN \leq 20$	Acceptable risk

Risk Control

Where risks are identified as unacceptable, risk control measures must be determined to reduce the risk prior to the process or system being implemented. A number of actions can be taken in order to further reduce risk including: (1) changing the design to reduce risk, introducing protective measures in the device or the manufacturing process, (3) inserting a warning statement into the instructions for use (IFU).

Risks scored as "investigate further risk reduction" should be examined to determine whether it is practicable to reduce the risk further. The risk should be reduced to as low as is reasonably practicable, (aka ALARP) taking into account the benefits of accepting the risk and the practicability of implementation. If risks classed as "investigate further risk reduction" are already at ALARP, no further risk reduction is necessary.

Residual Risk Evaluation

After risk control measures are applied, a new risk assessment will be carried out to determine residual risks. Residual risks will be assessed for acceptability using the same criteria as detailed in 6.5. If the residual risk is not judged acceptable then further risk control measures will be applied.

If the residual risk is not judged acceptable and further risk control is not practicable then the team may perform a risk/benefit analysis by evaluating data and literature on the medical benefits of the intended use to determine if they outweigh the risk. If this evidence does not support the conclusion that the medical benefits outweigh the residual risk, then the risk remains unacceptable. This analysis should be recorded and approved by both the risk management team and senior site management.

EN ISO 14971: 2009– Characteristics - Annex C

Annex C contains several questions that can be used to identify medical device characteristics that could impact upon safety:

1) What is the intended use and how is the medical device to be used?
2) Is the medical device intended to be implanted?
3) Is the medical device intended to be in contact with the patient or other persons?
4) What materials or components are utilised in the medical device or are used with, or are in contact with, the medical device?

EN ISO 14971:2009 - Annex E

Refer to EN ISO 14971:2009 Annex E - Examples of hazards, foreseeable sequences of events and hazardous situations.
Examples of hazards, foreseeable sequences of events and hazardous situations:

<u>Examples of energy hazards</u>

<u>Electromagnetic energy</u>
Line voltage
Leakage current
- enclosure leakage current
- earth leakage current
- patient leakage current
Electric fields
Magnetic fields
<u>Radiation energy</u>
Ionising radiation
Non-ionising radiation
Thermal energy

High temperature
Low temperature
Mechanical energy
Gravity

Examples of biological and chemical hazards

Bacteria
Viruses
Other agents (e.g. prions)
Re- or cross-infection
Chemical
Exposure of airway, tissues, environment or property, e.g. to foreign materials:
- acids or alkalis
- residues
- contaminates
- additives or processing aids
- cleaning, disinfecting or testing agents
- degradation products
- medical gases
- anaesthetic products

Biocompatibility
Toxicity of chemical constituents

Example of operational hazards

Function
Incorrect or inappropriate output or functionality
Incorrect measurement
Erroneous data transfer
Loss or deterioration of function

Use error
Attentional failure
Memory failure
Rule-based failure
Knowledge-based failure
Routine violation

Examples of information hazards

Labelling
Incomplete instructions for use
Inadequate description of performance characteristics
Inadequate specification of intended use
Inadequate disclosure of limitations

Operating instructions
Inadequate specification of accessories to be used with the medical device
Inadequate specification of pre-use checks
Over-complicated operating instructions

Warnings
Of side effects
Of hazards likely with re-use of single-use medical devices

Examples of initiating events and circumstances

Incomplete requirements

Inadequate specification of:
- design parameters
- operating parameters
- performance requirements
- in-service requirements (e.g. maintenance, reprocessing)

end of life

Manufacturing processes

Insufficient control of changes to manufacturing processes
Insufficient control of materials/materials compatibility information
Insufficient control of manufacturing processes
Insufficient control of subcontractors

Transport and storage

Inadequate packaging
Contamination or deterioration

Inappropriate environmental conditions

Environmental factors

Physical (e.g. heat, pressure, time)
Chemical (e.g. corrosions, degradation, contamination)
Electromagnetic fields (e.g. susceptibility to electromagnetic disturbance)
Inadequate supply of power

Inadequate supply of coolant

Cleaning, disinfection and sterilisation

Lack of, or inadequate specification for, validated procedures for cleaning, disinfection and sterilisation

Inadequate conduct of cleaning, disinfection and sterilisation

Disposal and scrapping

No information or inadequate information provided

Use error

Formulation

Biodegradation
Biocompatibility
No information or inadequate specification provided
Inadequate warning of hazards associated with incorrect formulations

Use error

Human factors

Potential for use errors triggered by design flaws, such as
- confusing or missing instructions for use
- complex or confusing control system
- ambiguous or unclear device state
- ambiguous or unclear presentation of settings, measurements or other information
- misinterpretation of results
- insufficient visibility, audibility or tactility
- poor mapping of controls to actions, or of displayed information to actual state
- controversial modes or mapping as compared to existing equipment
- use by unskilled/untrained personnel
- insufficient warning of side effects
- inadequate warning of hazards associated with re-use of single-use medical devices
- incorrect measurement and other metrological aspects
- incompatibility with consumables/accessories/other medical devices

slips, lapses and mistakes

Failure modes

Unexpected loss of electrical/mechanical integrity

Deterioration in function (e.g. gradual occlusion of fluid/gas path, pr change in resistance to flow, electrical conductivity) as a result of ageing, wear and repeated use.

3.5 Test Methods

This chapter introduces the design, execution and analysis of test method validation for medical devices. Test method validation involves the formal documentation of a test method used to capture and analyse data or information. The reason test methods need to be validated is to confirm that they are suitable and fit for the intended purpose. Secondly, and of equal importance is the need to verify that the test method performs to an acceptable level and is reliable and trustworthy over time. After all, test methods are used to assess product outputs such as dimensions, material strength and product functions. Getting this wrong will lead nowhere very quickly, so it is important to have confidence in the results of testing. Validation studies must demonstrate method capabilities in the testing environment. As a result, validation studies allow the formal documentation of the ruggedness of the test method in real-use conditions (i.e. demonstrating that the precision and accuracy limits are met with different technicians, different production batches and variable test equipment, etc.).

Where to Start?

The form and shape of any test method validation is influenced by several factors namely (1) in house requirements, these are internal procedures within a company that define the process and may provide or require the use of company templates or forms etc. (2) external requirements such as those relating to ISO and ASTM standards, (3) the type of test to be completed – visual inspection, manual contact measurement, non-contact (automated) measurement system, destructive tests and so on, (4) the type of equipment and status of equipment- simple equipment may not require equipment validation. However, complex equipment such as equipment that uses automation and software needs to be fully qualified. In addition, engineering studies may be require to test the robustness of the system, prior to measurement capability studies or formal test method validation. Also, existing equipment may be suitable for use and already qualified. Therefore, an assessment of the suitability for use must be completed to determine if all requirements arc met. (5) Product range and specifications. If a family of products are to be tested, the test method validation can be more complex. The product specifications is a critical input as the accuracy of equipment needs to be factored in, and this is dependent on the type of measurements required to be taken, and the upper and lower tolerances to be applied.

Example 1

A packaging company has a seal strength on the lid of a blister package. It wants to put in place a test method to test the seal strength of the package. This scenario would call for a test method validation.

Example 2

A medical device incorporates the use of a spring that is used to actuate a valve. The manufacturer of the device wants to develop a test method that examines the tensile strength of the spring on an ongoing basis. This scenario would also call for a test method validation.

Example 3

A contact lens manufacturer uses an optical comparator to measure the diameter of contact lenses during manufacture. The manufacturer must develop and validate a test method to facilitate the measuring of contact lenses.

What is test method validation?

Test method validation is the documented process of ensuring a test method is suitable for its intended use. The intended use of any system is normally documented and described in a User Requirements Specification.

Test method validation involves establishing the performance characteristics and limitations of a method and the identification of influences which may change those characteristics. TMV is an important element of quality control. Without validation there can be no assurance that the test results will be reliable and fit for the purpose. In some fields, validation of methods is a regulatory requirement. Generally, any method used to produce data in support of regulatory (e.g., FDA) filings or the manufacture of devices for human use should be validated. All are candidates for validation, though the process for each can vary. Most test methods exist as validated standards, methods developed by technical standard organizations (ANSI, ASTM, ISO…) to establish uniform methods and procedures for testing. But standard methods do not always fit the requirements of the tests to be performed.

When should methods be validated?

The risk associated with products an output (dimensions, features, chemical requirements etc.) often dictates the validation activity based on the potential level of patient harm weighed against the business risk of not performing the activities. The device risk index or harm classification dictates the minimum level of statistical confidence required. Higher risk requires more rigorous testing and higher levels of statistical confidence. In most cases, TMV is not mandated in the medical device industry (except ISO 11607). But demonstrating the safety and effectiveness of a device is difficult to do if the methods for establishing these parameters are not shown to be appropriate and reliable. Test Method Validation may be required for:

- A new method is developed
- A revision to established methods
- Methods that are moved or transferred

Take the case where a standard test method is established and in operation. However, a change to the system software is required. This type of change could impact the measured output. Therefore, the change needs to be considered for re-validation. Any other changes to the test procedure such as a change in handling of test specimens or the change, addition, removal or modification of equipment including fixturing can impact the measured output. It is important to note that validation of a test method is not required on each individual piece of equipment or fixturing, once replicate equipment or fixturing is assessed during the validation study. Some examples of changes not necessarily requiring any re-validation or change to a validation report etc. include:

- Clerical corrections to the test method that do not change the method or affect the measurement of the output.
- Removing of referenced supplies that do not impact test output, for example lint or cleaning agents.
- Movement of equipment does normally not merit re-validation of the test method, but a limited equipment qualification may be required.

Test method validations should be product and site specific. This means the site and product should be clearly defined in the scope of the test method validation documents. Before an already existing validated test method can be used with a new product or at a new site, the suitability of the existing test method must be documented.

The Code of Federal Regulations (CFR) Title 21 Part 820: Quality System Regulation (QSR) 21 does not clearly call out requirements on method validation. It does not actually mentions the words "method" and "validation" side by side. However, many warning letters have been issued to manufacturing on the subject since at least 2005. Method validation also protects the manufacturer from allowing defective product to circumvent inspection methods if not fit for purpose.

Regulating guidelines from a variety of sources covered in the sections below. The discussion, as it relates to method validation, is somewhat circuitous for medical devices. As stated previously, this is caused by the absence of the phrase method validation in FDA QSR documents. For this reason, some basic treatment relating to process validation is covered below even though this topic is covered in detail in a separate chapter especially because the actual CFR definitions are general enough to lump methods into the category of a process if a CDRH auditor sees fit to do so.

The first sentence of the 21 CFR 820 Quality System Regulation scope states:

"Current good manufacturing practice (cGMP) requirements are set forth in this quality system regulation. The requirements in this part govern the methods used in, and the facilities and controls used for, the design, manufacture, packaging, labeling, storage, installation, and servicing of all finished devices intended for human use."

In this first sentence, FDA has deemed the topic of methods as not excluded from the purview of FDA. This interpretation is evidenced by the warning letters and Form 483's issued to Medical Device companies described in later sections of this chapter.

Per FDA CDRH, the additional validation related definitions are:

Installation qualification: establishing documented evidence that process equipment and ancillary systems are capable of consistently operating within established limits and tolerances.

Process performance qualification: establishing documented evidence that the process is effective and reproducible.

Product performance qualification: establishing documented evidence through appropriate testing that the finished product produced by a specified process(es) meets all release requirements for functionality and safety.

Where process results cannot be fully verified during routine production by inspection and test, the process must be validated according to established procedures [820.75(a)]. When any of the conditions listed below exist, process validation is the only practical means for assuring that processes will consistently produce devices that meet their predetermined specifications:

Method validation as a requirement is not called out specifically; the FDA has issued warning letters and 483s in relation to the lack of "Method Validation".

W.H.O. Guidance

The World Health Organisation issued draft guidance for Test Method Validation of in vitro diagnostic medical devices in December 2016. Technical Guidance Series (TGS) for WHO:

Guidance on Test Method Validation of In Vitro diagnostic medical devices TGS-4

The guidance defines the terms verification and validation as follows:

"Verification is the documentary proof that particular specifications have been met. When designing and developing an IVD, relevant attributes such as cost, and those for performance such as precision, sensitivity and stability are identified and given numerical specifications in design input documentation. It is subsequently the role of the R&D department to design an IVD that will meet those specifications.

The R&D department consequently identifies valid test methods to demonstrate that the specifications have been met (verification) in the new design. Once design has been established, further numerical specifications are produced by the R&D department to ensure that the specifications of each attribute will be met consistently in routine production to ensure quality manufacturing. These new specifications are assigned to control critical production points and may include those for acceptance of raw materials, in-process materials, cleanliness of equipment, qualification of instrumentation and for the finalised IVD to verify its manufacture.

Again, it is also the role of the R&D department to identify appropriate test methods to monitor these specifications. An example of verification is related to incoming goods inspections; each time a raw material is purchased its properties will be verified against the specification using a validated test method."

Validation is the documentary proof that the particular requirements for a specific intended use can be consistently fulfilled. Validation is defined as "verification against needs for a specific use" (i.e. the specification for that use). Within this guide, consistency is essential: it is an expectation that every lot of an IVD will behave as all other lots and will continue to meet design inputs. To ensure this, it is necessary to have validated test methods for measuring and/or monitoring specifications that will consistently produce results fit for purpose. The test methods must be validated to ensure that the results of measuring and/or monitoring are meaningful. For example, the need for accurate measurement of a raw material weighed in micrograms will not be achieved by using a weighing device with tolerance measured in grams. A test method using such an instrument would not be valid for the intended use. Thus, for the example provided, a test method should be specified that has the necessary accuracy and precision for measuring such weights, and an instrument and procedure identified that will consistently achieve this requirement during its use. The test method is then validated to produce results fit for purpose.

Validation of a test method is distinct from its characterisation. Characterisation is documentation of some or all of the features of the method; validation is ensuring that the relevant characteristics are appropriate for the specific intended use. Validation of a method to be used widely, and for standard methods, often begins with complete characterisation. However, for each specific intended use it is likely that only a subset of the characteristics will be relevant and must be evaluated.

ISO 13485, Medical Devices Standard

ISO 13485 is the quality management standard of choice for manufactures of medical devices. Revised in 2016, ISO 13485:2016 "specifies requirements for a quality management system where an organisation needs to demonstrate its ability to provide medical devices and related services that consistently meet customer and applicable regulatory requirements."[1] The scope of the standard can apply to any organisation or company involved throughout the life-cycle of a product, including design and/or development, production, storage and distribution, installation, or servicing of a medical device and design and development or provision of technical or professional services.

The recent revision is designed to address recent developments in quality management and other updated regulations that relate to the industry. Improvements in the new version of the standard include broadening its applicability to include all organisations involved in the life cycle of the product, from the concept stage to end of life along with greater alignment with regulatory requirements and post-market surveillance including complaint handling.

ISO 13485:2016 is also used by suppliers or external vendors that provide QMS related management system-services. Requirements of ISO 13485:2016 are applicable to organisations regardless of their size and regardless of their type except where explicitly stated. Wherever requirements are specified as applying to medical devices, the requirements apply equally to associated services as supplied by the organisation. If any requirement in Clauses 6, 7 or 8 of ISO 13485:2016 is not applicable due to the activities undertaken by the organisation or the nature of the medical device for which the quality management system is applied, the organisation does not need to include such a requirement in its quality management system. For any clause that is determined to be not applicable, the organisation records the justification as part of their certification and quality management system.

Basic Definitions as defined in EU Annex IX of Directive 93/42/EEC)

Intended Purpose: Intended purpose means the use for which the device is intended according to the data supplied by the manufacturer on the labelling, in the instructions and/or in promotional materials. (Chapter I section 1 of Annex IX of Directive 93/42/EEC)

Transient: Normally intended for continuous use for less than 60 minutes.

Short Term: Normally intended for continuous use for not more than 30 days.

Long Term: Normally intended for continuous use for more than 30 days.

Invasive Devices: A device which, in whole or in part, penetrates inside the body, either through a body orifice or through the surface of the body.

Body Orifice: Any natural opening in the body, as well as the external surface of the eyeball, or any permanent artificial opening, such as a stoma.

Surgically Invasive Device: An invasive device which penetrates inside the body through the surface of the body, with the aid of or in the context of a surgical operation.

Implantable Device: Any device which is intended:

- to be totally introduced into the human body or,

- to replace an epithelial surface or the surface of the eye, by surgical intervention which is intended to remain in place after the procedure. Any device intended to be partially introduced into the human body through surgical intervention and intended to remain in place after the procedure for at least 30 days is also considered an implantable device.

Medical Device: means any instrument, apparatus, appliance, material or other article, whether used alone or in combination, together with any accessories or software for its proper functioning, intended by the manufacturer to be used for human beings in the:

- diagnosis, prevention, monitoring, treatment or alleviation of disease or injury.

- investigation, replacement or modification of the anatomy or of a physiological process.

- control of conception which does not achieve its principal intended action by pharmacological, chemical, immunological or metabolic means.

A medical device may be assisted in its function by the following means:

Active Medical Device: any medical device relying for its functioning on a source of electrical energy or any source of power other than that directly generated by the human body or gravity.

Active Implantable Medical Device: any active medical device which is intended to be totally or partially introduced, surgically or medically, into the human body or by medical intervention into a natural orifice, and which is intended to remain after the procedure.

Custom-Made Device: means any active implantable medical device specifically made in accordance with a medical specialist's written prescription which gives, under his responsibility, specific design characteristics and is intended to be used only for an individual named patient.

Device Intended for Clinical Investigation: any active implantable medical device intended for use by a specialist doctor when conducting investigations in an adequate human clinical environment.

Intended Purpose: means the use for which the medical device is intended and for which it is suited according to the data supplied by the manufacturer in the instructions.

Putting into Service: means making available to the medical profession for implantation.

Where an active implantable medical device is intended to administer a substance defined as a medicinal product within the meaning of Council Directive 65/65/EEC of 26 January 1965 on the approximation of provisions laid down by law, regulation or administrative action relating to proprietary medicinal products (6), as last amended by Directive 87/21/EEC (7), that substance shall be subject to the system of marketing authorisation provided for in that directive.

Where an active implantable medical device incorporates, as an integral part, a substance which, if used separately, may be considered to be a medicinal product within the meaning of Article 1 of Directive 65/65/EEC, that device must be evaluated and authorised in accordance with the provisions of this directive.

In Europe, EN ISO 13485:2013 helps companies meet the requirements of: Directive 93/42/EEC on medical devices. This harmonised standard gives companies the "presumption of conformity" to complying with directives. EN ISO 13485 was published in February 2013 and harmonised in August 2013 to cover the three directives:

90/385/ECC– The Active Implantable Medical Devices Directive (AIMDI)

93/42/ECC – The Medical Devices Directive (MDD)

98/79/EEC – In Vitro Diagnostic MDD (IVDMDD)

ISO 13485 has 8 Clauses or Sections which make up the structure of the standard.

Section 0 Normative References, Definitions and Terms

Section 1 Requirements of the Quality Management System (QMS)

Section 2 Normative References

Section 3 Terms and Definitions

Section 4 Requirements of the Quality Management System (QMS)

Section 5 Management Responsibility

Section 6 Resource Management

Section 7 Product Realisation

Section 8 Measurement, Analysis and Improvement

With regard to Test Method Validation, the relevant areas of ISO 13485 include:

(1)   Clause 7: Product Realisation- Section 7.3 Design and Development

(2)   Clause 8: Measurement Analysis

Clause 7: Product Realisation- Section 7.3 Design and Development:

Design and Development Verification and Validation ensure that the product is designed, developed and subsequently manufactured meeting all the customer requirements, regulatory requirements and business requirements. These requirements are classed as inputs to the design and development, and verification and validation ensure the inputs have been adequately taken into account. The design and development testing sometimes replicate the commercial applications of the medical device, hence providing a realistic challenge in order to have confidence in the medical device.

Design Control

Design control is a necessary practice that ensures good engineering principles are maintained throughout the design phase of a product. It also refers to the continual design and development of the product through its very lifecycle. The design and development files and history must be controlled and maintained, with any changes properly assessed, tested and documented.

Clause 8: Measurement Analysis:

Clause 8 includes:

8.1 General requirements

8.2 Monitoring and measurement

8.3 Control of nonconforming products

8.4 Analysis of data

8.5 Improvement

8.1 General Requirements

Measurement, analysis and improvement are the key themes of clause 8. As with all medical devices, inspection and testing both during manufacturing and post manufacturing is necessary to ensure products and services function as intended and without defects. With any type of measurement or inspection analysis, the method used to complete the testing is critical. The method must be fit for purpose, and the equipment must be suitable. This "method validation" typically is done during the design and development phase.

8.2 Monitoring and Measurement

Monitoring and measurement are dependent on the information or feedback provided from various sources. The most important feedback is the post-production feedback that is gathered from customers or the end user. Again, this occurs over the whole lifetime of the product or service in question. There are a number of methods that can be used to obtain feedback. Some examples include:

-Customer surveys

-Customer complaints

-Review of regulatory databases such as MAUDE.

-Repair and servicing information

8.3 Control of Nonconforming Product

Non-conforming product presents a risk to patients or users of medical devices. When a situation arises where non-conforming product is manufactured or detected through inspection processes, the product must be controlled and segregated to prevent unintended use or distribution.

Some examples resulting in non-conformance are:

- When a manufacturing process drifts outside its validation window or operating parameters.

- A certificate of analysis for a raw material is not provided by the supplier or the results are out of specification.
- In-process testing was not completed at the defined intervals.
- Training of personnel completing tests is not current or is inadequate.

8.4 Analysis of Data

In any engineering activity, data and the quality of the data is a key factor in making the right decisions. Provided the data collected is relevant and accurate, analysis of data can provide important insights into process performance, quality control and product functionality. Data should be collated in a consistent way and controlled by a procedure. When it comes to medical device manufacturing, the sources and types of data are multiple. Data can be generated from in-process testing and data can be generated from end of line testing aka finished product testing.

8.5 Improvement

ISO 13485 fosters a culture of continual improvement. As we have seen, each activity can be described as a process. For example, a manufacturing process, a procurement process, a complaints process. The set of processes that make up the quality management system need to be continually reviewed to ensure they are suitable and effective for the task at hand. Typical tools used to keep improvement in mind include:

- Review of the quality policy and quality objectives
- Frequency and category of corrective and preventative actions (CAPA's)
- Customer complaints
- Management review input

Test Methods - Definitions and Key Concepts

Attribute: is defined as the result of a property or characteristic. It is generally used with the terms pass or fail.

Accuracy: can also be defined as trueness. An expression of the closeness of agreement between the value that is accepted, either as a conventional true value or an accepted reference value and the value obtained. A system with low bias implies good accuracy and vice versa.

ANOVA (Analysis of Variance): a statistical method used to evaluate the significance of differences in means due to different factor-level combinations.

Bias: The difference between observed "average of measurements" and a reference value; also referred to as accuracy.

CQA (Critical-to-Quality): a property or characteristic with specific nominal value and appropriate limit and range providing a particular quality attribute.

Critical Process Parameter (CPP): a process parameter that has a direct impact on critical quality attributes.

Dichotomous Variable: an output with only two possible values. Also known as dummy or indicator variable.

Equipment Qualification: establishing documented evidence that the process equipment is suitable for the intended use and is capable of consistently operating within established limits and tolerances under normal operating conditions.

Process Validation: process validation is defined as confirmation via documented evidence that a particular process performs consistently to a high degree of assurance in accordance with predetermined specifications under anticipated conditions.

Measurement Capability Index (MCI): the Measurement Capability Index (MCI) represents the capability of the measurement system. It is used to evaluate the capability of the gauge to classify product against predetermined specifications.

MSA: a study to determine the degree of error involved in measuring the given parameter. The measurement system involves the combination of operations, procedures, gauges, instruments, environmental conditions, people and software.

Precision: the degree of agreement (scatter) between a series of measurements when a method is applied repeatedly to multiple samplings of a homogeneous sample or artificially prepared sample under the prescribed conditions. There are three types of precision; repeatability, intermediate precision and reproducibility.

Range: range is defined as the interval between the upper and lower measurements required. The minimum specified range should be within the equipment range and validated to operate at all points within the range.

Ruggedness (Intermediate Precision): variation on different days or with different analysts and equipment. The extent to which intermediate precision should be established depends on the circumstances under which the method is intended to be used.

Resolution: the smallest unit of measure that can be obtained reliably from a measurement device, also known as gauge discrimination.

Gauge R&R: represents the estimate of the measurement variation. The measurement variation has two components; repeatability or the precision under the same operating conditions (same operator, test method, sample, etc.) and reproducibility or the precision between operators when measuring the same sample with the same gauge.

Variable: is generally the output that is measured.

Validation: confirmation by examination and provision of objective evidence that the particular requirements for a specific intended use can be consistently fulfilled.

New Test Methods

A test method procedure should be created as early on as possible and trialed and examined for completeness and appropriateness. If new test methods are required, a revision controlled draft should be available for the purposes of the test method validation.

Changes to Existing Methods

If changes to existing test methods are required, a redlined version highlighting the changes should be made available for the test method validation.

Method Transfer

If an existing test method is suitable for the test method validation, a suitability report can be completed to document the suitability and show that all factors have been considered (see attachment 1). However, the test method should have been previously validated. The parameters at which the validation is to be conducted must be within the existing validated range. Equipment used in a test method must be assessed to ensure the process is within the equipment qualification. All validation testing must be done on qualified equipment. Equipment qualification is therefore a prerequisite of test method validation.

Test Method Ruggedness Study Protocols

Ruggedness refers to the variation, on different days or with different operators or equipment. The extent to which ruggedness (aka intermediate precision) should be established depends on the circumstances under which the method is intended to be used. An initial ruggedness assessment should be completed to understand the sources of variation. More formal ruggedness studies may be required which should be captured in a formal study protocol. The output of any ruggedness studies should detail any changes or modifications to the test method procedure. Generally, a scoring system is used to describe ruggedness which forms a ruggedness assessment. As a result of ruggedness studies and consequent updates to the procedure, the ruggedness assessment needs to be reassessed. This reassessment should be reflected in the final scores of a Ruggedness Assessment Matrix.

Accuracy

Accuracy is a measure of exactness of the test method output or another way of putting it is the closeness of agreement between a set of test results. For example, take a component that weighs exactly 4 kg according to an NIST traceable scale. If the weight of component is taken 10 times on the balance under study using the test method under study then calculate the mean weight of the 10 readings. The offset between the mean weight and the 4kg "accepted reference value" is a measure of bias.

A large bias = poor accuracy. A small bias = good accuracy.

It is important to note that accuracy does not address the variation between individual measurements.

Simply put, if the average is very close to 4kg, then the test method could have been declared to be very accurate.

It is advised that you consult any relevant standards (e.g. ISO, ASTM) to the product or feature being measured as standards often will call out an accuracy requirement. Generally, results should be accurate to ±1% of the measured value. Therefore, the equipment must be fit for the intended purpose or the measurements in mind.

Note: instrument or equipment accuracy can normally be found on calibration certs provided by the manufacturer or vendor.

Precision

The precision of a method is the degree of agreement among individual test results when the same test method or procedure is applied repeatedly to multiple samplings that represent a population. Precision can be a measure of either the degree of reproducibility or of repeatability of the method. Repeatability refers to the use of a method using the same operator/test person with the same equipment. Repeatability should be assessed using either a minimum of 9 determinations covering the specified range for the method (e.g. 3 concentrations /3 replicates each). Reproducibility refers to the use of the analytical method in different laboratories such as in a collaborative study.

Ruggedness

Intermediate precision (also known as ruggedness) expresses differences related to laboratory variation, as on different days, or with different analysts or equipment within the same laboratory. The extent to which intermediate precision should be established depends on the circumstances under which the method is intended to be used. The effects of random events on the precision of the analytical method should be established. The use of experimental design (matrix) may be used to study the effects of typical variation (dominance factors) on the analytical method (e.g. equipment, analyst, days).

Representative/Continuous Sampling

Representative sampling is used to determine overall process performance (e.g. Pp / Ppk), which is more applicable for processes known or suspected as less than stable or not in statistical control. Sampling in this way best determines overall spread, which includes within-time and time-to-time variation.

Below, some examples are given on how to sample representatively:

1.   Sampling over a given time-period: e.g. a tray of product is produced every 15 minutes, the period of interest is a 1 hour interval and the sample size is 40.

2.   Sampling a batch or product lot not assembled in any order: if the product is packed in a tray (without any grouping) then sample from various sections of the tray.

Consecutive Sampling

This type of sampling involves taking one sample immediately after each another for the subgroup or time period in question and is used to determine process capability (e.g. Cp / Cpk). Consecutive sampling is used in particular to create control charts where a process is sampled in time order by selecting a subgroup sample consecutively and repeating this sampling over a number of subgroups while in same time order. This method is typically used when the process is stable as there will be little or no causes of lot-to-lot variation.

Range

The range is defined as the interval between the upper and lower measurements required. The minimum specified range should be within the equipment range and validated to operate at all points within the range. If an existing test method or piece of equipment is to be used, it is important to determine if the method parameters for the new/modified test method are within the validated range of the equipment qualification. Remember, all validation testing must be done on qualified equipment. Typically, the equipment qualification assessment is documented in the test method validation protocol.

Resolution

We have previously defined resolution as the smallest unit of measure that can be obtained reliably from a measurement device or system. For example, a Vernier callipers may have different models with different resolutions. Some will have only two digits to the right of the decimal point (X.XX mm) and other models could read three digits to the right of the decimal point (X.XXX mm). The instrument resolution should be better than the resolution of the product specification. If the product specification is X.XXX, then at least a "four-digit" measurement device should be used.

Probability Of False Alarms P (Fa)

This signifies the likelihood of rejecting a conforming unit. This is typically an acceptance criterion for attribute tests. Refer to MSA template for further illustration.

Probability Of Misses P (M)

This indicates the likelihood of accepting a non-conforming unit. This also is typically an acceptance criterion for attribute tests. Refer to MSA template for further illustration.

What Can Impact the Accuracy of a Test Method?

Accuracy is influenced by both the instrument (scale) and the test method. If you drop the object on the scale and take a reading before the scale has stabilised, the accuracy is likely to be poorer than when using a test method that demands allowing the scale to stabilise. Examples include: Tensile strength at break - strength does not exist as a material property independent of the test method used to measure it.

For properties like time, distance, and mass, there are NIST traceable standards that can be measured. These standards have a generally accepted reference value that can be compared to the observed readings to assess accuracy (bias). No such reference sample exists for tensile strength at break, deflation time or implant radial strength. For tests without a reference value, the accuracy of the underlying sensor (e.g. load cell) used to determine the output should be addressed if possible.

Components of Test Method Validation

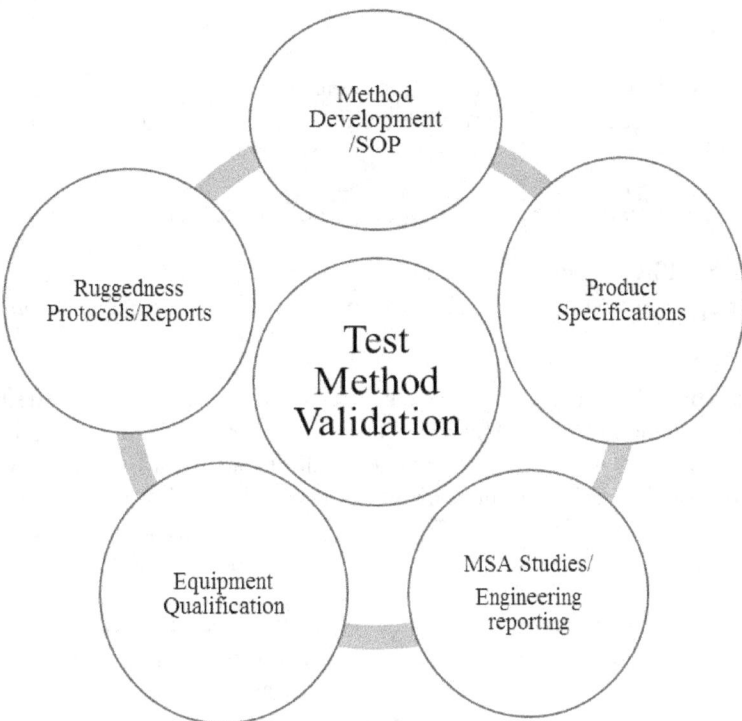

Figure: Typical components of a test method validation.

MSA Studies

A measurement system analysis (MSA) is an experimental design used to identify the elements that affect measurement variation. There are two types of data in which MSA studies can be completed i.e. variable data and attribute data. These terms are defined below. Variable data: data that can assume a range of numerical responses on a continuous scale. Most measurements yield variable data.

Attribute data: data that represents the absence or presence of a characteristic.

Non-destructive tests: test where the measured characteristic is not altered due to testing. Since the sample is not altered, multiple readings can be taken on the sample with the expectation of getting the same measured result.

Destructive tests: test where the measured characteristic is changed due to testing. Since the sample is changed, there is no expectation of getting the same measured result over multiple readings. So, in summary that makes up four types of MSA studies:

- Variable / Non-Destructive
- Variable / Destructive
- Attribute / Non-Destructive
- Attribute / Destructive Table

The following sections describe the requirements, measurement capability indexes and the typical acceptance criteria per MSA type.

General MSA Requirements

Test Environment Conditions - the test environment (i.e. temperature, humidity) should represent the conditions going forward. The effect of multiple environmental conditions can be evaluated if the study is properly designed and planned.

Sample Range - samples should cover the expected range of measurements.

Standard (for attribute MSA) - define the true answer (pass or fail). The standard is based on the inspection ratings of an expert opinion or a measurement system with known better inspection capability than the one under evaluation.

Measurement Instructions/ Training - follow the inspection instructions as defined in the controlled documents or redlines included with the protocol. Do not minimise variability by adding special instructions not defined in the controlled documents or redlines included with the protocol. Reference the controlled documents in the protocol. Special instructions are allowed when using pseudo samples provided that the variability is not minimised due to the instructions. Testers should have a high degree of skill and experience. Do not use new personnel or inexperienced people to conduct measurement studies.

Equipment Qualification and Calibration – The equipment must be calibrated prior to conducting the study. Evidence of the calibrated state should be documented in the report (e.g. calibration certificates etc.). It is important not to re-calibrate the equipment during the study as results can be different due to the calibration effect. The effect of calibration can only be evaluated if the study is properly designed.

Randomisation –

1.   Assign the samples to the first operator in random order. Operator measures the parts.

2.   Assign the samples to the second operator in random order. Operator measures the parts.

3.   Assign the samples to the third operator in random order. Operator measures the parts. Repeat the process described in steps 1 to 3 with the operators for a second and third trial.

Data collection - when documenting the results of a trial, the operator should not have access to the results from the previous trials. A different data collection sheet must be provided for each operator involved in each trial. In lieu of a different data sheet, a data recorder may be used to blind the data recording operator to the test data of previous runs.

Variable MSA Studies

Non Destructive/Variable Msa Studies

The key requirements for non-destructive and variable MSA studies include:

No. of Operators – at a minimum, 3 operators should be used during the study. More operators are also recommended if human/operator interaction is a source of measurement error.

Sample Size – a minimum of 10 units is recommended.

Trials - a minimum of 3 trials should be completed.

Destructive/Variable Msa Studies

If a test is destructive in nature, repeated measurements cannot be taken as the sample is damaged or destroyed as part of the test. One solution is to adopt standardisation of units where homogeneous samples are created by standardising the material or manufacturing process.

- No. of Operators – 3
- Sample Size – 10 units
- Trials – 3 trials

This equates to 90 measurements in total. If standardisation is not feasible, the use of non-destructive pseudo-samples can be used. However, equivalence should be demonstrated between the pseudo sample and "true" units.

Attribute MSA Studies

Non destructive

The recommended and minimum sample size requirements for attribute/non-destructive MSA studies are shown below:

Recommended Minimum Sample Size Requirement

- Operators - 3
- Sample size - 25
- Trials – 3

Destructive

When the test is destructive, repeated measurements cannot be taken as the sample is destroyed or altered. Some approaches are outlined below in order to quantify the measurement variability for destructive tests.

Standardisation Approach: homogeneous and representative samples are created by standardising the method of sample preparation, or material.

Sub-samples: cut each sample into three sub-samples to represent the three trials.

Pseudo-samples: create non-destructive pseudo-samples, documenting a rationale justifying the equivalence of the pseudo samples to the true samples.

Measurement Capability Index

The Measurement Capability Index (MCI) is calculated to assess the capability of the measurement system. The MCI is calculated as a % tolerance.

Measurement Capability Index acceptance criteria:

This index is used to evaluate the capability of the gauge to classify product against the specifications.

The index represents the % of the tolerance (upper specification limit (USL) and the lower specification limit (LSL) that is consumed by the measurement system variation. Figure 9 shows a graphical representation for this index.

Suitability for Use Reports

If an existing test method can be used with no or minor changes, a Test Method SFU Report can be used to document the test method validation.

Suitability for use report is appropriate only if the new product test method parameters are within the existing validated range.

If the test method parameters for the new product are outside of the validated range, the test method must be re-validated. Examples of cases which can utilise such suitability for use reports include:

Test method transfer to a new manufacturing site.

- o New product where the product specifications fall within the validated output range.
- o Minor changes in component material which do not impact the validated test. Examples of changes that require full validation include:

New products:

- o Extension of product sizes that fall outside the validated range.

Suitability for Use

If an existing test method can be used with no or minor changes, a Test Method SFU Report can be used to document the test method validation.

A suitability for use report is appropriate only if the new product test method parameters are within the existing validated range.

If the test method parameters for the new product are outside of the validated range, the test method must be re-validated. Examples of cases which can utilise such suitability for use reports include:

- Test method transfer to a new manufacturing site.
- New product where the product specifications fall within the validated output range.
- Minor changes in component material which do not impact the validated test. Examples of changes that require full validation include:
- New products.
- Extension of product sizes that fall outside the validated range.

Test Method Validation Plan -Example

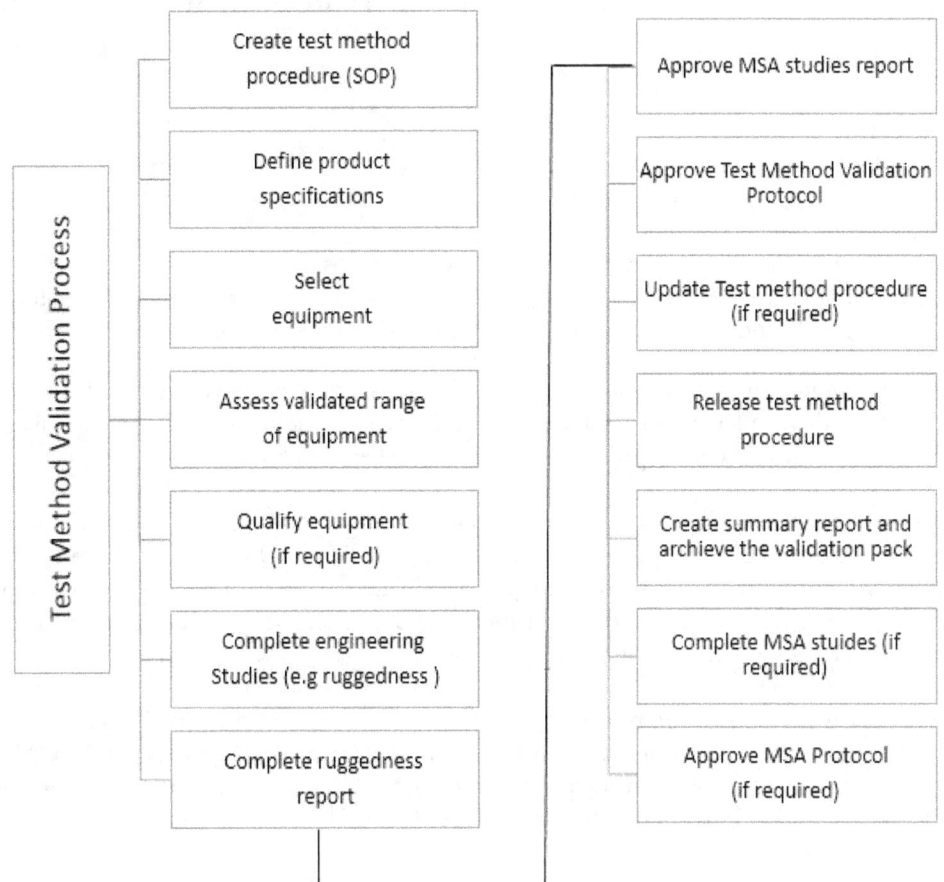

Part 4
- PHARMACEUTICALS AND BIOPHARMACEUTICALS -

4.1 Facilities

Facilities and utilities qualifications are typically prerequisites to the validation of manufacturing equipment and systems. Much of the activity that deals with establishing a facility or building that is *fit for purpose* is managed under the broad heading of commissioning and qualification (C&Q). The terms C&Q are often used interchangeably and in practice some overlap in activity is expected.

Commissioning can be defined as the planned, documented, and managed engineering approach to the start-up and handover of facilities, systems and equipment to the end-user. It must deliver a safe and functional environment that meets the predefined design and user requirements.

In strict terms, qualification is more concerned with the confirmation and documentation showing that equipment or systems are properly installed and functional. Qualification forms part of validation, but the individual qualification steps do not equal a validated process. The establishment of a user requirements specification (URS) and detailed design specifications ensure that the building or facility will meet end-users' needs and that it is fit for the intended purpose.

It also provides a level of protection to the contracting company responsible for the project or facility construction. Post-URS approval requires an approved Design Qualification (DQ). This provides verification and a documented record that the proposed design is suitable for the intended purpose. Further verification including IQ/OP/PQ should be applied as required based on the system impact and criticality of facilities/utilities.

Risk and Impact Assessment

A risk-based qualification process should assess the potential of a system to impact the product quality. The boundaries of any system (HVAC, compressed air supply etc.) should be identified in order to help establish the scope of any system and determine if it has a direct, indirect or no impact on product quality.

Direct Impact: a system that can directly impact product quality.

Indirect Impact: where a system is not expected to directly impact the product quality but supports or is ancillary to a direct impact system.

No Impact: a system that does not directly impact product quality and does not support a direct impact system.

Example of HVAC System Boundaries

For a HVAC System supplying a classified area, only once the air enters the room must the air quality meet the classified designation. The Critical Quality Attributes (CQAs) are routinely monitored through the Environmental Monitoring Programme and the Critical Process Parameters (CPPs) should be monitored through the calibrated and validated Environmental Monitoring System (QBMS). The direct impact (level 1) for the HVAC systems are indicated on the boundary diagram shown below.

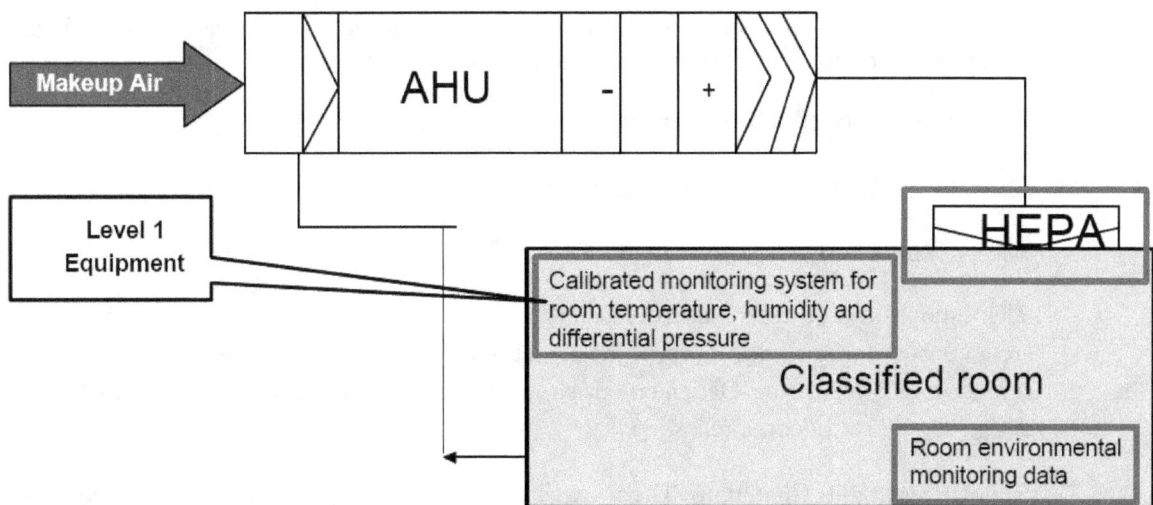

Figure : HVAC System Boundary Diagram (Level 1). Each individual system is represented by a green box. Separate qualifications should be performed for each one. The room environmental monitoring system is typically called a Building Management System (BMS). The *calibrated* monitoring system for room temperature, humidity and differential pressure is called a QBMS (where the "Q" stands for quality indicating the system is used to monitor critical parameters).

Qualification Levels

Qualification levels are often used within companies to classify the criticality of equipment or systems. Level 1 requires the highest level of verification.

Level 1: a system where an **undetected change** in system performance poses a significant risk to the product and product safety. Level 1 systems require the highest degree of qualification and validation. This should include URS/DQ/IQ/ OQ/ PQ.

Level 2: a system where a change that **may be detected** in system performance poses a significant risk to product and product safety. These systems require a level of qualification including IQ, however, OQ and PQ testing may not be required. This should be based on the intended use of the system, impact on product quality and overall risk.

Level 3: All other systems.

Typically IQ or equivalent testing is sufficient. **Note:** other requirements or qualifications should be based on risk.

The level of qualification and validation testing required for any system should be based on a risk assessment, examining the criticality of the system and environment. Risk assessments should consider the following points:

- Building design and construction features
- System boundaries and complexity
- Potential product impact
- Environmental controls and monitoring systems
- Potential impact to operator safety
- Type of qualification/validation (e.g. prospective, concurrent, or retrospective)

Controlled-not-Classified (CNC) environments, utilities, and facility control systems also require adequate qualification/validation. Again, the impact on product quality should be determined in order to shape any validation. Routine monitoring test locations as well as alert and action levels should be determined in advance of any validation for environmental monitoring or utility systems.

Clean Room Design Considerations

Air Handling Unit (AHU) -Air Intake Quality

Seasonal Variations

All locations on earth except latitudes near the equator experience seasonal temperature changes. The changes are a consequence of Earth's orbital motion about the sun, coupled with the tilt of its axis of rotation with respect to its orbital plane. Design criteria should be based on published temperature data. The HVAC system design should consider the following:

Standard Operating Conditions: These are climatic conditions against which the systems must be designed to operate, control, and maintain required conditions. (These may be based on published data, which are only exceeded 2.5% or 1% of the time).

Extreme Operating Conditions: These are climatic conditions against which the systems must be designed to operate, without manual intervention, and without damage to the systems or the facility. Based on product / process risk assessments, extreme or standard conditions shall be used for HVAC design for dedicated areas.

Location

Based on the building layout, footprint and design intent, a suitable and adequate space must be identified for HVAC location. This must include provision of chilled water, heating systems, ducts and drainage. HVAC plants must be accommodated in designated HVAC plant rooms or interstitial areas.

Air Intake

During the design phase, the air intake locations should be selected to ensure air is in the best environmental condition. The below considerations help to achieve a strong starting point:

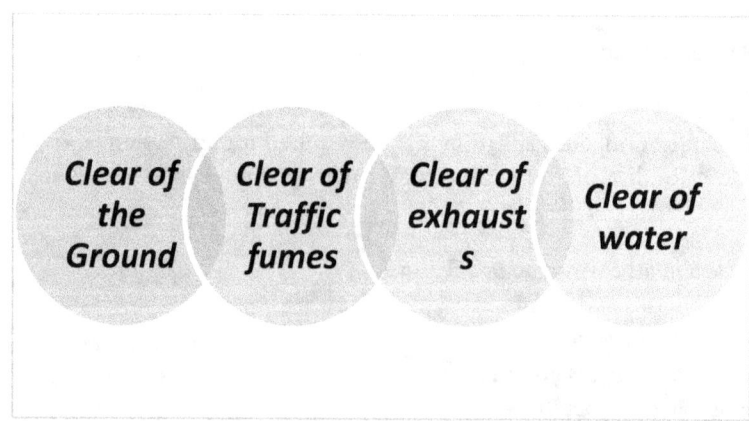

Figure : Air intake considerations

Thermal Load

Thermal load can be defined as the amount of heat energy to be removed from an inner environment by equipment (HVAC) used to maintain that environment at the design temperature when worst case external temperature(s) are being experienced. The thermal load requirement should be calculated for the following:

- ➢ Max summer conditions
- ➢ Minimum winter conditions
- ➢ High rainfall
- ➢ Standard operation
- ➢ Extreme operating conditions

Room Recovery Time

Room recovery time to return to the required pressure differential and cleanliness

Specification should be minimised.

Dust, Vapour, or Fume Control

Highlight areas requiring dust, vapour, gas and/or fume control on the room data sheet. These areas must be controlled to remove the possibility of product contamination and to ensure the safety of the operator and environment. Areas requiring 100% fresh air or extraction to atmosphere may require greater airflow or other measures within the room to maintain environmental conditions.

In order to meet the appropriate level of cleanliness, HVAC systems require sufficient filtration to provide "clean" air to prevent contamination of the product. Pre-filters and main filters are normally suitable for most operations; however, HEPA filters are required to prevent particulate or microbial contamination for higher-classification areas

Air Change Rates

The air change rates for each room must be calculated to be sufficient for clean-up to achieve specified particulate conditions "at rest" in static conditions after a maximum of 20 minutes from completion of operations. The actual air change rate must be chosen to satisfy the most stringent requirements including GMP, GLP, heat gain, ventilation requirements and/or occupancy, including an appropriate safety factor.

The air change rate must be optimised for energy savings; however, specific attention must be paid to air locks where a greater air change rate must be applied. Air changes can be reduced (e.g. setback modes) in some circumstances ("at rest" mode, with no production activity and no personnel interventions).

Room Environmental Conditions

Other environmental conditions to be controlled, such as temperature and relative humidity, depend on the product and nature of the operations carried out in those areas. These parameters should not interfere with the defined cleanliness standard.

Temperature Requirement

The normal operating temperature requirement for each classification. Temperature and humidity must be appropriate to the product and process. Consideration should be made for specific product and process requirements.

<u>Humidity Requirement</u>

The normal operating humidity requirement.

<u>Particulate Levels</u>

Particulate levels are specifically defined for each room classification "at rest" and "in operation". The levels are controlled though air filtration, facility design, gowning requirements, and decontamination

<u>Room Exhaust</u>

Where there is a risk of active compounds being present in extracted air, filters should be fitted, preferably in the room, to prevent contamination of ductwork and the environment. The filters must be selected based on the particle size distribution of the products to be handled.

HVAC System Design

The HVAC system must be appropriately selected using the specific design requirements as outlined above. The system must be able to provide clean, conditioned air to the specified areas to meet all of the quality requirements. The most important precursor to HVAC design is the comprehensive definition of the function and performance required followed by the selection of an appropriate system. A poor selection can lead to unnecessarily high-energy consumption, and operational deficiencies. All-air systems rely on the movement of large quantities of air through a central air handling unit to control room conditions, as well as provide for ventilation requirements. They have the advantage of being relatively simple with most of the unit situated in one location; however, they are very space consuming. All-air systems tend to be relatively inflexible and not ideal for areas that are likely to need environmental alteration on a regular basis.

These HVAC systems are used for areas that have a lot of small zones, each with slightly different thermal loads but which requires constant ventilation. These systems can have poor energy efficiency if a lot of reheat is required. These are typically used in large manufacturing areas, and laboratories with many small rooms.

<u>Dust Extraction and Collection</u>

It is essential to capture dust as close as possible to the point of generation without affecting the process. In most cases dust capture should be within 100mm from the point of release. Air velocity is the key parameter in dust capture.

Pharmaceutical and chemical applications have specific collection requirements as any dust build-up in the system is likely to be of a pharmacologically active nature, sensitising, toxic and/or corrosive. It is vital to maintain transport velocities and minimise any potential for cross contamination.

A typical system should have a minimum transport velocity of 18 m/s, but this may need to be higher if heavy particles are to be collected. This velocity must be maintained throughout the system to prevent dust from dropping out in the ducts. The dust collection must be configured with the hazardous nature of the dust in mind. A clearly defined disposal procedure for the collected dust (e.g. bag-in / bag-out system for filter and dust bin) needs to be understood at the design stage. HVAC unit shall meet EN 1886 and EN 13053 requirements.

<u>Fans</u>

Certified performance curves are required to verify correct fan operation. Fans that may be subjected to high temperatures, humidity, corrosive fumes or other hazardous atmospheres should be constructed using non-reactive, non-corrosive, suitable and approved materials (such as epoxy painting). Whenever H2O2 or other disinfection application is planned, material compatibility certificates shall be supplied by the vendor.

Fans must be selected to supply the design volume, taking into account the assumption that filters are half clogged, except for the terminal filter which shall be considered to be fully clogged according to EN 13053. If the terminal filter is HEPA, clogging shall be considered according to EN 1822 and the target volume is 80% of the given maximum clogged specified value.

Filtration

Face-fitting filters shall be used in all cases, as slide-in filter elements never give a good seal. The installation must be such that the airflow pushes the filter against the seal. The face velocity across the filter section shall not exceed 2 m/s. For ventilation and air conditioning applications, two minimum filtration stages are required. For certain applications, return air filtration will be required to contain highly active materials (e.g. viruses or potent compounds). Normally, these filters should be changed from the room side. However, since those filters must be integrity tested, it is recommended to place one filter in the main return duct before the exhaust fan and design return duct network, in order to ensure tightness of the duct between the room and the filter (bag-in / bag-out filter change systems should be provided for BSL-3 areas). In case of live biological agent biocontainment, decontamination up to the filter must be proven. The grade of filter and technical solution must be selected based on the product particle size distribution and occupational exposure band (OEB) level.

HEPA filters and Dehumidification

For most HVAC applications, dehumidification is best achieved by the use of cooling coils. It should be noted that dehumidification is a very high consumer of energy and should only be used if there is a real process need. When areas are not in use, the dehumidifier should be turned off, if possible.

When room humidity must be maintained below 50% during warm weather, an absorption dryer may be necessary unless the room temperature can be increased within specification to compensate.

Normal practice is to optimise size and efficiency of the absorption dryer by first removing as much moisture from the air as possible by cooling. The design of absorption dryers is normally based on a slowly rotating desiccant wheel.

Air is passed through the wheel and dried by the desiccant coating (guidance: lithium chloride especially if the wheel is not used frequently and silica gel if used permanently and with low humidity target). It is not normally necessary to size a dryer to handle the entire air volume. Drying a proportion of air and re-mixing to achieve the desired moisture content is usually sufficient.

Air humidification may be necessary during cold weather when introducing fresh air to spaces that require humidity control. When air humidification is necessary, humidifiers should be selected on the following basis:

> direct steam injection using steam
> direct steam injection using self-generative electric or gas steam humidifier.

Humidifiers should be located before the fan and the final filter which will remove any particulate generated. At least 300 mm clearance should be allowed upstream and 1 m downstream between humidifier manifolds and coils, attenuators etc. (general recommendation to be confirmed through calculation note provided by the vendor). A single manifold or multiple manifolds in parallel may be used to meet the humidification requirements as per manufacturer's recommendations.

Sound Attenuators

Sound attenuators should be provided as necessary, to achieve the specified noise levels within occupied spaces. To minimise external noise nuisance, assessment can confirm the necessity to use acoustic media (enveloped in polyester film), that is inert and corrosion-resistant at normal operating conditions. Material quality shall be equivalent to that specified for HVAC unit or ducts.

Sound attenuators should be installed in the air handling unit or ductwork. The use of sound attenuators in the air supply and air return should be based on requirements for fresh air inlet and air exhaust, and according to external noise levels that might need to be maintained at or below the ambient site noise levels.

<u>Dampers</u>

The provision of sufficient dampers is essential for proper control. To minimise noise transmission into the room, these should be mounted as far as possible from the diffuser.

Carefully evaluate the space-by-space pressure control that will be used in the design. Static pressure control via hard balance or dynamic control via air terminal control units are both appropriate. Consideration should be given to the overall project size, the complexity of the facility and the project budget.

Automatic volume controllers are recommended for regulating air volume independently of supply pressure. They can be selected for constant volume, variable volume or dual duct mixing applications. Automatic low-leakage fresh air and exhaust air shutoff dampers are strongly recommended to isolate the HVAC network. Fresh air dampers shall be Class 3 minimum (maximum leakage preventing coil freezing). Whenever fumigation is performed shutoff damper shall ensure Class 4 leakage rate. Where dampers are required to provide modulating control of airflow, they must be selected to provide an appropriate level of control authority. This will normally mean a damper smaller than the duct size.

Heating and Cooling

Heating mode: Low pressure hot water (LPHW) is the preferred heating medium for HVAC applications and should be used whenever practicable. Electrical heating should be avoided due to fire risk and should be limited to low power coil and in locations where no other energies are available. Hazard operability analysis (HAZOP) must be conducted if electrical heating is being considered. Cooling mode: Chilled water is the preferred cooling medium for HVAC applications and should be used whenever practicable.

The direct expansion of refrigerant in coils is an acceptable method of cooling, particularly on small isolated plants, or where lower temperatures are needed for dehumidification or for cold room. This system, however, does not normally give close control. Direct expansion coils should only be used with extreme care on variable air volume systems (if speed driver available on compressors).

<u>Heating Coils</u>

The face velocity of air across heating coils should not exceed 2 m/s. Coils should be made of material suitable for applicable constraints. Drains shall be located outside the casing of the HVAC unit. Coils shall be removable.

<u>Cooling Coils</u>

Cooling coils have been identified as potential sources of microbial contamination; therefore, careful design is required to prevent water carryover and to ensure that drain pans do not retain water. Double tube, non-welded units are recommended. The face velocity of air across cooling coils should not exceed 2 m/s. Where necessary, stainless steel or plastic eliminator blades should be provided to prevent any moisture carryover. Where provided, these must be removable for cleaning.

<u>Ductwork</u>

For most applications, galvanised steel ductwork will be the most appropriate form of construction; however, stainless steel or plastic construction may be necessary where there is a higher risk of corrosion due to moisture or fumes (exhaust ducts usually). Where operating pressures above 2,000 Pa are necessary, fully welded construction is recommended. For contained ducts (e.g., exhaust duct before bag-in / bag-out filter), air tightness Class C shall be followed (EN 12237). For BSL-3, fully welded construction should be considered.

Generally ductwork should be constructed to an appropriate local standard, suitable for the maximum design pressure (positive or negative), such as those published by Sheet Metal and Air Conditioning Contractors' National Association (SMACNA) in the USA, Building and Engineering Services Association (B&ES) in the UK . Where flexible connections are proposed these must be designed for the same pressure as the ductwork. Solid ducted connections are preferred for final connections to terminal HEPA filter housings. For applications where flexible connections to diffusers are used, these should be no longer than 500 mm and nominally straight.

Special consideration must be given to fume extract ducts where these pass through fire barriers. Using fire dampers should be avoided where the loss of extraction could make a fire situation worse. An alternative design, such as the use of fire-rated ductwork, may be necessary in these cases. A thorough risk assessment must be conducted.

Simple Representation of HVAC system

Position	Description
1	Fresh air intake (°C, %RH, flow rate)
2	Dampers
3	Filter creating a differential pressure
4	Filter creating a differential pressure
5	Control valves for cooling fluid
6	Exhaust fan
7	Steam flow rate
8	Supply fan
9	Filter creating a differential pressure
10	Controlled room/ area
11	Extraction

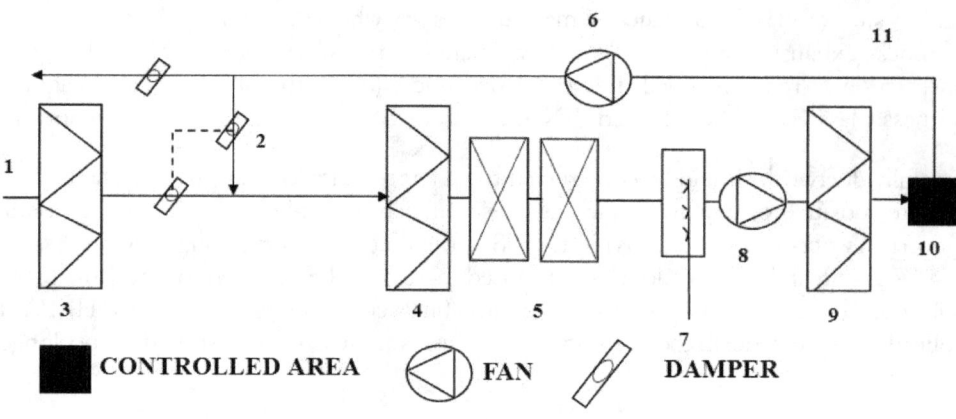

Figure: Simple HVAC diagram

Parameter	Description
Temperature	The HVAC must be capable of operating over a range of temperatures and accurate to a tolerance. Temperature probes/detectors must be placed at various points to provide feedback and control.
Relative Humidity	Relative humidity must be monitored continuously. Typically, humidity sensors should be effective over an operational range (e.g. 5-95% R.H.) Accuracy should also be no less than $\pm3\%$
Air flow/ Air Pressure	Air flow is proportional to the square root of the differential pressure.
Dampers	Dampers are used to control inlet and outlet airflow.
Valves	Ball valves or globe valves control the flow of air. Valves are designed with specific safety features to meet the intended use (e.g. CLOSED without energy supply-cooling valve, OPEN without energy supply-heating valve.

Environmental Monitoring

An environmental monitoring programme is required for GMP controlled areas. The purpose of such programmes is to document, define and describe parameters to be monitored, monitoring frequency and methods. Environmental monitoring is a regulatory requirement. It also demonstrates that the GMP areas are been controlled and are fit for purpose.

Key Requirements of Environmental Monitoring

Identification and classification of environmental areas that require monitoring
Test methods and sampling procedures
Defined testing frequencies
Sample locations based on Risk
Microbial monitoring of personnel
Monitoring of non viable particles
Monitoring of temperature, relative humidity and differential pressures
Defined alert and action levels for each environmental area
Trending of Enviromental data
Change Control

Figure: Key requirements of environmental monitoring

Other parameters such as those controlled by the HVAC system (air changes/hour etc.) should also be verified according to a defined schedule.

<u>Grade A, B and C</u>

➤ Viable and non-viable particles monitored under operational conditions
➤ Risk-based approach to sampling points that represent high risk/critical positions

<u>Grade D</u>

➤ Non-viable particles must be measured at-rest conditions
➤ Viable particles measured under operational conditions

Building Management Systems

A Building Management System (BMS) is an automated control system that is used to manage building and facilities heating, ventilation and air conditioning, security, fire protection systems and so on. It is made up of many different input / output subsystems, controller(s), server(s) and workstation(s) communicating over a control network to control, monitor, alarm and trend equipment. . BMS systems are also referred to as a Facilities Management/Monitoring System (FMS), Energy Management System (EMS), Building Automation System (BAS) or other equivalent.

Environmental Monitoring Systems (EMS) are automated control systems consisting of input / output subsystems, controller(s), server(s) and workstation(s) communicating over a control network to monitor, alarm and trend environmental critical process parameters such as temperature, humidity, differential pressure, conductivity, cooler / refrigerator status amongst others. Suggested classification of BMS based on intended use:

Building Management System (BMS)	
System Classification	GxP.
Data Usage	Data not used for GxP impacting decisions Engineering use only.
Monitoring	No critical process parameters are monitored by the system.
Controls	No GxP equipment.
System Boundaries	Up to the point of use of the system or equipment.
Validation	Not required.
Environmental Monitoring System (BMS)	
System Classification	Non GxP.
Data Usage	Data may be used to make quality decisions and product release decisions. Data is used to determine compliance.
Monitoring	Critical process parameters are monitored by the system.
Controls	Critical alarm limits are controlled.
System Boundaries	From the point of use.
Validation	Not required.

Contamination Control

The philosophy of containment control requires it to be applied across all inputs that make up a facility, equipment, processes, utilities and so on. Containment is primarily concerned with keeping things in by preventing product or processing agents from egressing into the surrounding atmosphere. Ensuring adequate containment protects personnel who interact with the process, equipment and systems. Aseptic processing often deals with biological agents or compounds that may be harmful to operators or technicians. A secondary concern of containment is protection of the environment. Containment also complements efforts in contamination prevention. As with aseptic processing the risk to the patient and product must be at the forefront of activity. Risk-based approaches and tools should be used to identify potential risks and put in place adequate controls and mitigations. Any assessment should take into account all the following systems:

- ➤ Facility layout
- ➤ Drainage systems
- ➤ HVAC requirements
- ➤ Location and adequacy of utilities
- ➤ Personnel flow and procedures for entering and leaving
- ➤ Behavioural requirements of personnel in the clean room
- ➤ Flow of materials and products to prevent cross-contamination and mix-ups between products and between dirty and clean or sterile and non-sterile equipment and products
- ➤ Design to avoid cross-contamination when manufacturing live biological agents, e.g. local exhaust air HEPA filtration, dedicated air handling units.

Material Flow

The design and layout of any manufacturing area should facilitate the effective flow of materials. This is a fundamental requirement no matter what the industry, e.g. medical devices, pharmaceuticals, bio pharmaceuticals and even non-regulated engineering companies that assemble, machine or fabricate products. However, the manufacture of medicinal products that are required to be sterile imposes a greater level of control and thought. With regard to aseptic processing facilities, material flows do not only require efficient and effective flow of materials; the activity should also support the requirements of aseptic processing while minimising any risk of contamination. Identifying critical processing zones is a crucial step in ensuring the right building design and controls are implemented. Isolators and aseptic filling require the highest classification with strict environmental controls. Secondary packaging operations such as cartonning are often completed in areas controlled and operated to a lower classification.

Design and layout of facilities should:

- ➤ Maintain microbiological integrity of the identified critical processing zones
- ➤ Prevent or minimise contamination from outside critical processing zones
- ➤ Control the flow of materials by restricting access to trained and authorized personnel

Material Transfer

Material transfer from the outside of clean rooms to the inside is completed via material air locks or hatches. Material air locks and hatches ensure that there is clear separation between controlled clean areas and less clean areas. Many suppliers provide products that are double bagged. This provides an added level of control when transferring materials. The outer bag can be removed within the air lock thus providing a clean inner product. Material air locks also allow the sanitisation of products. Tools and other items must be clean and dirt free.

Controls that prevent personnel from the clean area and less clean area being present in the material air lock at the same time. This can be achieved by training and educating staff on the importance of contamination control. A simple visual check of the air lock to confirm it is vacant can be done in order to avoid mixing of personal from different zones. Decontamination procedures are necessary to ensure materials or tools entering the controlled area are decontaminated.

Material Air Lock Considerations:

> ➤ Interlocked doors
> ➤ Access control
> ➤ Sanitation/cleaning procedure
> ➤ Double or triple bagged products
> ➤ Dedicated trolley for air locks

Disinfection and Cleaning Agents

When materials are being transferred via an air lock, consideration must be given to the status of materials and products. As a rule, no cardboard or unnecessary paper should enter a clean room. Wooden pallets are not acceptable as they can carry dirt and microorganisms and wood cannot be sanitised due to its porous nature. Soft fabric cases often used to carry tools should also be avoided as the material can carry dirt and grease. Cleaning and disinfecting agents should be tested and approved prior to their use onsite. The choice of agents should be backed up with studies that demonstrate the effectiveness of disinfectants and cleaning agents.

Gown-Up Areas

Gowning rooms are designed in order to minimise contamination and facilitate the orderly change over from street clothes to scrubs and/or gowns. Hand washing facilities help reduce the risk of humans carrying unwanted microorganisms into the aseptic processing area. The design of the room should result in clear separation between the less clean side and the clean side. This can be achieved with a step-over segregating the two areas.

Other features of gowning rooms should include:

> ➤ Storage lockers for street clothes
> ➤ Gown and garment storage
> ➤ Body-length mirrors
> ➤ Hand washing /drying and disinfection facilities

GMP Zoning

Selecting a suitable classification for a room or manufacturing facility depends on several factors. Firstly, it can be said that sterile products require a more stringent set of criteria than non-sterile products. However, there is an extensive range of products and medical devices that are sterile but are used in different ways and consist of different materials and technology. Some sterile products are single-use only and used for short term purposes and then disposed of. Other sterile products are used subcutaneously for longer periods or even require implantation. Therefore, the design of a facility along with its HVAC specification must be appropriate to the product being manufactured. High-risk products require greater control. The goal of facilities and HVAC systems is to minimise contamination and the associated risks. Using a "sterile versus non-sterile" rule of thumb is not adequate when classifying a room or facility. Standards including EN ISO 14644-1 and guidelines such as EU cGMP Guidelines EudraLex volume 4 Annex 1 (2008) should be consulted in order to fully understand the requirements of each ISO classification and grade of room.

ISO classifications do not specify room occupancy states but when a designation is applied, the occupancy state must be stated in the relevant documentation or procedure. The most relevant European guideline (Annex 1 of the EU cGMP Guideline) lists four classification grades and their associated particulate limits in the 'at-rest' and 'in-operation' conditions. In general, for the sterile and non-sterile products, similar classes are applied, but in non-sterile production the producer could assign their classes, having similar particulate concentration, temperature, pressure etc. but lower air-change rate could be used.

Types of Contamination:

- cross contamination (of a product/material with another product/material)
- non-microbial particulate contamination (non-viable particles)
- biological/microbiological contamination (viable particles/micro-organisms)

Factors Influencing Contamination Cleanliness Levels in the Manufacturing Processes:

- process
- air cleanliness
- personnel hygiene and clothing
- work practices
- material design (material of construction, surface finishes, room finishes, equipment, open system/enclosed system, utensils, etc.)material cleanliness

Room Air Classification (By Limits of Particulate Contamination)

ISO CLASS	FDA	cCMP	Permissible particle number in 1 m3					
			0,1 μm	0,2 μm	0,3 μm	0,5 μm	1 μm	5 μm
1			10	2				
2			100	24	10	4		
3	1		1,000	237	102	35	8	
4	10		10,000	2,370	1,020	352	83	
5	100	A	100,000	23,700	10,200	3,520	832	29
6	1,000	B	1,000,000	237,000	102,000	35,200	8,320	293
7	10,000	C				352,000	83,200	2,930
8	100,000	D				3,520,000	832,000	29,300
9						35,200,000	8,320,000	293,000

Figure 2: Table showing ISO classes and EudraLex Grades A-D.

| Grade | Maximum permitted number of particles per m³ equal to or greater than the tabulated size | | | |
| | At rest | | In operation | |
	0.5 μm	5.0 μm	0.5 μm	5.0 μm
A	3 520	20	3 520	20
B	3 520	29	352 000	2 900
C	352 000	2 900	3 520 000	29 000
D	3 520 000	29 000	Not defined	Not defined

Figure 3: maximum permitted airborne particle concentration for each grade. Showing both "at-rest" and "in-operation" conditions (EU V4 Annex 1). The EU guidance given for the maximum permitted number of particles in the "at-rest" column corresponds approximately to the ISO classifications.

Room Air Classification (By Limits of Microbial Contamination)

The HVAC systems help maintain the viable (microbial) limits within a specific area. These limits are defined in Annex 1 of the EU GMP Guide as shown below.

| Grade | Recommended limits for microbial contamination (a) | | | |
	air sample cfu/m³	settle plates (diameter 90 mm) cfu/4 hours (b)	contact plates (diameter 55 mm) cfu/plate	glove print 5 fingers cfu/glove
A	< 1	< 1	< 1	< 1
B	10	5	5	5
C	100	50	25	-
D	200	100	50	-

Figure 4: Recommended limits for microbial contamination

Environmental Grade A (Aseptic)

Grade A is reserved for critical processes in manufacturing sterile products, product components or product contact. This is generally achieved using isolator technology which maintains a barrier to the background environment or surrounding room.

Grade A Operations include:

➤ Aseptic processing of sterile ingredients
➤ Filling of sterile products not for terminal sterilisation
➤ Stopper insertion
➤ Crimp capping

Environmental Grade B

Grade B is used for supportive work for aseptic processing corresponding to ISO 14644 (Part 1) Class 5 ("at-rest") and Class 7 (when "in-operation"). Grade B areas typically serve as the background environment of Grade A areas for aseptic processing.

Environmental Grade C

Suitable for non-critical processing steps, Grade C corresponds to ISO 14644 Part 1 Class 7 ("at-rest") and Class 8 ("in-operation"). Grade C operations include:

> ➤ Clean side of material air locks and gowning rooms
> ➤ Filling of products that are to be terminally sterilised

Environmental Grade D

Grade D at least corresponds to ISO 14644 Part 1 Class 8 ("at-rest" / no definition for "in-operation").

> ➤ Clean section of material air locks and final compartments of gowning rooms
> ➤ Dispensing of raw materials and excipients and preparation of solutions for sterile products to be sterile filtered and terminally sterilised
> ➤ Background environment for transfer and crimp capping of stoppered containers with sterile products

Compliance Tests for GMP Zones

Test	Requirements
Particle count test	Test covers verification of cleanliness. Dust particle counts to be carried out and result printed. The number of readings and positions of tests should be defined in accordance with ISO 14644-1 Annex B5.
Air pressure difference	This test is used to verify non cross-contamination. Log of pressure differential readings to be produced or critical plants should be logged daily, preferably continuously. A 15 Pa pressure differential between different zones is recommended. Refer to ISO 14644-3 Annex B5.
Airflow volume	To verify air change rates. Airflow readings for supply air and return air grilles to be measured and air change rates to be calculated. Refer to ISO 14644-3 Annex B13.
Airflow velocity	To verify unidirectional flow or containment conditions. Air velocities for containment systems and unidirectional flow protection systems to be measured. Refer to ISO 14644-3 Annex B4.
Filter leakage tests	To verify filter integrity. Filter penetration tests to be carried out by a competent person to demonstrate filter media, filter seal and filter frame integrity. Only required on HEPA filters. Refer to ISO 14644-3 Annex B6.
Containment leakage	To verify absence of cross-contamination. Demonstrate that contaminant is maintained within a room by means of: • airflow direction smoke tests • room air pressures. Refer to ISO 14644-3 Annex B4.
Recovery	To verify clean-up time. Test to establish time that a clean room takes to recover from a contaminated condition to the specified clean room condition. Should not take more than 15 minutes. Refer to ISO 14644-3 Annex B13.
Airflow visualisation	To verify required airflow patterns. Tests to demonstrate air flows: • from clean to dirty areas

	• do not cause cross-contamination • uniformly from unidirectional airflow units Demonstrated by actual or video-taped smoke tests. Refer to ISO 14644-3 Annex B7.

Further reading

ISO 14644-1: International Organisation For Standardisation – Cleanrooms and Associated Controlled Environments. Part 1: Classification of Air Cleanliness.

ISO 14644-3: International Organisation For Standardisation – Cleanrooms and Associated Controlled Environments. Part 3: Test Methods.

ISO 14644-4: International Organisation For Standardisation Cleanrooms and Associated Controlled Environments: Part 4: Design, Construction and Start-Up.

EudraLex, Vol 4, Annex 1: EU Guide to Good Manufacturing Practice (EU GGMP) Governing Medicinal Products for Human and Veterinary Use, Annex 1 – Manufacture of Sterile Medicinal Products.

EN 1822:2009: European Standard For HEPA Filter Classification.

US FDA CFR 211: Code of Federal Regulations Food and Drug Administration Title 21 Part 211 – Current Good Manufacturing Practice for Finished Pharmaceuticals – Section 211.46 Ventilation, Air Filtration, Air Heating and Cooling.

ICH Q7: International Conference on Harmonisation - Good Manufacturing Practice Guide for active Pharmaceutical Ingredients – Section 4.21 and 4.22 – Utilities.

US FDA: Food and Drug Administration - Guidance for Industry "Sterile Drug Products Produced by Aseptic Processing – Current Good Manufacturing Practice".

4.2 Utilities

Introduction

The term "clean utilities" in the life science industry refers to utilities that have to fulfil regulatory requirements. The most common utility is water, which can be supplied in different pharmaceutical grades of purity. Purified water (PW or PUW), highly purified water (HPW) and water for injection (WFI) are the most common. Water quality specifications can be found in the pharmacopeias, e.g. the US Pharmacopeia. Other clean utilities can also include clean compressed air, clean gases (e.g. nitrogen, argon and oxygen), and clean steam.

Key Definitions

Alert Limit: a value reached when the normal operating range of a critical parameter has been exceeded, indicating that corrective measures may need to be taken to prevent the action limit being reached.

At-Rest: a condition where the installation is complete with equipment installed and operating in a manner agreed upon by the customer and supplier, but with no personnel present.

Clean Room: an area (or room or zone) with defined environmental control of particulate and microbial contamination, constructed and used in such a way as to reduce the introduction, generation and retention of contaminants within the area.

Containment: a process or device to contain product, dust or contaminants in one zone, preventing it from escaping to another zone.

Contamination: the undesired introduction of impurities of a chemical or microbial nature, or of foreign matter, into or onto a starting material or intermediate, during production, sampling, packaging or repackaging, storage or transport.

Point Extraction: air extraction to remove dust with the extraction point located as close as possible to the source of the dust.

Pressure Cascade: a process whereby air flows from one area, which is maintained at a higher pressure, to another area at a lower pressure.

Relative Humidity: the ratio of the actual water vapour pressure of the air to the saturated water vapour pressure of the air at the same temperature expressed as a percentage. More simply put, it is the ratio of the mass of moisture in the air, relative to the mass at 100% moisture saturation, at a given temperature.

Turbulent Flow: turbulent flow, or non-unidirectional airflow, is air distribution that is introduced into the controlled space and then mixes with room air by means of induction.

Identifying Critical Utilities

The process of identifying critical utilities can be done with the application of direct impact, indirect impact and no impact definitions (see previous section "*risk and impact assessment*"). Risk assessments, CQAs and CPPs should also help identify critical utilities. When critical utilities are required as part of manufacturing and processing, the following points should be examined during the requirements and design stage:

- Materials of construction
- Internal surface finishes
- System sizing
- Flow rates, dead legs, drainage etc.

The process of identifying critical utilities can be done with the application of direct impact, indirect impact and no impact definitions (see previous chapter). Risk assessments, CQAs and CPPs should also help identify critical utilities. When critical utilities are required as part of manufacturing and processing, the following points should be examined during the requirements and design stage:

➢ Materials of construction
➢ Internal surface finishes
➢ System sizing
➢ Flow rates, dead legs, drainage etc.

Compressed Air

Compressed air is used for valve actuation, instrument air and process air to name but a few applications. Only the point-of-use filtration and the gas quality instrumentation should be classified as level 1. When flow or pressure is a CPP, the measurement/monitoring should be performed by the system into which the gas is flowing. Additionally, the CQAs and CPPs should be routinely monitored through the calibrated monitoring system. For compressed air, the potential CPPs are listed below. For the physical system being evaluated, the use and the application of the compressed air will determine which (if not all) CPPs are needed to ensure the system produces product of the desired quality.

➢ Hydrocarbons
➢ Moisture
➢ Particulates
➢ Temperature

It is important that each point of use has appropriate sterile filters in place. If the filter is not placed directly at the point of use, control and counter measures should be implemented to address any risk of contamination downstream of the filter. Compressed air for bio-pharmaceutical use must be generated using oil-free compressors with appropriate temperature controls in place.

Water Systems

Water supply and the associated water systems in biotechnology and pharmaceuticals are vital components of the manufacturing process. They are used to clean equipment and vessels, to cool or heat processing pipes and systems, and in many circumstances certain grades of water are components of the finished product (e.g. water-for-injection). Various grades of water service a particular purpose. Some common types include:

➢ Potable water
➢ Soft water
➢ Purified water
➢ Water-for injection

Water used in-process and in-cleaning should be pure and free from microbial and chemical impurities. As the water gets easily contaminated by environmental conditions, diligence in the design is essential. Typically water systems are supplied on a continuous loop with recirculation.

CPPs typical for a water system include:

➢ Pressure

- ➢ pH
- ➢ Conductivity
- ➢ Level
- ➢ TOC
- ➢ Flow
- ➢ Temperature
- ➢ Resistivity

Water-for-Injection

The use of WFI is twofold. Firstly, it can be used for critical processing steps such as washing and rinsing . It can also be used in injectable products. WFI is a key raw material for sterile intravenous and intradermal products. WFI is produced by Multi Column Distillation Plant (MCDP), and must meet the microbial requirements of regulated bodies.

Clean-in-Place (CIP) / Sterilise-in-Place (SIP) System

The cleaning of equipment, vessels and process piping is a critical activity. Any residue from a previous production batch needs to be removed in order to avoid cross contamination. CIP and SIP skids are often utilised to allow efficient switchover between batches and/or products.

Clean steam

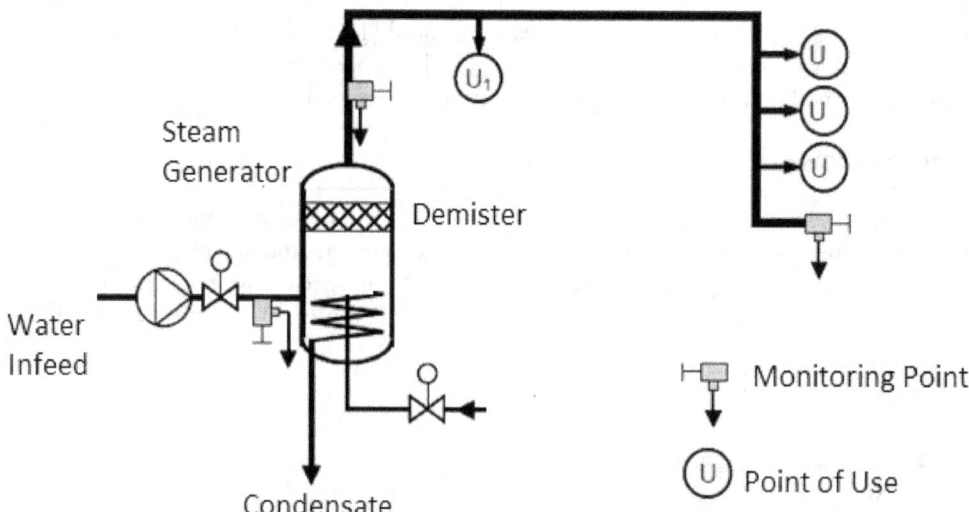

Figure: Simple Clean Steam Generation Piping and Instrumentation

Pure Steam is used in pharma and biotech for sterile application, autoclave sterilisation etc. Distribution piping of clean steam is a critical aspect. Improper sizing of pipes may impact the production process and lead to loss of time during sterilisation.

Clean steam, also referred to as "pure steam", and gases used in manufacturing operations must be of a quality suitable for their intended purpose. The intended use of clean steam and gases must be understood in order to determine any risks to the patient or product. For example, gases that end up being part of the product must fulfil the regulatory requirements. Preventative maintenance and on-going monitoring must be implemented for clean steam systems.

> ➢ Routine inspection and maintenance
> ➢ Frequency of filter change
> ➢ Frequency of the sterilisation for the gas distribution system, if applicable
> ➢ Frequency for integrity testing of the sterile filter

Water systems for purified water, de-ionised water and water-for-injection (WFI) must provide a consistent and reproducible output. Where there is moisture, there is always a risk of microbial contamination. Therefore, the design of water systems should mitigate against such risks. Good engineering practices such as using circulation loops, no dead legs and polished surface finishes all work to provide an effective and safe system. The design should also take into account ease of sampling at the point of use. The removal of endotoxins is a requirement for WFI. On-going sampling to monitor the quality of water is particularly important where water systems are concerned. Procedures should be in place to ensure effective monitoring and testing is maintained. Action limits and acceptance criteria should be clearly documented in approved SOPs or equivalent. Failure to meet limits or acceptance criteria should initiate an investigation. The potential CPPs are listed below for clean steam systems:

> ➢ Conductivity
> ➢ Flow
> ➢ Level
> ➢ Pressure
> ➢ Resistivity
> ➢ Temperature

HVAC

Heating, ventilation and air-conditioning (HVAC) plays an important role in ensuring the manufacture of quality products. Furthermore, HVAC systems also provide comfortable conditions for operators based in the manufacturing environment. HVAC system design influences the layout of airlock positions and doorways. In turn, airlocks, entrances and exits have an effect on room pressure differential cascades and cross-contamination control. The prevention of contamination and cross-contamination is an essential design consideration of the HVAC system. In view of these critical aspects, the design of the HVAC system should be considered at the concept design stage of a manufacturing plant.

Temperature, relative humidity (RH) and ventilation should not adversely affect the quality of products during their manufacture and storage, or the proper functioning of equipment. CPPs for HVAC systems include:

> ➢ Temperature
> ➢ Humidity
> ➢ Particle count (viable and non-viable)
> ➢ HEPA filter certification/leak test/air flow rates
> ➢ Room differential pressures

The Displacement Concept (Low Pressure Differential, High Airflow)

This concept is commonly found in production processes where large amounts of dust are generated. Under this concept the air should be supplied to the corridor, flow through the doorway, and be extracted from the back of the cubicle. Normally the cubicle door should be closed and the air should enter the cubicle through a door grille, although the concept can be applied to an opening without a door. The velocity should be high enough to prevent turbulence within the doorway resulting in dust escaping. This displacement airflow should be calculated as the product of the door area and the velocity, which generally results in relatively large air quantities.

Note: This method of containment is not the preferred method as the measurement and monitoring of airflow velocities in doorways is difficult.

Pressure Differential Concept (High Pressure Differential, Low Airflow)

The pressure differential concept may normally be used in zones where little or no dust is being generated. It may be used alone or in combination with other containment control such as a double door airlock. The high pressure differential between the clean and less clean zones should be generated by leakage through the gaps of the closed doors to the cubicle. The pressure differential should be of sufficient magnitude to ensure containment and prevention of flow reversal, but should not be so high as to create turbulence problems.

In considering room pressure differentials, transient variations, such as machine extract systems, should be taken into consideration. A pressure differential of 15 Pa is often used for achieving containment between two adjacent zones, but pressure differentials of between 5 Pa and 20 Pa may be acceptable. Where the design pressure differential is too low and tolerances are at opposite extremities, a flow reversal can take place. For example, where a control tolerance of $\pm$ 3 Pa is specified, the implications of rooms being operated at the upper and lower tolerances should be evaluated. It is important to select pressures and tolerances such that a flow reversal is unlikely to occur. The pressure differential between adjacent rooms could be considered a critical parameter, depending on the outcome of risk analysis.

The limits for the pressure differential between adjacent areas should be such that there is no risk of overlap in the acceptable operating range, e.g. 5 Pa to 15 Pa in one room and 15 Pa to 30 Pa in an adjacent room, resulting in the failure of the pressure cascade, where the first room is at the maximum pressure limit and the second room is at its minimum pressure limit. Low pressure differentials may be acceptable when airlocks (pressure sinks or pressure bubbles) are used to segregate areas.

The pressure control and monitoring devices used should be calibrated and qualified. Compliance with specifications should be regularly verified and the results recorded. Pressure control devices should be linked to an alarm system set according to the levels determined by a risk analysis. Manual control systems, where used, should be set up during commissioning, with set points marked, and should not change unless other system conditions change. Airlocks can be important components in setting up and maintaining pressure cascade systems and also help to limit cross-contamination. Airlocks with different pressure cascade regimes include the cascade airlock, sink airlock and bubble airlock.

4.3 Materials Management

Introduction

The key theme of effective materials management is control of materials from the incoming stage through the manufacturing process. Specifications and testing support the control of materials ensuring they are meeting the key quality requirements to allow consistent manufacturing and quality products.

Table: GMP Materials Management

Overview of GMP Materials Management					
Item	**PICS /s**	**Eudr aLex**	**FDA**	**WHO**	**ICH**
Reference	GMP for Medial Products Part I, Chapter 5 Production	EU GMP V4 Part 1, Chapter 5 Production 5.27	CFR - Code of Federal Regulations Title 21, Part 211	Annex 2, Section 14.0	ICH, Q7, Section 7, Materials management
Key Headings	Starting materials Processing operations-intermediate and bulk products Packaging materials Packaging operations Finished Products Rejected,	Starting Materials Processing operations: intermediate and bulk products Packaging materials Finished products Rejected, recovered and returned materials Product shortage due to manufacturin	Subpart E-- Control of Components and Drug Product Containers and Closures	General Starting materials Packaging materials Intermediate and bulk products Finished Products Rejected, recovered, reprocessed and reworked Recalled product Return goods	General controls Receipt and quarantine Sample and testing Storage Re-evaluation

	recovered and returned materials	g constraints			
Key Words/ Themes	Specifications CofA	Active substances Excipients	Specifications Retesting Approved materials CofA	Specifications CofA	Specifications Sampling CofA Storage

Starting Materials

The designated name of the product and the internal code reference where applicable;

- ➢ Manufacturers batch number
- ➢ the status of the contents (e.g. quarantined, on test, released)
- ➢ the expiry date or a date beyond which retesting is necessary

Packaging Materials

The purchase, handling and control of primary and printed packaging materials should be as for starting materials. Particular attention should be paid to printed packaging materials. They should be stored in secure conditions so as to exclude the possibility of unauthorised access. Roll feed labels should be used wherever possible. Cut labels and other loose printed materials should be stored and transported in separate closed containers so as to avoid mix ups. Packaging materials should be issued for use only by designated personnel following an approved and documented procedure.

Intermediate

Intermediate products can be simply described as raw materials that may have been mixed and processed to some degree or other. Intermediate and bulk products should be kept under appropriate conditions and must be used within specified dates and according to specifications.

Finished Product

Finished products should be held in quarantine until their final release, after which they should be stored as usable stock under conditions established by the manufacturer.

For the approval and maintenance of suppliers of active substances and excipients, the following is required:

Active substances

Supply chain traceability should be established and the associated risks, from active substance starting materials to the finished medicinal product, should be formally assessed and periodically verified. Appropriate measures should be put in place to reduce risks to the quality of the active substance.

The supply chain and traceability records for each active substance (including active substance starting materials) should be available and be retained by the EEA based manufacturer or importer of the medicinal product. Audits should be carried out at the manufacturers and distributors of active substances to confirm that they comply with the relevant good manufacturing practice and good distribution practice requirements.

The holder of the manufacturing authorisation shall verify such compliance either by himself or through an entity acting on his behalf under a contract. For veterinary medicinal products, audits should be conducted based on risk.

Further audits should be undertaken at intervals defined by the quality risk management process to ensure the maintenance of standards and continued use of the approved supply chain.

Excipients

Excipients and excipient suppliers should be controlled appropriately based on the results of a formalised quality risk assessment in accordance with the European Commission 'Guidelines on the Formalised Risk Assessment for Ascertaining the Appropriate Good Manufacturing Practice for Excipients of Medicinal Products for Human Use'. For each delivery of starting material the containers should be checked for integrity of package, including tamper evident seal where relevant, and for correspondence between the delivery note, the purchase order, the supplier's labels and approved manufacturer and supplier information maintained by the medicinal product manufacturer. The receiving checks on each delivery should be documented.

Prevention of Cross-Contamination

Cross-contamination should be prevented by attention to design of the premises and equipment. This should be supported by attention to process design and implementation of any relevant technical or organisational measures, including effective and reproducible cleaning processes to control risk of cross-contamination.

A Quality Risk Management process, which includes a potency and toxicological evaluation, should be used to assess and control the cross-contamination risks presented by the products manufactured. Factors including; facility/equipment design and use, personnel and material flow, microbiological controls, physico-chemical characteristics of the active substance, process characteristics, cleaning processes and analytical capabilities relative to the relevant limits established from the evaluation of the products should also be taken into account. The outcome of the Quality Risk Management process should be the basis for determining the necessity for and extent to which premises and equipment should be dedicated to a particular product or product family. This may include dedicating specific product contact parts or dedication of the entire manufacturing facility.

Suggested Technical Measures

> Dedicated manufacturing facility (premises and equipment)
> Self-contained production areas having separate processing equipment and separate heating, ventilation and air-conditioning (HVAC) systems. It may also be desirable to isolate certain utilities from those used in other areas
> Design of manufacturing process, premises and equipment to minimise opportunities for cross-contamination during processing, maintenance and cleaning
> Use of "closed systems" for processing and material/product transfer between equipment
> Use of physical barrier systems, including isolators, as containment measures
> Controlled removal of dust close to source of the contaminant e.g. through localised extraction
> Dedication of equipment, dedication of product contact parts or dedication of selected parts which are harder to clean (e.g. filters), dedication of maintenance tools
> Use of single-use disposable technologies
> Use of equipment designed for ease of cleaning
> Appropriate use of air-locks and pressure cascade to confine potential airborne contaminant within a specified area
> Minimising the risk of contamination caused by recirculation or re-entry of untreated or insufficiently treated air

➢ Use of automatic clean in place systems of validated effectiveness
➢ Common general wash areas, separation of equipment washing, drying and storage areas

Suggested Organisational Measures

➢ Dedicating the whole manufacturing facility or a self-contained production area on a campaign basis (dedicated by separation in time) followed by a cleaning process of validated effectiveness
➢ Keeping specific protective clothing inside areas where products with high risk of cross-contamination are processed
➢ Cleaning verification after each product campaign should be considered as a detectability tool to support effectiveness of the Quality Risk Management approach for products deemed to present higher risk
➢ Depending on the contamination risk, verification of cleaning of non-product contact surfaces and monitoring of air within the manufacturing area and/or adjoining areas in order to demonstrate effectiveness of control measures against airborne contamination or contamination by mechanical transfer
➢ Specific measures for waste handling, contaminated rinsing water and soiled gowning
➢ Recording of spills, accidental events or deviations from procedures
➢ Design of cleaning processes for premises and equipment such that the cleaning processes in themselves do not present a cross-contamination risk
➢ Design of detailed records for cleaning processes to ensure completion of cleaning in accordance with approved procedures and use of cleaning status labels on equipment and manufacturing areas
➢ Use of common general wash areas on a campaign basis

Rejection

Intermediates and components failing to meet established specifications should be identified as such and quarantined according to a procedure. These items can be reprocessed or reworked as described below.

Reprocessing

Reprocessing by repeating a manufacturing step or a chemical or physical process of an established manufacturing process is generally considered acceptable. Reprocessing should involve evaluation to ensure the quality of product and must not adversely impact the safety of the finished product.

Recovery of Materials and Solvents

Recovery (e.g. from mother liquor or filtrates) of reactants, intermediates, or the API is considered acceptable, provided that approved procedures exist for the recovery and the recovered materials meet specifications suitable for their intended use.

Returns

Records of returned intermediates or APIs should be maintained. For each return, documentation should include:

− Name and address of the consignee

− Intermediate or API, batch number, and quantity returned

− Reason for return

− Use or disposal of the returned intermediate or API

Testing of Materials

The tests performed should be recorded adequately. EU GMP V4 Part 1 Chapter 6: Quality Control recommends the following information as a minimum:

- ➢ Name of the material or product and, where applicable, dosage form
- ➢ Batch number and, where appropriate, the manufacturer and/or supplier
- ➢ References to the relevant specifications and testing procedures
- ➢ Test results, including observations and calculations, and reference to any certificates of analysis
- ➢ Dates of testing
- ➢ Initials of the persons who performed the testing
- ➢ Initials of the persons who verified the testing and the calculations, where appropriate

Sampling Checklist

The sample taking should be done and recorded in accordance with approved written procedures that describe:

- ➢ The method of sampling
- ➢ The equipment to be used
- ➢ The amount of the sample to be taken
- ➢ Instructions for any required sub-division of the sample
- ➢ The type and condition of the sample container to be used
- ➢ The identification of containers sampled

➢ Any special precautions to be observed, especially with regard to the sampling of sterile or noxious materials
➢ The storage conditions
➢ Instructions for the cleaning and storage of sampling equipment

4.4 Sterile Manufacturing Operations

Introduction

Sterile manufacturing operations depends on several factors including the right design and operation of facilities, utilities and equipment. Sterility assurance must be demonstrated to be in control within a manufacturing setting. This is achieved by:

- Qualification and validation of the processes, facilities, utilities, equipment, cleaning methods and sterilisation operations
- Qualified personnel for aseptic handling in conventional clean rooms or by barrier systems
- Control of critical aspects and critical parameters via the application of change management, change control and a suitable quality management system
- Environmental monitoring
- Routine Maintenance
- Analytical method validation

Sterility

The impact of contaminated injectable products can result in serious illness or death to patients. Many injectable treatments sustain life and bio-chemical processes or genetic conditions. While there is always residual risks or acceptable risks, it is important to mitigate against any risks throughout the manufacturing process. Furthermore, a risk-based approach to operations and in particular changes to the process must be maintained throughout the life cycle of a product. Contamination can be caused by particles or microbes. Where appropriate and technically permissible terminal sterilisation is the preferred point of sterilisation. Terminal sterilisation is when the final sealed product in its container is sterilised at the end of the process.

Unit Operations

Bioprocesses treat raw materials and generate useful products. Unit operations are the individual steps in the process that modify materials and their properties at each step of the process. Each unit operation comes together to create a complete process. The term unit operation usually refers to processes that cause physical modifications to materials such as a change in phase or component concentration. Chemical or biochemical changes are the subject of reaction engineering.

Bioreactor Engineering

The design and manufacture of bioreactors is yet again an area within bioprocessing that depends on scientific and engineering expertise. It should be pointed out that there is no standard design procedure for the design of reactors. However, knowledge of bioprocess reactions and kinetics is a key element. Other knowledge such as mixing, mass transfer and heat transfer also contribute to the design process. Key aspects of bioreactor design include:

Reactor size: What is the capacity of the reactor? This is generally driven by the expected production volumes.

Reactor configuration: Is the reactor air driven, stirred, agitated etc.?

Operating configuration: Is it a continuous operation or a batch driven operation?

Process Requirements: Refer to the required operating temperatures, pH that needs to be maintained in the vessel.

Stirred Tank

A conventional tank involves mixing and bubble dispersion done via mechanical agitation. This requires a high energy input per unit volume. Headspace is an important consideration when filling tanks. Typically, only between 60% to 80% of the tank volume is used. This headspace is important especially if foaming of the broth occurs. Some tanks are designed to take account of foaming issues with the addition of a foam. Chemical means of reducing or preventing foam formation can also be employed. However, these chemicals can impact the process (reduction in rate of oxygen transfer). Temperature modulation is typically controlled using coils.

Bubble Column Bioreactor

A bubble column is a type of bioreactor. Bubble columns offer an alternative to stirred reactors, having no mechanical means of stirring. Mixing and aeration is done by gas sparging by the use of a gas sparger placed at the bottom end of the vessel. This type of reactor requires a lot less energy to mix compared to mechanical stirring.

Airlift Reactors

Airlift reactors are similar to bubble columns as neither require mechanical mixing. A key difference between bubble columns and airlifts is that the air is channelled through a riser in the airlift, which allows more control of the bubble patterns. Airlift reactors can be categorised into either internal loop or external loop configurations.

Aseptic Operation

With the exception of food and beverage fermentations, cultures used in the treatment of medical conditions frequently require sterile conditions.

This is especially important for slow growing cultures that can be quickly compromised by unwanted contaminates. Typically, up to 5% of fermentations in industrial settings are lost as a result of failings in sterilisation. Slow growing cells would have a higher rate of contamination due to sterility issues. Antibiotics by their nature have a higher resilience to this type of loss.

Industrial fermenters are designed to allow in-place steam sterilisation under pressure.

For effective steam sterilisation, the vessel must be fully purged of air. Dead legs, stagnant areas or crevices should be avoided during the design phase as these can be a point of microbial contamination. Polished welded joints with a high surface finish are desired.

Valves

Valves control the introduction of liquids to the vessel and their removal when required. Valves therefore, can be a potential entry point for contaminants. Traditional gate and globe valves do not suffice for aseptic operations.

Pinch and diaphragm type valves are more commonly used as they do not contain any dead spaces within their assembly. The closing mechanics also provide isolation from the liquid or product contents.

Materials of Construction (MOC)

Fermenters are made of materials that are suited to the use of steam sterilisation techniques and regular cleaning. These materials can be classed as both non-reactive and non-absorptive surfaces. Most large-scale reactors are made of high-grade stainless steel. Cheaper classifications of stainless steel can be used for jacketing and other non-product contact areas.

All interior product contact surfaces should be polished to a "mirror" finish. Welds also need to be finished in a similar manner. Electro polishing provides a better quality surface finish than mechanical polishing.

As with any chemical reaction, factors such as temperature, pH and oxygen concentration can impact the performance and yield. To ensure the optimum conditions are maintained, it is important to monitor and control such parameters and factors. By far the most common these days is automatic control of systems and equipment with automatic feedback and adjustment.

Downstream Processing

In fermentation processes, raw materials are altered most significantly by the reactions occurring in the fermenter. In addition, physical changes after fermentation are also important in order to extract and purify the desired product from the culture broth. Any treatment completed after fermentation is referred to as downstream processing. In most instances, downstream processing only requires physical modification of material rather than any chemical or biochemical processing. Although the product will dictate the downstream processing model, there are general major steps which are detailed below.

CELL REMOVAL: this step involves the removal of cells from the fermentation liquor. If the cells are the product itself, little downstream processing is required. Typical unit operations for cell removal include filtration, microfiltration and centrifugation.

CELL DISRUPTION/CELL DEBRIS REMOVAL: If the product desired is located within the cell itself, then these unit operations are required in order to open the cells and release their contents. Such a unit operation includes high-pressure homogenisation. The cell debris is then separated from the desired product via filtration unit operations.

PRIMARY ISOLATION: The purpose of primary isolation is to remove components that differ from the product. Unit operations such as solvent extraction, precipitation and ultrafiltration are typically used.

PRODUCT ENRICHMENT: The purpose of product enrichment is to separate the product from impurities with properties that are close to those of the product. Chromatography is usually used at this stage of the process.

FINAL ISOLATION: The method used for final isolation depends greatly on the product in question. Ultrafiltration is used for liquids and drying for solid products.

Removal Operations

One of the first steps in downstream processing is the removal of cells from the culture liquid. The major process options for cell removal are filtration, microfiltration and centrifugation.

In general, filtration and microfiltration use particle size as the principle of operation, whereas centrifugation relies on particle density. Other factors such as viscosity of broth and surface charge contribute to the performance of these removal operations.

Filtration

Basic filter design involves solids being retained in the filter cloth, while the liquid passes through the cloth/membrane. However, the liquid filtrate that passes through typically contains a small portion of solids. It should be noted that large scale filtration is expensive and difficult to perform under sterile conditions.

Microfiltration

Microfiltration uses microporous membranes to recover cells. Unlike filtration, microfiltration generally does not require preconditioning (heating or addition of agents to reduce viscosity). Microfiltration allows cell recovery of typically 100%, so it is therefore very efficient. Microfiltration can also be done under sterile conditions. It is also typically less expensive than filtration and centrifugation.

Centrifugation

Centrifugation is ideal for cell recovery if the cells or product is too small to filter using conventional filtration. Centrifugation of fermentation broths results in a thick cream-like sludge. Another advantage of centrifugation systems is that many of them are steam-sterilisable. Centrifugation must always take place under sterilised conditions.

Precipitation

Precipitation is a method that is frequently used for the recovery of proteins from culture broths. It also has applications in downstream processing for products such as antibiotics. Typically, precipitation is used in the early steps of downstream processing as it facilitates the reduction in liquid volume, which therefore makes the preceding steps less costly and easier to manage. Precipitation is achieved by the adding of *precipitants* such as salts, solvents and polymers to work to change properties such as pH, ionic strength or temperature of the solution. These effects reduce the solubility of the product which forces it to precipitate out of the solution in particles which are insoluble. The precipitated solid particles can then be recovered by filtration, microfiltration or centrifugation.

Proteins and Precipitation

The most common application of precipitation is recovery of proteins. Proteins treated using precipitation include enzymes used for medical treatments, food proteins from plants and animals and also recombinant proteins manufactured using genetically engineered organisms.

Precipitation Methods

Examples of precipitation methods include:

> Salting out
> Isoelectric precipitation
> Organic solvent precipitation

In summary, the goal of precipitation is to reduce solubility of materials, hence inducing precipitation, which forces solid formation (particles of product). For protein recovery, precipitation is aimed at separating the protein from solution, without causing damage or changes that cannot be reversed.

Salting Out

High salt concentrations facilitate the aggregation and precipitation of proteins. The salt causes the water surrounding the protein to move into the bulk solution. This creates a "hydrophobic zone" on the protein surface. In simple terms, "hydrophobic" means water repelling — a surface that does not take up water due to the hydrophobic zones allowing sites of attraction between the protein molecule and other proteins within the solution. The success and applicability of salting out depends on the hydrophobicity of the protein.

Proteins that have few hydrophobic zones tend to remain in solution — even at high salt concentrations.

Isoelectric Precipitation

Isoelectric precipitation works by reducing the electrostatic forces to near zero, allowing the proteins to precipitate out. A benefit of isoelectric precipitation compared to salting out is that desalting of the precipitate is not required.

Organic Solvent Precipitation

The addition of solvents such as ethanol or acetone to aqueous protein solutions generally causes the protein to precipitate. Organic solvents have a lower dielectric constant than water which means they store less electrostatic energy when compared to water. Therefore, in the presence of the solvent, oppositely charged groups of proteins experience greater attractive forces which cause protein aggregation and precipitation.

Membrane Filtration

Membrane filtration is another type of unit operation that is used in downstream processing. It can be applied in order to separate, concentrate or purify a product. Applications include:

- Cell removal
- Cell debris removal
- Desalting
- Removal of viruses
- Recovery of precipitates

Membrane filtration has a number of advantages compared to other unit operations used to concentrate products:
- Low process energy requirements
- Membrane filtration can be done aseptically
- Does not need harsh chemicals

Membrane filtration can be categorised according to the size of the particles that are retained by the membrane:

Microfiltration: used to remove particulate such as cells and cell debris ranging in size from 0.2 to 10μm from broths. Typical membranes have a nominal pore size diameter of 0.05 to 5μm.

Ultrafiltration: Membranes for ultrafiltration have pores typically of a nominal size between 0.001μm to 0.1μm.

Raw Materials

Raw materials and components used in the manufacturing process should be properly sourced and approved through a supplier quality programme. This ensures that the vendor or supplier of raw materials has the necessary regulatory status and quality controls in place. A robust supplier approval process ensures materials are provided consistently to pre-approved specifications. Raw materials for sterile products must be tested for their bioburden and when necessary for bacterial endotoxin levels to determine acceptability of their use.

Upstream Processing

Most aseptic filling processes are made up of a number of key steps. However, it must be noted that the compounding processes required in order to supply the active product used in filling can be complex in nature. Filling and closing operations tend to be more similar in nature across different companies and different products. The first step normally involves some format of a glass container such as a vial or bottle being processed through a washer. Utilising ultrasonics, heated WFI baths, WFI spray and process air blowing, components are washed to remove any dirt or debris. At the outfeed of the washer, components then travel into a depyrogenation tunnel where they are dried, depyrogenated and sterilised. While the washing of containers and vials is an important step in many manufacturing processes, it does not clean or sterilise components. Its role is to remove any particles or debris. Typically no detergents are used for vials destined to deliver intravenous products. As no detergents or other chemicals are used, cleaning does not occur. The term "cleaning" applied to the biopharmaceutical industry refers to the removal of soils or greases. Vial washers using ultrasonic systems, heated WFI and process air blowers are not designed to "clean" in this regard.

Filling Operations

Suspensions and solutions that are filled in glassware such as vials provide lifesaving and sustaining medical treatments for millions of patients worldwide. When the product reaches the filling unit operation, it has been through many unit operations. The product and components must be sterile at this point. Transfer of product to individual vials or containers may be facilitated by employing piston valves, pressure control and peristaltic pumps.

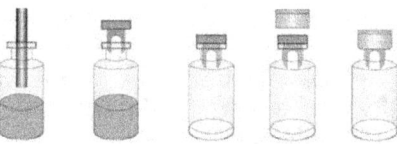

Figure: Representation of filling and closing of vials.

Once the required quantity of solution or suspension has been filled, the next unit operation required is container closure achieved by the insertion or application of a stopper or cap. Key consideration for filling and closing operations include:

➢ Design and function of filler heads
➢ Design and function of filler needles
➢ Fill accuracy and fill weight

The filling of Biotechnology Derived Products (BDP) into ampules or glass vials presents similar problems as with the processing of conventional products. Attempting to develop a site, prove clinical effectiveness and safety, as well as the validation of sterile operations, equipment, processes and systems often necessitates a lengthy process to achieve success for a start-up BDP facility.

The batch size initially produced by a BDP is likely to be small. Because of the small batch size, filling lines may not be as automated as for other products typically filled in larger quantities. Thus, there is more involvement of people filling these products. This can present more chances of contamination meaning any operation or involvement must be controlled and monitored.

Problems that have been identified during filling include inadequate attire, deficient environmental monitoring programmes, hand-stoppering of vials, particularly those that are to be lyophilised and failure to validate some of the basic sterilisation processes. Because of the active involvement of people in filling and aseptic manipulations, the number of persons involved in these operations should be minimised, and an environmental programme should include an evaluation of microbiological samples taken from people working in aseptic processing areas.

Another concern about product stability is the use of inert gas to displace oxygen during both the processing and filling of the solution. As with other products that may be sensitive to oxidation, limits for dissolved oxygen levels for the solution should be established. Likewise, validation of the filling operation should include parameters such as line speed and location of filling syringes with respect to closure, to ensure minimal exposure to air (oxygen) for oxygen-sensitive products. In the absence of inert gas displacement, the manufacturer should be able to demonstrate that the product is not affected by oxygen.

Typically, vials to be lyophilised are partially stoppered by machine. However, some filling lines have been observed that utilise an operator to place each stopper on top of the vial by hand. The concern is the immediate avenue of contamination offered by the operator. The observation of operators and active review of filling operations should be performed. Another major concern with the filling operation of a lyophilised product is assurance of fill volumes. A low fill would represent a sub-potency in the vial. Unlike a powder or liquid fill, a low fill would not be readily apparent after lyophilisation, particularly for a product where the active ingredient may be only a milligram. Because of the clinical significance, sub-potency in a vial can potentially be a very serious situation. A common method of filling vials consists of a two-step filling process. Generally, the first step fills up to 90% of the vial, with the second more accurately filling the remaining amount. The following parameters must be maintained to achieve the same fill volumes at each filling cycle:

- ➢ Viscosity of the product
- ➢ Product temperature
- ➢ Pressure in the dosing vessel
- ➢ Level in the dosing vessel
- ➢ Needle/filling head properties
- ➢ Properties of the hose material

Container Closure Integrity

Upstream processes need to take into account the many requirements that aim to produce products that are safe and effective for patients. Operations such as dispensing and compounding apply GMP principles from the very beginning of the manufacturing process. When the product has been manufactured and is ready to be filled and closed, so too the container closure methods must ensure that sterility and integrity of the product is preserved. Therefore, sterile product container closure systems (or closing systems) must be designed, qualified, and controlled in accordance with international and local regulatory requirements and GMP guidance.

Regulatory Requirements

FDA 21 CFR Part 600. PART 600 -- BIOLOGICAL PRODUCTS: GENERAL Subpart B--Establishment Standards, h)

h) Containers and closures.

"All final containers and closures shall be made of material that will not hasten the deterioration of the product or otherwise render it less suitable for the intended use. All final containers and closures shall be clean and free of surface solids, leachable contaminants and other materials that will hasten the deterioration of the product or otherwise render it less suitable for the intended use. After filling, sealing shall be performed in a manner that will maintain the integrity of the product during the dating period. In addition, final containers and closures for products intended for use by injection shall be sterile and free from pyrogens. Except as otherwise provided in the regulations of this subchapter, final containers for products intended for use by injection shall be colourless and sufficiently transparent to permit visual examination of the contents under normal light. As soon as possible after filling final containers shall be labelled as prescribed in 610.60 et seq. of this chapter, except that final containers may be stored without such prescribed labelling provided they are stored in a sealed receptacle labelled both inside and outside with at least the name of the product, the lot number, and the filling identification."

<u>Quality Requirements</u>

Product containers and closure systems must be capable of being sterilised and depyrogenated before the product is filled. A simple example would be glass vial and stopper components used for various injectable products undergoing sterilisation as part of the process (depyrogenation and sterilisation tunnels for glass components, and stopper sterilisation using vessels and moist heat).

During the product development stage, the type of container closure system must be developed based on the intended use, physical and chemical requirements of the medicine, storage requirements, delivery methods and shelf life. A detailed testing strategy should be developed as early as possible to test for the suitability of the container components. Test strategies can be best developed using a risk-based approach along with knowledge and experience of personnel. In addition to the components used and the size and shape of components, engineering studies are also required in order to define the critical parameters to be used during container closing operations. Depending on the methods of closure, some critical parameters may include:

- ➢ Crimping force (in crimp caps are used)
- ➢ Closure torque
- ➢ Stopper position
- ➢ Stopper force
- ➢ Closure

The parameter selection must ensure the integrity of the container closure system is not compromised. Principles of quality management must be applied to container closure systems when validated and should address the following:

- ➢ Approved container closure components
- ➢ Control srategy for critical parameters
- ➢ Finished product release testing
- ➢ Changes to the container closure system managed under formal change control procedures

<u>Testing</u>

Integrity testing is the most critical test with regard to closed containers. Depending on the container closure format, the followings tests may be used:

- ➢ Dye bath test
- ➢ Vacuum test
- ➢ Headspace analysis
- ➢ Electronic spark test
- ➢ Bubble testing
- ➢ Leak testing
- ➢ Pressure decay
- ➢ 100% visual inspection
- ➢ Presence of stopper or closure cap
- ➢ Presence of tip/cap
- ➢ Presence of cracks, sealing issues or evidence of crimping deficiencies e.g. crimp height.

<u>Electronic Spark Test</u>

This test allows the detection of very small pinholes or cracks that may cause leaks or effect the container integrity. It is used to assess ampoules, vials and glass cartridges. By placing high-voltage through the item and measuring impedance, any cracks or closure issues can be identified in samples.

Vacuum Test

This is completed by applying a vacuum for a defined time and then allowing it to reach ambient pressure. This is a common test used, however it can lack sensitivity.

Head Space Analysis

For containers that are filled using nitrogen or some other inert gases such as argon, head space analysis provides an accurate and repeatable test method. The O2 levels can be detected by the increase in oxygen content. A non-destructive way of testing for head space analysis is to use laser measurement.

Aseptic Process Simulation

Validation of aseptic processing for products must include simulating the process using aseptic process simulations. For simulations of final product filling, the number of containers filled should be representative of the projected batch size and be sufficient to enable a valid evaluation, including all routine operator interventions.

Isolator Barrier Systems

An isolator is a complex barrier system designed to support aseptic processing and manufacturing. The supplied air to such systems is generally supplied through a microbially-retentive filtration system. High efficiency particulate air (HEPA) filters are capable of removing particles as small as 0.3μm making them an integral part of isolator technology.

Figure : Photograph of a typical isolator showing isolator doors fillted with glove ports.

HEPA filters should be capable of achieving Grade A (ISO Class 4.8) at-rest and in-operation. Some exceptions are permitted, such as powder filling, however, risk assessments should mitigate risk to patients. The isolator is a sealed enclosure where there is no direct opening to the external environment or room. Transfer of materials or utensils is done in a controlled manner using a decontaminated interface.

Isolator Interfaces

Depending on the design considerations and individual vendor designs, isolators can have a number of operation interfaces. The term "interface" refers to the ability of an operator or process technician to interact with the machine. The primary method of intervention utilizes glove systems. Four part glove systems consisting of a gauntlet, glove, cuff-ring and sleeve. When used properly and by trained personnel, glove systems support critical line interventions required during aseptic processing and manufacturing.

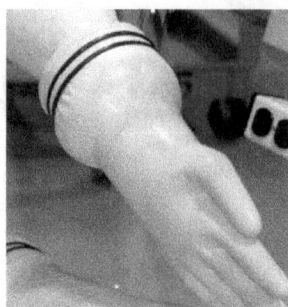

Figure: Isolator glove inflated (undergoing glove integrity test).

Figure: Rapid Transfer Port (RTP). Another means of transferring materials and tools.

The gasket of RTP systems has been identified as a potential source of contamination in isolators since there may be a small contact surface around the gasket that may not be exposed to the decontaminating agent. A risk analysis should be done to evaluate potential contamination risks with the gaskets and the need for maintenance programmes. Transfer of material into and out of the isolator is also a potential source of contamination.

Furthermore, isolators may also be designed in combination with smaller enclosures associated with them to allow the continuous ingress of materials through the smaller isolator into a main isolator.

Classification of Isolator Rooms

The surrounding room of an isolator should have limited access to staff (ensuring only the presence of authorised personnel), adequate space around the isolator and temperature/humidity under control for the effective utilisation of decontamination technologies (e.g. vapour phase hydrogen peroxide systems).
Regulatory authorities require background environments of aseptic production isolators to be classified at minimum in zone (Grade) D (ISO 8 at-rest). However, there is a general consensus that sterility testing isolators need not be placed in a classified clean room, but it is important that such isolator surrounding rooms impose restricted access.

Isolator Decontamination

The purpose of bio-decontamination is to remove viable bioburden on exposed surfaces inside the isolator; a decontamination process should be performed using sporicidal chemical agents associated with decontamination equipment such as gas/vapour phase decontamination systems using hydrogen peroxide (e.g. VHP) or the equivalent. A decontamination cycle is an automated machine cycle that is controlled and monitored during each stage of the cycle. Cycles can be divided into four stages:

-Dehumidification

-Conditioning
-Decontamination
-Aeration

Dehumidification: The dehumidification stage (also known as pre-conditioning) is designed to ensure that the isolator enclosure has a predefined humidity value (< 20 % RH) to ensure a proper concentration of decontaminating agent.

Conditioning: Depending on the complexity of the system, at a minimum, the isolator must have a tightly controlled temperature range, positive pressure and air velocity control. During this initial stage, the isolator doors and ports must be closed and sealed. Any defects in the barrier system should result in an alarm and abort the cycle. During conditioning, an automated leak test should be initiated to detect any breaks in the barrier system (e.g. defective gloves or seals). Heating of VHP delivery pipework also occurs. The conditioning stage is when the decontaminating agent shall reach the minimum concentration required to achieve the desired microbial reduction.

Decontamination: At this stage the VHP is maintained in the isolator according to the dosing rate contained in the recipe or cycle settings. The time and total amount of VHP must result in a kill in BIs placed within the isolator. Generally a 6 log reduction is required for a cycle to be deemed a success.

Aeration: During the aeration stage the amount of residual decontaminating agent must fall to safe levels. (< 1ppm). This is done by blowing the hydrogen peroxide carrying air out of the barrier system using fresh air.

Recommended Critical Process Parameters	Typical Units
Amount of H2O2 during conditioning	(g)
Dosing rate (conditioning)	(g/min)
Time for conditioning	(mins)
Amount of H2O2 during decontamination	(g)
Dosing rate decontamination	(g/min)
Time for decontamination	(mins)
Aeration time	(mins)

Decontamination Agents

Decontamination of isolators is achieved by the supply of gaseous sporicidal agents. These agents must be capable of killing both bacterial endospores and fungal spores. The system typically turns liquid agents into a gaseous vapour. The decontamination agent typically used in industry is hydrogen peroxide. Other agents include formaldehyde, peracetic acid and chlorine dioxide. The rationale for selecting a particular agent should be based on technical data, sporicidal efficacy and the materials and products that come into contact with such agents. Often the starting point when selecting an agent is the manufacturer's recommendations. Manufacturers of equipment trains are best positioned to understand interactions with seals and surfaces etc. In many cases, the equipment is designed with a particular type of decontamination agent in mind. Another source of information is the datasheets provided by the agent manufacturers. Datasheets also give an insight into the suitability of a chemical based on its purity, concentration and safety.

The below factors should be considered with regards to biodecontamination:

➢ Ensure as much surface area as possible of components are exposed.
➢ Minimise loads in order to limit the bioburden levels prior to the cycle starting.
➢ For filling and closing machines, design automation to ensure parts are moving during the cycle to facilitate exposure to the agent.
➢ Ensure all areas are dry and free of foreign objects and debris.

Containment Bioreactor systems

Containment bioreactor systems designed for recombinant microorganisms require not only that a pure culture is maintained, but also that the culture be contained within the systems. Both GLSP and biosafety levels are detailed in this section. A GLSP (Good Large-Scale Practice) level of physical containment is recommended for large-scale research of production involving viable, non-pathogenic and non-toxigenic recombinant strains derived from host organisms that have an extended history or safe large-scale use. The GLSP level of physical containment is recommended for organisms such as those that have built-in environmental limitations that permit optimum growth in the large scale setting but limited survival without adverse consequences in the environment.

BL1-LS

A BL1-LS (Biosafety Level 1 - Large-Scale) level of physical containment is recommended for large-scale research or production of viable organisms containing recombinant DNA molecules that require BL1 containment at the laboratory scale.

BL2-LS

A BL2-LS (Biosafety Level 2 - Large-Scale) level of physical containment is required for large-scale research or production of viable organisms containing recombinant DNA molecules that require BL2 containment at the laboratory scale.

BL3-LS

A BL3-LS (Biosafety Level 3 - Large-Scale) level of physical containment is required for large-scale research or production of viable organisms containing recombinant DNA molecules that require BL3 containment at the laboratory scale.

No provisions are made at this time for large-scale research or production of viable organisms containing recombinant DNA molecules that require BL4 containment at the laboratory scale.

Steam Sterilisers

Autoclaves or steam sterilisers are used to sterilise items such as tools, fixtures and utensils used in aseptic processing. Modern systems are designed to fulfil the requirements of FDA and EU regulatory requirements. DIN 58950/58951 is a standard in which many manufacturers design and build steam sterilisers to fulfil the requirements set out in the document. Conformance to this standard ensures autocalves comply with the FDA and GMP directives. Industrial steam steriliser systems used in biotechnology companies comprise the following main components:

➢ Pressure container for sterilisation

> ➤ Vacuum pump
> ➤ PLC controller
> ➤ Human Machine Interface (HMI)
> ➤ Cycle software

The sterilisation process can be divided into three distinct stages:

> ➤ **Pre-treatment Stage:** during this stage the autoclave begins to heat up and the air in the chamber is replaced by a mixture of steam and air.
> ➤ **Sterilisation Stage:** the purpose of this stage is to kill any harmful microbes by using steam sterilisation. The temperature and pressure of the chamber is held at predefined settings for a specific period of time.
> ➤ **After-treatment Stage:** cooling, decompression and drying occurs in this stage of the cycle.

The steriliser can be loaded with the help of a loading trolley manufactured with suitable materials or an automatic loading and unloading system. The steriliser can alternatively be equipped with trays for accommodating the goods to be sterilised.

Air Leakage Test (Vacuum) Test

The purpose of air leakage testing is to verify that the chamber is vacuum-tight and can maintain the vacuum over a period of time. To avoid loose interpretations, a formal definition of vacuum-tight should be documented. The British standard EN 285+A2 "Sterilisation. Steam sterilisers. Large sterilisers" provides definitions, guidance and a framework for testing steam sterilisers. The air leakage test should result in the chamber maintaining a predetermined pressure over a set period of time e.g. ten minutes.

Steam Penetration (Bowie Dick) Test

Steam penetration is tested using a Bowie Dick test kit. To verify the consistency of the process, this is typically done three times for a recipe or cycle.

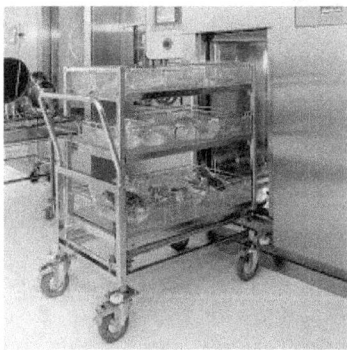

Figure: An autoclave trolley loaded with parts for sterilisation.

Pressure Leak Testing

This test is used to ensure the chamber does not leak. During the course of the test, the pressure is trended. The pressure drop over the test must be within specification. For example, the pressure decrease should be less than 100mbar during the course of the cycle (e.g. 10 minutes).

4.5 Depyrogenation

What Is Depyrogenation?

Depyrogenation is a thermal process that involves the removal of pyrogens from components (e.g. vials or containers) that are used for injectable pharmaceuticals and biopharmaceuticals. A pyrogen is defined as any substance that can cause a fever. Bacterial pyrogens include endotoxins. Later on we shall see that endotoxins are used to challenge depyrogenation tunnels. Depyrogenation tunnel design varies depending on the manufacturer, however, they usually consist of the following components:

> ➢ Infeed and preheating
> ➢ Heating zone
> ➢ Cooling zone
> ➢ Outfeed and transport to next unit operation
> ➢ Automatic emptying

Pyrogens

Pyrogens are fever inducing proteins of low-molecular-weight proteins. Pyrogens of external origin are referred to as exogenous pyrogens. Modern injection and delivery systems are largely safe, yet adverse reactions are still reported. If a treatment or medication administrated via hypodermic needle is contaminated with toxins such as pyrogens fever can be induced which can lead to some death in some cases. It was known in the latter part of the 19th century that some parenteral solutions caused a marked rise in body temperature. The fever producing agents were not known, and hence described in general terms such as "injection fever," "distilled water fever," and "saline fever". Bacterial pyrogens are responsible for many of those early fevers and for many of the other biological effects described incidental to parenteral therapies. The route of administration of a drug allows a pyrogen, if present, to bypass the bodies primary defences. The host's response is mediated through the leukocytes (white blood corpuscles) which in turn release their own kind of pyrogen (endogenous pyrogen) and this in turn initiates a fever like response and other biological reactions.

Bacterial Toxins

There are two general kinds of bacterial toxins: (1) endotoxins and (2) exotoxins. Endotoxins can be extracted from a wide variety of gram-negative bacteria. The term "endotoxin" is usually interchangeable with the term "pyrogen" although not all pyrogens. Higher doses of endotoxin are required to produce a lethal effect in the experimental animal than are required for exotoxins. The effects produced by endotoxins on the host are systemic such as fever and general body reactions, rather than strictly neurological effects, as is the case with most exotoxins. Endotoxins are found in the gram-negative bacteria mostly and are obtained subsequent to the death and autolysis of the cells. The endotoxins are extracted from and associated with the cell structure (cell wall). Good examples of pyrogen producing bacteria are S. typhosa, E. coli, and Ps. aeruginosa.

Exotoxins are produced during the growth phase of certain kinds of bacteria and are liberated into the medium or tissue. Exotoxins are protein in nature and their reactions are specific. For example, Clostridium botulinum produces an exotoxin of unusual potency which affects only neurological tissue. Other well-known examples of exotoxins are tetanus toxin, Shiga toxin, and diphtheria toxin.

Properties of Pyrogens

Pyrogens are:

> ➤ Known to consist biochemically of a lipid-polysaccharide-peptide substance
> ➤ Heat stable at the temperature of boiling water
> ➤ Demonstrate a low order of immune response
> ➤ Produced from persistent gram-negative bacteraemia which could have a 50% mortality rate

Bactericidal procedures such as heating, filtration, or adsorption techniques do not eliminate pyrogens from parenteral solutions. All ingredients must be kept pyrogen-free in the first place. For this assurance, the manufacturer carries out comprehensive pyrogen screening tests on all parenteral drug ingredients and sees to their proper storage prior to use. Ideally, the manufacturer recognises the critical steps in the manufacturing operations that could allow growth of pyrogen producing bacteria and monitors these areas routinely. For example, the water in the holding tanks would be tested for pyrogens and the manufacturer would insist on minimum holding times so that only pyrogen-free water is used. Pyrogen-free water, as "water-for-injection" outlined in the USP, is the heart of the parenteral industry.

Pyrogen Assay - Limulus Amoebocyte Lysate

Many laboratories conduct pyrogen assays by means of the limulus amoebocyte lysate (LAL) test method. The LAL method is useful especially for screening products that are impractical to test by the rabbit method. Products best tested for endotoxins by LAL techniques are: radiopharmaceuticals, anaesthetics, and many biologicals. Essentially, the LAL method reacts hemolymph (blood) from a horseshoe crab (limulus polyphemus) with an endotoxin to form a gel. The quantity of endotoxin that gels is determined from dilution techniques comparing gel formation of a test sample to that of a reference pyrogen, or from spectrophotometric methods comparing the opacity of gel formation of a test sample to that opacity of a reference pyrogen. The LAL test is considered to be specific for the presence of endotoxins and is at least a hundred times more sensitive than the rabbit test. Even picogram quantities of endotoxins can be shown by the LAL method. Although LAL is a relatively new pyrogen testing method, it has produced a wide variety of polysaccharide derivatives that give positive limulus test results and also show fever activity. It is also a fact that some substances interfere with the LAL test even when pyrogens are present.

Some firms use the LAL test for screening pyrogens in raw materials and follow up with pyrogen testing on the final product by means of the USP rabbit assay. The LAL test for pyrogens in drugs requires an amendment to the NDA on an individual product basis. LAL test reagents are licensed by the Bureau of Biologics. For devices, a firm must have its protocol approved by the Director Bureau of Medical Devices, before it can substitute the LAL assay for the rabbit. What is certain is that pyrogens remain a potential source of danger with the use of parenteral therapy.

Endotoxins and Depyrogenation

Endotoxins are used to challenge the effectiveness and consistency of depyrogenation tunnels. Endotoxin challenge vials must be processed through a depyrogenation process that must demonstrate a ≥3 log reduction in endotoxin. Typically, endotoxin challenge vials are placed in close proximity to thermocouples. Using this approach, the temperature profile of the position can be obtained during a cycle. Endotoxin challenge testing is often done during SAT and process development.

Endotoxin challenge testing is typically a requirement of validation, however, no commercial product can be used during a depyrogenation tunnel performance qualification using endotoxins as the product would be potentially contaminated with endotoxins. Therefore, performance validation of depyrogenation processes results in the discarding of the vials or ampules. Depyrogenation tunnels generate tremendous amounts of heat and can operate up to temperatures as high as 320°C (depending on design and operational constraints). Air handling is also a key function of the depyrogenation tunnel. Tunnels should not allow non-sterile air from the room into the sterile air inside tunnel zones. This is done through the use of HEPA filters and an overpressure cascade approach of the tunnel compared to the surrounding room or environment. Air flow must be laminar in nature to ensure the tunnel can maintain the correct pressures and temperatures. Most tunnels are divided into two sections:

-Hot zone (depyrogenation)

-Cool zone (sterilisation/cooling)

The depyrogenation section typically operates at higher temperatures in excess of 270°C which is the recognised depyrogenation temperature. Depending on the technical specification of the components, set-points of 290°C, 300°C or 320°C can be used. Components move slowly through the depyrogenation stages of tunnels. The "sterilising cooling section" operation mode sterilises the cooling sections. Sterilisation cycle consists of the following steps:

1. Pressure drop
2. Draining heat exchanger
3. Heating up of cooling sections set value of temperature (e.g. 240°C)
4. Sterilisation cooling sections: keeping temperature at recipe set value for recipe set value of time
5. Cooling down without heat exchanger: until temperature reaches < 95°C
6. Cooling down with heat exchanger: until temperature reaches < 25°C

The key requirement of cool zone sterilisation is that the temperature within the zone is maintained at a minimum of 170°C for a period of no less than two hours. This gives a very high degree of assurance that the zone is sterile and suitable for sterile manufacturing operations to occur. In summary, the endotoxin challenge must be sufficient to demonstrate a ≥3 log reduction in endotoxin.

Biological Indicators for Dry Heat

Biological Indicators (BIs) (most commonly Bacillus atrophaeus) are used to demonstrate the efficacy of cool zone sterilisation in depyrogenation tunnels. Using a known indicator population and D-value, the delivered lethality needed to obtain an SAL of at least 10-6 can be determined.

The lethality of a cycle can be calculated using the below equation:

Lethality, $F_{(h)} = \Delta T \times \Sigma L$
$L = 10(t-to) / Z$
$Z = 20$ constant

$to = 170$ the base temperature (°C)

$t =$ actual temperature (°C)

$\Sigma L =$ cumulative sum of time

$\Delta T =$ time differential (scan time)

Control of Materials

Items intended for sterilisation or depyrogenation should be prepared and maintained under conditions that will ensure that pre–sterilisation or depyrogenation levels of bioburden, particulate and pyrogen contamination are minimised. Items that will come into contact with sterile dosage forms, filling equipment, containers and closures after sterilisation or depyrogenation in a dry heat oven should be packed for sterilisation in an appropriate clean environment. An appropriate standard would be environmental grade C or D under local protection by HEPA-filtered air.

Contamination Considerations

Protection of items against contamination before sterilisation or depyrogenation is not generally an issue when washers and tunnels are integrated. However, components should in all cases be received in packaging that minimises contamination risk (e.g., from fibres) and handled in such a way as to minimise contamination Items should be clearly identified and controlled to avoid mix-ups between sterile and non-sterile items. Chemical indicators may be attached to containers or placed within loads.

These indicate, through a colour change, that items have been exposed to steriliser conditions but cannot be taken as proof of the adequacy of sterilisation cycles. However, if chemical indicators do not change colour they should be interpreted as confirming sterilisation failure. Items that are not dried immediately after cleaning should be sterilised as soon as possible (no longer than eight hours and preferably within four hours of cleaning) to minimise the risk of microbial proliferation and eventually pyrogen formation between cleaning and sterilisation.
A maximum storage time before re-sterilisation should be specified in case the equipment is not used immediately. Adequate cleaning, drying, and storage of equipment provide for control of bioburden and prevent contribution of endotoxin load.

Start-up Conditions

In sterilising ovens, following any drying phase, the load is typically heated up by closing the dampers to the fresh air supply. Air within the oven is continually recirculated over heating elements and through HEPA filters. In modern sterilisers the cycle is usually under automatic control. If the steriliser requires manual intervention for adjustments (e.g. dampers), then this should be very clearly and precisely defined in the operating SOP and details recorded on the record of each sterilisation cycle. In tunnels, the heating occurs as the components progress into the heating zone.

Control and monitoring should be independent and operate from different temperature sensors. Normally, temperature control and routine monitoring is by fixed position chamber sensors. The relationship to load temperature is established by the validation. If there are movable permanently installed temperature sensors, then these should be placed within the oven chamber and within the most difficult to heat position of the load as determined during validation.

This should be very clearly and precisely defined in the operating SOP and details recorded on the record of each sterilisation cycle. Where only data from fixed position chamber sensors are available, the chamber sensor should be positioned in the same position as used in the validation, generally the most difficult to heat position of the chamber. An appropriate allowance for lag phase should be included in the standard cycle (e.g. to set the steriliser timer). This approach is used to compensate for load lag times (the time difference between chamber probes and the load cold point reaching sterilising temperature) as established during validation as part of performance qualification. Note that this correction for lag should be part of the standard cycle as defined by validation and incorporated into the operating SOP, included in the automatic or manual control (as applicable) and included as part of the master process record (mpr) or acceptance criteria used to assess the cycle.

Tunnel control and monitoring should be independent, and operate from different temperature sensors. The control and routine monitoring of tunnels is by fixed position sensors. The relationship between tunnel and load temperature is established by the validation. The tunnel sensors should be positioned in the same position as used in the validation. The acceptance criteria for the cycle are set on the basis of the validation data such that the tunnel load receives the correct heat input. For tunnels, the sterilisation process is continuous and so the temperature record is of a set temperature. Thus, in order to verify heat treatment of components, the belt speed should be confirmed and, if adjustable, recorded either continuously or intermittently (at least at start and end during each day of operation).

In-Process Controls

Process parameters that are essential to sterility assurance should be verified and documented for every load processed. Other less critical process parameters that may be indicative of actual or potential steriliser failure should be verified at a lower frequency. Periodic checks:

- Confirmation of instrument calibration and performance of any applicable calibration checks
- Data to be obtained, documented and verified for each cycle
- Identification of the contents of the load
- Confirmation of compliance with validated loading pattern (ovens only)
- Confirmation of correct sealing of doors
- Confirmation of correct differential pressures
- Continuous record of the temperature, time, belt speed where applicable, throughout each cycle from at least one sensor

Cooling

Oven loads are generally cooled by switching off the heating elements and opening the fresh air dampers, which allows cool HEPA-filtered air to circulate around the load. The rate of cooling should be a compromise between rapidity and the need to avoid product damage. In particular, glass components may be adversely affected by internal stresses caused by rapid and uneven cooling. Note that the cooling phase is established in the validation and fixed as part of the standard cycle.

Failure of Depyrogenation

The checks on cycle records are vital as a failure of sterilisation or depyrogenation cycles may not be readily detectable in the product testing as there are no visible or practicable non-destructive means of testing for sterility.

The assurance of sterility is thus very heavily based upon the validated process conditions being consistently reproduced during routine operation. It is essential that any failures are promptly detected and that there is a clearly defined course of action in the appropriate operating procedure. Any cycle that does not meet any of its acceptance criteria should be thoroughly investigated.

Materials processed through such a cycle cannot be released solely on passing a test for sterility. Any abnormal or unusual occurrences should be formally recorded on the appropriate site documentation and notified to production management and quality management (even if the occurrence is not formally part of acceptance criteria). They should then be assessed for impact on the sterilisation or depyrogenation and on the functionality of the unit. Procedures should be in place to address such situations (e.g. containment measures). There must be a formal, thorough and fully documented investigation of all cycle failures under the site failure investigation procedure. Possible causes of sterilisation or depyrogenation failure(s) include but are not restricted to:

- Components held for insufficient time during sterilisation
- Too low of a depyrogenation temperature in the hot zone

This may happen in the event of the load lag time being longer than expected due to use of unapproved load patterns, over-loaded ovens, inadequate drying etc.

- Ingress of non-sterile air due to inadequate over-pressure, faulty door seals or filter failure
- Dampers failing to operate correctly
- Excessive residual water in containers (from washing stage)

Part 5
-VALIDATION-

5.1 Validation Planning

Introduction

Validation planning plays a key role in the qualification and validation of new equipment and processes. It also has a role in established processes and is used to plan and manage the ongoing validation requirements within a company. So why the need for validation plans? Firstly, the requirement for validation within medical device and pharmaceutical companies is a legal and regulatory one. The Food and Drug Administration (FDA) stipulates validation as a regulatory requirement of Good Manufacturing Practices (GMP) for both pharmaceuticals (21 CFR 211) and medical devices (21 CFR 820). Validation plans act like a qualification plan that can be used to document strategies, technical rationales and key deliverables. They are a regulatory requirement for medicinal products manufactured in the United States and Europe.

Although not stated in 21 CFR Part 820 (Medical Devices), validation planning is an important activity that helps to document the validation strategy and is commonplace with medical device manufacturers. In Europe, EudraLex (V4 GMP) is the collection of rules and regulations governing medicinal products in the European Union which also require validation planning.

All equipment, processes, facilities and utilities that are GxP impacting need to be qualified. To facilitate the validation efforts, a Validation Plan (VP) creates a roadmap and structure to meet the validation requirements. For simple processes or simple equipment qualifications, a stand-alone validation plan may not be required and can be captured within a protocol or change control. The requirements of validation plans can be driven by a procedure which may be local to a site or factory, or may be corporate and applicable to multiple sites. Consistency of requirements can also be managed by the use of an approved validation plan template.

Generic Benefits

Apart from regulatory or procedural requirements to create a validation plan, there are many other beneficial reasons to complete one. A validation plan acts as a top level document that can pull together the many references, protocols, reports and rationales that make up a project. It is also a powerful asset to introduce new staff and team members to a project or process who need to get up to speed quickly and comprehensively. Validation plans are often the first documents an auditor will request to see in relation to a new process or new product introduction. They force the various stakeholders to sit down and agree upon the strategy and any technical rationales required to deliver a successful project.

Types of Validation Plans

Validation plans can be divided into three different types or configurations. Depending on the validation activity or the project, a validation plan may take the form of a (1) Site Validation Plan (aka Site Master Validation Plan) (2) Master Validation Plans (MVPs) or (3) Individual Validation Plans (VPs).

From the outset, it is important to highlight that different companies may adopt different terminology with regard to validation planning. Typically, large companies will have a site validation plan aka site master validation plan or equivalent document.

A site MVP details the products, processes and associated validation protocols and reports for a manufacturing site/factory. It is the overarching validation plan. Typical components of a site MVP include: description of products and processes, test methods (analytical, physical), specifications, an up-to-date list of utility qualifications, equipment qualifications and process validations.

An MVP encompasses all aspects of a validation strategy and may include multiple processes, multiple pieces of equipment/machines that require validation. MVPs are common for new product introductions. Although not stated in 21 CFR Part 820, MVPs are useful in documenting the validation status.

An individual validation plan generally details the validation strategy of one piece of equipment or machine, therefore, an individual validation plan tends to be limited to a handful of pages. Nonetheless, it is valuable in documenting the validation approach and is a central document that can detail process development reports, specifications and so on.

Matrix or Family Approaches

A family or matrix approach to validation can be used where similar products are produced using the same equipment and processes. A particular product size or product configuration may be selected to represent the "worst-case" product. Therefore, by qualifying the worst case, all of the other products within the family are considered validated. Matrix or family approaches must be clearly documented with technical rationale provided in advance of any qualification activities. This can be addressed in a validation plan or within a protocol. Alternatively, a technical report or product development report can be created and referenced. Taking a family of products, the worst case product might be, the smallest, the largest, the heaviest, or the product requiring the greatest precision and so on.

Changes to the Validated State

Revalidation may be necessary under the following conditions:

- change(s) in the actual process that may affect quality or its validation status
- change(s) in the product design which affects the process
- transfer of processes from one facility to another
- change of the application of the process
- change in materials
- change in a manufacturing agent (cleaning agent, oils, greases, coolant, detergents etc.)

The need for revalidation should be evaluated and documented. Evaluation needs to consider historical results from quality indicators, product changes, process changes, changes in external requirements (regulations or standards) and other such circumstances, as applicable. Revalidation may not be as extensive as the initial validation if the situation does not require that all aspects of the original validation be repeated.

Changes Impacting Operational Qualification - OQ

For changes made to a qualified process, it is necessary to evaluate whether worst case conditions exist. The VOSA may be used to facilitate this evaluation. For changes impacting worst-case conditions, conduct an OQ study in order to challenge those outputs and related inputs at worst-case conditions. If no worst-case conditions exist for the affected outputs, then an OQ of the change is not required. Rationale for not conducting an OQ must be documented in the PQ report.

Narrowing or tightening process parameters within a qualified range does not typically require qualification. Take a scenario that seeks to widen process parameters outside of the qualified range. Suggested minimum validation activity would include:
- Both operational and performance qualification runs to qualify the new range.
- All product outputs impacted by the change should be tested and challenged as would happen in an initial validation.

Changes Impacting Performance Qualification - PQ

If changes are proposed to a qualified process, the impact of the changes should be assessed and documented. Changes to critical process parameters may require a full re-qualification depending on the level of change that is proposed. Supporting studies may be required to support the proposed change such as engineering studies or testing.

What Factors Should Determine the Content of a Validation Plan?

An initial barrier to developing an effective validation plan is the availability of information. When a project is in the concept stage, decisions may still be required on the direction of the project, e.g. what technology will be used? How many new machines will be purchased? What is the timeline?

At the project level, it is likely that a project charter or some scoping document is available in draft copy or is being developed. These documents can help create a validation plan and give it an initial framework. It might also suggest minimum requirements for particular disciplines. An analytical VP would differ to a VP for a new manufacturing facility. In addition, a formal or approved template will prompt the author to include the right information and content.

Content of a Site Validation Master Plan

A site VMP tends to be quite top level in nature with mostly generic and non-specific content. It is not used to describe what the strategy involves for a new project or modification to equipment or process. Some suggested headings include:

Products: a list of products and a brief description of them. Some background on their uses.

Equipment and Processes: a description of the different equipment, technologies and processes used in the manufacture of the different products.

Facilities/Utilities: a brief description of the facilities and utilities onsite.

Overview of Process Validation: a general explanation of the validation policy e.g. risk-based etc.

Summary of Validation Reports for Site: a listing or summary of the completed validations.

Notes: section and headings that are generally common across all validation plans are listed in Appendix 1 at the end of this book.

Content of an Individual Validation Plan

Depending on the type and scope of validation, the following headings should be considered for inclusion (as applicable):

Products: as above.

Equipment and Processes: as above.

Facilities/Utilities: as above.

Overview of Process Validation: as above.

Validation Inputs: inputs include URSs, design history files, project charters etc.

Validation Deliverables: typically, a list of both activities and documents to be approved/completed can be included here.

Test methods and Specifications: a list of relevant test methods used during process testing or product testing.

Process Development: if process development studies are to be completed, this section can be populated. Studies can be referenced here with rationale for the validation strategy outlined.

Risk Management: if risk assessments are completed for the equipment or process, the critical items can be listed here along with the risk references.

FAT(Factory Acceptance Testing): this section can document the need for a FAT or not e.g. "No FAT is required for this equipment/process as it is an off-the-shelf item" OR "The FAT will be completed prior to shipping".

SAT (Site Acceptance Testing): this section can document scope and approach to the SAT.

Validation Plans and FMEA

A Process FMEA (pFMEA) involves assessing each of the process steps, documenting the things that can go wrong at each process step and determining potential consequences for the product. In turn, this information is used to adopt the appropriate risk mitigation and control measures. These mitigations may be applied up front (redesign of a jig or fixture), during the process (e.g. monitoring and controlling a critical process parameter), or after the process at inspection.

Qualification Plan

A Qualification Plan (QP) describes all the qualification measures and at which stage of the qualification the verification will be completed. A qualification plan typically contains detailed descriptions of the necessary test measures and a description of the interdependencies of the individual tests. References to other test documents such as FAT or SAT and a description of the deviation management may also be integrated into the qualification plan. In some instances, there may not be a need or a requirement for a qualification plan. A validation plan can also serve to detail the qualification strategy. A simple table or matrix can be used to map out the requirements and qualification activities.

The FDA provides clear definitions on the four types of validation which include:

(1) Prospective validation
(2) Concurrent validation
(3) Retrospective validation
(4) Revalidation

These terms are defined in the next chapter - *Equipment and Software Validation*. A validation plan should identify what type of validation is to be conducted, e.g. prospective or concurrent.

5.2 Equipment and Software Validation

Introduction

The principles outlined in this chapter can be applied to manufacturing equipment and systems, test equipment and lab equipment. Equipment used in any manufacturing, testing or activity that has the potential to impact product quality needs to be properly commissioned and installed to a full validated state. The definition of the word "validation" should not be underestimated as it is the key driver of why processes and systems must be validated within regulated industries. It is evident as one progresses through the validation life-cycle that the definition and its key terms provide the framework to achieving a validated state. So here it is:

Validation is *"Establishing documented evidence that provides a high degree of assurance that a specific process will consistently produce a product meeting its pre-determined specifications and quality attributes".*

The Goal of Equipment Qualification

The ultimate goal of Equipment Qualification is to ensure that equipment is fit for its intended use. Therefore, equipment is validated to confirm it functions as intended and meets all requirements to manufacture product safely and consistently. FDA requires that "Each manufacturer shall ensure that all equipment used in the manufacturing process meets specified requirements and is appropriately designed, constructed, placed and installed to facilitate maintenance, adjustment, cleaning and use". In other words all manufacturing equipment, support facilities, measuring and test equipment must be "qualified". (FDA 21 CFR 820.70 (G))

Equipment qualification protocols are developed to document this testing and hence provide evidence on the functionality and consistency of the equipment. There are two distinct parts within the scope of equipment qualification, installation qualification and operational qualification. Often these subparts are abbreviated to IQ and OQ. Other combinations such as IOQE and IQ/OQ can be encountered within industry. This is often defined in a company's procedure or SOP relating to equipment validation.

A User Requirements Specification (URS) is often used to document the "specified requirements" of a particular piece of equipment. A URS can then be used as an input document when equipment qualification is required. While a URS document can be extensive covering areas such as equipment functionality, utility requirements, safety features, software specs etc. not all requirements documented in a URS will need to be verified or validated. Critical requirements should be identified early and should always be verified.

In short, equipment qualification is confirmation via documented evidence that the particular requirements for a specific intended use can be consistently fulfilled under anticipated conditions.

Often referred to as the three "Cs" of validation; confirmation, consistency and conditions (anticipated) are key themes that validation must address. Confirmation is addressed by the process of completing a formal validation. When it's done, it is documented and available for review to auditors. To assess consistency, there must be a number of batches or "runs". Typically, there are minor batch-to-batch differences or variations between batches. These differences can be as a result of setup or raw material differences. Process validation must ensure that despite minor changes, there is consistency between batches, with product meeting specifications. Controlled or anticipated conditions are the machine or process settings that are known, documented and controlled during the manufacture of products.

What Is Equipment Installation and Operational Qualification IQ/OQ?

I.   Establishing documented evidence that all key aspects of the process equipment installation adhere to the manufacturer's approved specifications and any recommendations of the supplier of the equipment are suitably considered.

II.   The process/equipment operates as intended and all user requirements are adequately fulfilled.

Inputs to a User Requirements Specification (URS)

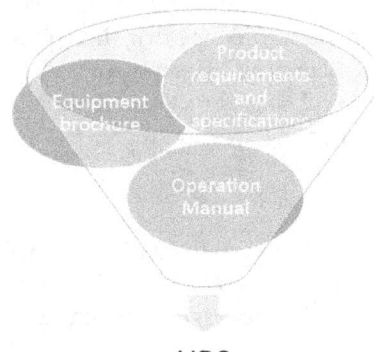

URS

When creating a URS it is important to:

- Specify hardware requirements
- Describe software requirements
- Determine the scope of operating parameters
- Identify safety requirements
- List calibration requirements
- Itemise materials of construction (if required)
- Stipulate the sequence of operation as required

URS documents should be reviewed by appropriately experienced personnel familiar with the product, process and equipment. This ensures critical functions and operations will be identified.

Equipment Classification

Custom: Custom equipment is where the design and requirements are specific to a customer's requirements. Often they are complex as "off-the-shelf" equipment does not meet the customer's needs. With customisation, the potential risk, or potential for manufacturing errors to occur increases.

Off-the-Shelf: Equipment which is available from suppliers that is ready made and does not require customisation.

Risk/Impact Assessment

Prior to any equipment qualification activities a risk or impact assessment should be completed. This is to identify all risk factors which may impact the product quality, performance of the process and ultimately the application of the product (with or by a patient). A risk-based approach to equipment qualification is best practice. The most commonly used process is Failure Mode Effect Analysis Methodology (PFMEA). It is also useful to categorise the potential risks into customer impact and process impact.

Installation Qualification

We have previously defined Equipment Qualification (EQ) and the two components to it (IQ and OQ) as "Establishing by documented evidence that all key aspects of the equipment installation meets the manufacturer's specification"). IQ is required in order to ensure that the equipment is installed, positioned and sited in a manner that is safe and in-line with manufacturer's recommendations. Once a piece of equipment is sited, it must be integrated into the utilities that are required to operate the equipment. An example of typical checks is listed on the following page.

Operational Qualification

The second element of equipment qualification now must be considered; equipment-operational qualification. This is "Establishing by documented evidence that the equipment operates per specifications and over the required ranges and to required tolerances". Equipment is also tested to ensure alarms and controls operate as required and intended. Some typical checks included in an equipment-operational qualification are testing of alarms, control system testing, utility failures and functional and operational testing.

The Validation Lifecycle

The Validation lifecycle refers to the requirement to control and document all validation activities from conception and URS stage to the retirement of equipment or a process. The lifecycle approach ensures compliance throughout the life of the process/equipment while maintaining a validated state throughout the application of change control.

Types of Validation

With most equipment, systems and processes it is best practice to complete all qualification and validation activities in advance of the manufacture of any products for sale, commercial use and use in certain trials. The FDA provides clear definitions on the four types of validation which are explained below.

Prospective Validation

Establishing documented evidence **in advance** of process implementation that a process or system operates as intended. This is the preferred approach and is most common when new products must be validated before commercial manufacturing.

Concurrent Validation

Establishing documented evidence that a process operates as intended, based on information generated during process implementation. Concurrent means that the outputs and performance of the system are monitored at the time of manufacturing which can include commercial lots.

Retrospective Validation

Retrospective validation is used for facilities or processes that have not completed formal validation. Historical data or a retrospective review can provide the evidence that the process or facility is operated as intended. This type of validation is uncommon.

Revalidation

Revalidation involves the re-execution of validation activities in order to maintain a validated state. This can be a result of substantial changes to product attributes, specification or changes to the manufacturing process itself. Other reasons a partial or full revalidation may be required involve instances where product quality issues have increased.

Requirements Specifications

While quality system regulations state that design input requirements must be documented, and specified requirements must be verified, the regulations do not further clarify the distinction between the terms "requirement" and "specification." The URS is the starting point for any piece of equipment or machine. The

URS drives the project to conclusion and should take into account all of the factors relating to the product and process.

A requirement can be any need or expectation for a system or for its software. Requirements reflect the needs of the customer, and may be: (1) market-based, (2) contractual, or (3) statutory, as well as a (4) company's internal requirements. There can be many different kinds of requirements (e.g., design, functional, implementation, interface, performance, or physical requirements).

For example, take the scenario where a company wishes to purchase a thermal oven. From a statutory perspective, the company will ensure that the equipment is CE marked and is built in accordance with all EU directives as this is a fundamental requirement for equipment used within the EU. Similarly, validation is a statutory requirement for all medical devices and pharmaceuticals. Thus, it will be the responsibility of the customer to validate the equipment and the process. Internal requirements may include the need for suppliers to be certified to a QMS, or that they operate as a limited company.

Software requirements are typically stated in functional terms and are defined, refined, and updated during the development phase. Success in accurately and completely documenting software requirements is a crucial factor in successful validation of the resulting software. A specification is defined as "a document that states requirements." It may refer to or include engineering drawings or other relevant documents *21 CFR 820.3(Y).

There are different kinds of written specifications:

- User requirements specifications
- System requirements specification
- Software requirements specification
- Software design specification
- Software test specification
- Functional design specification

All of these documents establish "specified requirements" and are design outputs for which various forms of verification or validation are required. The URS must also define non-software requirements and hardware. Non-functional requirements such as maintainability and usability can also be included. There should be a clear distinction between mandatory regulatory requirements and optional features. The URS should be understood and agreed by both the user and supplier.

Factory Acceptance Testing (FAT)

An FAT or Factory Acceptance Test is an engineering activity. The purpose of the FAT is to verify the equipment or system meets the requirements of the URS. From the validation engineers' perspective, it can be a learning activity and an opportunity to gather data, documentation and supporting design documents that will prove valuable during the equipment and process validation of the equipment. SAT is an engineering activity that is completed at the site of the vendor or equipment manufacturer, post FAT.

Equipment Qualification (EQ) Protocols

Protocol Preparation

Thorough and careful preparation is critical in order to successfully complete a qualification without deviations. In the preparation of the protocol, the URS is a key document. Often quotations, design documents, vendor drawings and owner manuals can contribute to the test and verifications to be completed during EQ.

Protocol Approval

Approval is always required prior to executing an equipment qualification protocol. Approvers should be aware that they are signing for the accuracy and content of the whole document. It is strongly advised that prior to final approval and execution of a protocol, a dry-run or trial is completed to ensure the test methods and acceptance criteria are accurate.

Post Execution Review

Upon execution of a protocol, timely review is advised in order to catch any errors or omissions. The person reviewing the protocol should not be the same person who performed the test. This review is best completed by a quality engineer; however, each organisation should identify personnel responsible for EQ reviews.

Some points to remember when completing protocols:

- Ensure the protocol is fully approved prior to execution
- Ensure personnel are trained on the protocol (if required) and trained to the specific work instructions
- Ensure all team members, contractors etc. sign the signature log
- Observe safety precautions and wear PPE as required
- Ensure other employees that may be impacted by qualifications are aware that a qualification is in progress
- Ensure that any test product required in support of the qualification is identified, segregated and stored to internal standards
- Check that accurate work instructions are available (some companies may allow redlined copies to be used)
- Complete all tests in the protocol
- Always use indelible ink
- Carefully check each result against any acceptance criteria
- Ensure all test equipment is validated within calibration prior to use
- Handwritten comments should be signed and dated per GDP
- Deviations should be written up at the time of observation
- Data records and attachments should be identified with the protocol number, signed, paginated and dated
- When data is transcribed it should be verified by a second person. The source of the data should also be recorded
- All product manufactured should have relevant batch documentation as per normal production conditions

Equipment Qualification Reports

On completion, equipment installation and operational protocol reports are required. The format of any report largely depends on company specific procedures. If the protocol is an executable document (results are hand written in) then the executed version can be deemed the report. A summary report may be required but this depends on the requirements within the company or organisation.

The typical requirements of a completed EQ validation include:

- Equipment qualification protocol
- Equipment qualification protocol (executed)
- Raw data
- Attachments (examples of attachments include: material certs, calibration certs, CE Certs and MSDS.

Software Validation

Where there is potential to affect product conformance to requirements or where software or IT systems provide support to aspects of quality management, validation is required.

Most companies categorise software validations to account for the different applications of software and IT systems. For example, enterprise systems, such as the drawing package SolidWorks, would be validated in a different manner to manufacturing systems that contain software (a.k.a. embedded software).

"Embedded" software is where the software is integrated into the manufacturing equipment. Embedded software is typically validated during the equipment qualification stage, process validation stage or test method validation. Enterprise software falls outside of equipment or process validation but does require validation if it impacts product quality or is used to make quality decisions. Standalone systems such as ERP (Enterprise Resource Planning) systems also require validation.

Software Validation and GAMP

Good Automated Manufacturing Practice (GAMP) is a set of guidelines for manufacturers and users of automated systems in regulated industries. These guidelines are particularly important for the medical device, pharmaceutical and biopharmaceutical industries. The application of GAMP and validation of automated systems in manufacturing helps ensure that regulated medical devices and medicinal products have the required quality and are manufactured according to good practices, meet regulatory and legal requirements and ensure patient safety. GAMP ensures quality is in-built into each stage of the manufacturing process. Therefore, GAMP has a place in all aspects of automation and production, including the handling of raw materials, control of facilities and equipment etc.

Key Terms

Automated System: Term used to cover a broad range of systems, including automated manufacturing equipment, control systems, automated laboratory systems, manufacturing execution systems and computers running laboratory or manufacturing database systems. The automated system consists of the hardware, software and network components, together with the controlled functions and associated documentation. Automated systems are sometimes referred to as computerised systems.

Commercial Off-the-Shelf (COTS): Configurable programs and stock programs that can be configured to specific user applications by "filling in the blanks", without (COTS) altering the basic program.

Computer System Validation: A process that confirms by examination and provision of objective evidence that the computer system conforms to user needs and intended uses. System validation is a process for achieving and maintaining compliance with GxP regulations and fitness for intended use by adoption of life cycle activities, deliverables, and controls.

GAMP 5: Is a set of guidelines that offers a risk-based approach to ensuring the compliance of GxP-impacting computerised systems.

V- Model: Is a development process which sets out a roadmap of stages and deliverables during a project.

21 CFR Part 820: FDA requirements pertaining to medical devices.

User Requirement Specification, URS: The URS is a critical document that defines the requirements of the computerised system and agreement to the requirements.

Software Requirement Specification, SRS: An SRS can be written to interpret the requirements of a URS and how they relate to the requirement or how the requirement is met in practical terms regarding software.

Functional Design Specification, FDS: A functional design specification is a document that specifies how particular requirements are met – this can be a combination of how the equipment/process operates mechanically/automatically etc. An FDS is typically written in response to a URS.

Computer System Validation Life Cycle

The Computer System Validation Life Cycle refers to all activities from initial concept to retirement of a computer system. The life cycle of the system includes the defining of, and performance of activities in a systematic way from conception, requirements, development or configuration, testing, release and operational use.

The four GAMP life cycle phases include:

Concept

Planning and project stage

Operation

Retirement

The concept stage is concerned with understanding the need or the problem to be addressed. We will see that the User Requirement Specification (along with other specifications) and the initial risk assessment help to drive a project forward in a systematic manner. The most common life cycle approach for computerised and automated systems is the V-Model. The GAMP based V-model lays out a roadmap which facilitates the validation of equipment and automated systems.

The planning and project stage involves the planning of the validation effort required to implement the system into the business area(s) based on identification and approval of system concept. This phase includes assessments of the regulatory and system risks, supplier assessment, development of validation strategies, identification of deliverables that will be generated, definition of the business process the system will support as well as the user requirements which the system will fulfil. Design, development and configuration of the hardware and software are also required to meet the system requirements as per specifications. In case of custom software components, this effort could also include detailed software design and developmental testing to ensure readiness for verification testing.

The verification stage confirms that specifications have been met and releases the system for use. This phase will involve multiple stages of reviews and testing depending on the system type, the development method applied and its use. Once verification activities have begun any changes to the system must be captured through change control. On successful completion of the verification activities, the system is then released for effective use. The test strategy and other verification activities will vary widely between simple equipment and more complex customised/configurable systems. The verification and validation approach is typically agreed and detailed in the validation planning stage. The VP can be updated accordingly as the project develops with more detail being added. Alternatively, a test strategy document or matrix could be written to provide more specific test plans.

Verification deliverables vary based on the complexity and level, or customisation of the system in question. Corporate or company specific procedures also shape the required activities to be completed and reported. Some generic deliverables are listed below:

- Approval, execution and review of test protocols
- Writing and approving SOPs for operation and maintenance of the system
- Traceability matrix
- Completion of any risk mitigations (e.g. updates to FMEA etc.)
- Validation summary report(s)

Validation reporting requirements vary depending upon the scope of the system and should also be driven by a procedure and template. The validation plan can also outline the deliverables and what needs to be addressed in the report. A Validation Summary Report (VSR) must be written which summarises the results of executing the VP. The documents created for the validation activities summarise (or point to summaries) of the testing performed. Finally, the VSR indicates the acceptance of the system/equipment by the user and by the project team stating that the equipment is released for commercial operation/production.

The operation phase supports the need to maintain compliance and fitness for intended use after the system is released for normal use. It is important to ensure the system remains within a continued validated state. All proposed or necessary changes to the system must be assessed and controlled as part of a change control process. Once the system has been accepted and released for use, the operation phase begins. This phase consists of maintaining the system's compliant state and fitness for intended use through the control of the procedures supporting the system's operational use.

During the operation phase the below activities are typically completed:

- Ongoing training
- Preventative maintenance
- Service management and performance monitoring
- Change control
- Periodic review
- Maintaining system security
- Records management
- Calibration

The retirement phase involves the planning and proper management of activities relating to the removal of systems from service (shutdown). The retirement should take into account the storage of any data and any data migration that needs to occur prior to retirement. The retirement plan, if needed, will outline the retirement strategy from the roles and activities that will be conducted to the removal of the system for use. A Retirement Summary Report is produced that documents the results of the activities defined in the retirement plan including:

- Retirement Plan and timelines.
- Summaries of any data migration activities.
- Identification of the storage location of documentation relating to the system.
- Obsoleting of SOPs.

It must be stressed that GAMP is a set of principles, a set of guidelines that aims to achieve compliant computerised systems that are fit for intended use. GAMP Guidelines differ to 21 CFR QSR regulations as they are not legal or statutory requirements. However, they represent industry best practice and compliment the validation efforts that are legal requirements and statutory requirements.

Regulatory Review

Software validation is a requirement of the quality system regulation, 21 Code of Federal Regulations (CFR) Part 820. Validation requirements apply to:

(1) software used as components in medical devices
(2) software that is itself a medical device
(3) software used in production of the device or in implementation of the device manufacturer's quality system.

Note: EU GMP Annex 11, provides information on the inspection of 'computerised systems'.

In addition, computer systems used to create, modify, and maintain electronic records and to manage electronic signatures are also subject to the validation requirements. Such computer systems must be validated to ensure accuracy, reliability, consistent intended performance and the ability to discern invalid or altered records. The regulated user should be able to demonstrate through the validation evidence that they have a high level of confidence in the integrity of both the processes executed within the controlling computer system and in those processes controlled by the computer system within the prescribed operating environment.

Specification Hierarchy

An equipment URS can define the requirements of a computerised or automated system along with the operation, function, process and safety mandated by the customer. If the system is bespoke and complex, an SRS may be written to more clearly detail the software requirements, automation and functionality. Similarly, an FDS (functional design specification) can be written to address how mechanical or physical processing occurs.

URS-to-SRS

Scenario: a URS is written to specify the requirements for an automated blister packaging line in a medical device company.

The URS details the following:

> URS R1.0 – The machine shall be capable of operating in various modes to allow the manufacture of product and other debugging activities.

In turn, an SRS can be written to interpret the URS, for example, the SRS shall have the following modes:

> SRS 1.1 Run empty mode - in this mode the equipment does not accept any new product.
>
> SRS 1.2 Production mode – every station operates within the machine.
>
> SRS 1.3 Bypass mode - where any operation can be disabled.

Another two examples of a URS requirement being transposed to an SRS requirement are shown below:

Example 1: URS Requirement

URS R1.0: In the event of E-Stop activation, all sequencers shall be maintained and shall retain the sequence step that they were in at the time of E-Stop activation.

SRS Requirement

SRS 1.0 The lot count and lot integrity must be maintained after E-stop activation.

Example 2:URS Requirement

URS R1.0 The system shall use password protection.

SRS Requirement

SRS 1. 1 Basic machine functionality (cycle start/stop, fault reset, manual operations) require no security.

SRS 1.2 As required, user IDs will be assigned to security groups for authentication. Authentication will be via active directory authentication against B+L domain accounts.

SRS 1.3 An auto-logoff feature shall be incorporated in the design.

Examples of Security Requirements:

- Three levels of access required, operator, and engineer and maintenance.
- Engineer - access to all screens, to modify process settings.
- Maintenance - access to functions required to perform machine maintenance activities.
- Operator - restricted access, does not have access to change process settings.
- Different access levels will require different passwords.

- No security will be required for basic operations (start/stop).
- A user auto-logoff feature shall be incorporated in the design. The auto-logoff time shall be configurable.
- A soft copy of program settings must be provided with delivery of the equipment.

Examples of HMI Requirements:

- The reject count and yield must be displayed on the HMI screen.
- Real-time readings for all critical parameters shall be visible on the HMI Screen.
- All critical parameters shall be adjustable via the HMI Screen.
- The status of each door should be visible on the HMI screen.

Examples of EHS Requirements:

- Activated E-Stops shall be clearly displayed on the HMI screen with a suitable alarm message generated.
- Activated E-Stops shall result in no further movement of the system until the E-stop is reset and all alarms are cleared.

System Categorisation

GAMP 5 makes provision for four categories of software in order to distinguish the level of customisation/configurability that exists across software serving different functions.

GAMP Software Category 1, Operating systems
GAMP Software Category 2, Non-configured software
GAMP Software Category 4, Configurable software packages
GAMP Software Category 5, Custom software

GAMP Software Category 1, Operating Systems

Category 1, operating systems, covers established, commercially available operating systems.

These are not subject to validation themselves; the name and version of the operating system must, however, be documented and verified during Installation Qualification (IQ). Application software hosted on operating systems need to be validated.

GAMP Software Category 2, Non-configured Software

Category 3 covers commercially available, standard software packages and "off-the-shelf" solutions for certain processes. The configuration of the software packages should be limited to adaptation to the runtime environment (for example network and printer connections) and the configuration of the process parameters. The name and version of the standard software package should be documented and verified in an Installation Qualification (IQ). Special user requirements, such as security, alarms, messages, or algorithms must be documented and verified in an Operational Qualification (OQ).

GAMP Software Category 4, Configurable Software Packages

GAMP Software Category 4, Configurable Software Packages Category 4 covers configurable software packages that allow special business and manufacturing processes. This involves configuring predefined software modules. These software packages should only be considered as belonging to Category 4 if they are well-known and mature. Normally, a supplier audit is necessary. If this is not available, the software packages should be handled as Category 5. The name, version, and configuration should be documented and verified in an Installation Qualification (IQ). The functions of the software packages should be verified in terms of the user requirements in an Operational Qualification (OQ). The validation plan should take into account the life cycle model and an assessment of suppliers and software packages.

GAMP Software Category 5, Custom Software

GAMP Software Category 5, Custom Software Custom/Bespoke Software (GAMP Software Cat 5) is software that contains custom code designed or modified specifically for a particular customer. As the code is custom it presents a greater risk. This risk must be mitigated with the right approach to the validation.

GAMP Considerations

Correctly assigning a GAMP software category to equipment, a system or process is an important activity that should be completed early on in the planning stage of a project. There must be some degree of familiarity with the equipment or system. The manufacturer or vendor can be a source of information that may help the designation. In many cases, companies create tools or processes that help determine what GAMP software category applies. These have different names such as questionnaires, screening tools, planning tools etc.

Risk Assessments

A risk assessment process should be applied to cGxP computerised systems in order to identify and mitigate potential risks to (1) patient safety, (2) product quality and (3) data integrity. Results identified through a risk assessment help to determine the validation strategy, the effort and time required, and allow better targeting of the validation activities to the highest risks.

The Risk Assessment should be revised during the Software Development Life cycle (SDLC) if the functionality, requirements or intended use of the system changes. The risk assessment activity should also be evaluated during system build-up as well as when implementing changes. Risk assessment tools for cGxP computerised systems are typically completed during the planning stage, specification stage and post qualification if a change or update is required.

Planning Stage

Initial Impact/Risk Assessment – during the planning phase to identify the level of impact and GxP relevance of the system/equipment. (Tools used: High-Level Risk Assessment).

Specification Stage

Functional or Quality Risk Assessment – during the specification phase – identify potential risks and possible mitigations to be to be introduced to the process. (Tools used: Quality Risk Matrix, (p)FMEA).

Changes to the System

Impact Assessment of Changes – as part of the change control process in the system operational phase. The following diagram defines the risk assessment steps within the system life cycle (Tools used: Impact Assessment Checklist, Change Control Procedures).

Quality Risk Matrix

A QRM is a risk assessment that identifies and manages the risk to patient safety, product quality and data integrity that relates to the systems processes. Risk scenarios or potential causes should be developed for each identified function or process step and then assessed for the impact on patient safety, product quality or data integrity. Risk mitigations and controls should then be introduced to address both medium and high levels of risk. The QRM requires three "assessments" in order to produce an estimation of overall risk (low, medium, high)

Assess Likelihood
Assess Detectability
Assess Severity

Traceability Matrix

A Traceability Matrix should be prepared as required in accordance with company and internal policy. It is also recommended by GAMP guidelines, ASTM E2500 and ISPE risk-based approach to validation. The matrix links the user requirements and specifications to the testing and validation activities. A traceability matrix illustrates that all user requirements are traceable to the verification/validation activity or vendor documents as relevant (FDS if applicable, design specifications etc.). Generally, individual organisations will have an approved template to work from. However, the URS structure can form the basis of the template, with additional columns added to document the test/verification method, reference documents (such as FDS and vendor specifications and design documents)

General Requirements

Configuration Identification

Software and hardware packages should be identified by a unique product identifier and a version number. For the software end-user, the parts of an automated system that are subject to configuration management should be clearly identified. The system should therefore be broken down into configuration items. These should be identified at an early phase of development so that a complete list of configuration items is defined and maintained. The application-specific items should have a unique name or version ID. The depth of detail when specifying the elements is decided by the needs of the system, and the organisation developing that system.

Requirements for the User ID and Password

User ID: The user ID of a system should have a minimum length agreed with the customer and should be unique within the system.
Password: A password should always consist of a combination of numeric and alphanumeric characters. When setting up passwords, the number of characters and a period after which a password expires should be stipulated. The structure of the password is normally selected to suit the specific customer. The configuration is described in the section ***Security Settings of Password Policy***.
Criteria for the structure of a password are as follows:
  Minimum length of the password
  Use of numeric and alphanumeric characters
  Case sensitivity

Audit Trail

The audit trail is a control mechanism of a system that allows all data entered or modified to be traced back to the original data. A reliable and secure audit trail is particularly important in conjunction with the creation, change or deletion of GMP relevant electronic records. In this case, the audit trail must archive and document all the changes or actions made along with the date and time. Typical contents of an audit trail must be recorded and describe the procedures "who changed what and when" (old value/new value).

Uninterrupted Power Supply

An uninterruptible power supply (UPS) is a system for buffering the main power supply. If the power supply fails, the battery of the UPS supplies the required power. When the power supply returns, the UPS battery stops supplying power and is recharged. Some UPS systems provide the option of main power supply monitoring in addition to the buffering function. They guarantee an output voltage at all times without interference voltages. UPS systems are necessary so that process and audit trail data can continue to be recorded during power failures. The design of the UPS must be agreed with the system user and must be specified in the URS, FS or DS. The following points must be considered:

- Energy requirements of the systems to be supplied
- Power of the UPS
- Required duration of UPS buffering

The energy requirements of the systems to be buffered decide the size of the UPS. A further selection criterion is the priority of the systems. Systems with high-priority include:

- Automation system (AS)
- Archive server
- Operator station (OS) server
- Operator station (OS) clients
- Network components

Field devices that generally have relatively high energy requirements may also be included in the buffering depending on the power of the UPS. This must be decided in consultation with the system user and related to the classification of the process. Whatever is decided, it is important that the systems for logging data are included in the buffering. The time at which the power failure occurred should also be recorded. The use of UPS systems involves the installation of software. This should be installed and configured on the PC-based computers of the process control system to be buffered. The setup should also account for:

- Configuration of the power failure alarms
- Stipulation of the time before the PC is shut down
- Stipulation of the time during which UPS buffering is provided

The automation systems (AS) must be programmed so that the process control system changes to a safe state after a selectable buffer time if a power failure occurs.

Types of UPS

Due to the different requirements of the various devices involved, three classes have established themselves as stipulated by the International Engineering Consortium (IEC) in product standard IEC 62040-3 and the European Union EN 50091-3:

Offline UPS

The simplest and least expensive UPS systems (according to IEC 62040-3.2.20, UPS class 3) are standby or offline UPS systems. They protect only against power outages and brief voltage fluctuations and peaks. Undervoltage and overvoltage are not compensated for. Offline UPS systems switch to battery supply automatically if there is overvoltage or undervoltage.

Line-interactive UPS

The way in which line-interactive UPS systems (according to IEC 62040-3.2.18, class 2) function is similar to standby UPS systems. They protect against power outage and brief voltage peaks and can compensate voltage fluctuations continuously using filters.

Online UPS

Double conversion or online UPS systems (according to IEC 62040-3.2.16, Class 1) count as genuine power generators that continuously generate their own line voltage. Connected consumers are therefore supplied permanently with line power without restrictions. At the same time, the battery is charged

Software Source Code Review

For GAMP Software Categories 4 and 5, source code review is advised unless the supplier has evidence of the same available for review. As part of Good Automated Manufacturing Practices, reviews should be completed as part of the development life cycle. If a source code review is not completed a justifiable rationale should be documented in an applicable document such as a validation master plan.

Calibration

A key part of any qualification is to confirm that the equipment is fit for the intended purpose. Each piece of equipment will have a defined operating range. For example an oven may have an operating range of 20°C to 100°C ±5°C. However, the process window may only require a temperature range of 30°C to 60°C. In this instance a calibrated range of 20°C to 70°C would suffice. However, if the process window or the temperatures at which product was manufactured ranged from 20°C to 100°C this would present a problem as it falls outside of the equipment qualification range when the calibration tolerance is taken into account.

Deviations

A deviation can be simply described as an unintended event which causes a test or verification to fail to meet expected acceptance criteria. Each company or organisation should have a procedure detailing the management of deviations. It is critical that all deviations are identified, investigated and evaluated for their impact on product quality, the risk/impact to the patient and the impact on the qualification or validation. The basic components to a deviation are listed below:

- Deviation Description - provides the page and section of the deviation and an overall description e.g. document generation error, operator error, machine crash etc.
- Potential impact on product – does the deviation impact the product?
- Potential impact on validation/qualification – will the validation have to be repeated in part or in full?
- Investigation – DMAIC, RCA, fishbone diagram, 5Ws
- Root Cause- what is the concluding root cause?
- Planned Resolution- what actions are required to be implemented?
- Deviation Resolution (Actions completed) – were all the actions in the planned resolution implemented? What is the final result? Have the actions been effective?

Requalification

Over the lifetime of a piece of equipment, the need to requalify may arise. Therefore, any proposed change to equipment or a process must be assessed to see if the validated state will be impacted. It is therefore critical to understand clearly the nature of the change(s). Some scenarios where requalification of equipment may be required include:

- Major equipment repairs
- Moving equipment
- Changes to the upper and lower operating limits of the equipment
- Upgrading of software
- Hardware upgrades or changes
- Changes in performance and/or defect levels

After assessing any proposed changes based on the reasons listed above, a determination of the level of requalification is required. This may be limited to a partial requalification (addendum) or it may require a full requalification.

5.3 Process Validation

Introduction

This chapter provides an introduction to process validation for medical devices. Process validation is a statutory and regulatory requirement for the manufacture of medical devices. Per FDA 21 Code of Federal Regulations process validation is a regulatory requirement of Good Manufacturing Practices (GMP) for both pharmaceuticals (21 CFR 211) and medical devices (21 CFR 820). In addition to the regulatory drivers, process validation is a requirement in order to obtain certification to international standards issued by many notified bodies. (E.g. ISO 13485 Medical Devices – Quality Management Systems, ASTM E2500-Standard Guide for Specification, Design, and Verification of Pharmaceutical and Biopharmaceutical Manufacturing Systems and Equipment etc.)

Traditional and New Approaches to Validation

Historically, process validation involved the testing and verification of all aspects of a process. While this may seem appropriate, it must be understood that in order to test/verify all aspects of a process, and for it to hold weight, this activity must be documented and recorded. In this respect, an "all aspects" approach to process validation can be burdensome to resources. The traditional approach largely used the V-Model which set out a sequence of deliverables that should be completed. The use of risk assessments were limited as all requirements of a system were tested and qualified.

In recent years, a risk-based approach has been increasingly endorsed by regulatory authorities and hence adopted by medical device manufacturers. One such standard is the ASTM E2500. As the title suggests, it is primarily used within pharmaceutical and biopharmaceutical industries; its principles and core approach can be adopted by medical device manufacturers also. ASTM E2500 was designed to make the implementation process for GMP systems and validation more cost-effective. It aims to achieve this based on scientific and risk-based principles, focusing on the risk to the patient. However, at just a five-page document, ASTM E2500 lacks the detail required in order to meet regulatory expectations. While different terminology and philosophies exist, they do not change the regulatory expectations relating to validation.

Both approaches exhibit common elements which include:

- Good engineering practices
- Planning
- Requirements definition (URS etc.)
- Design review
- Change management
- Documented testing and inspection

While many manufacturers may predominantly choose a particular approach, it is common to see elements of both approaches (traditional and risk based). Each individual company will shape its internal validation procedures to best suit the business needs of the company.

What Is Process-Operational Qualification (OQ-P)?

The ability of a process to produce product in accordance with pre-determined specifications under worst case conditions. PQ is only required if no worst-case conditions are evident.

What Is Process-Performance Qualification (PQ)?

The ability of a process to consistently produce product in accordance with predetermined specifications under anticipated conditions (normal/routine conditions). Before considering process validation in further detail, it is important to look at the prerequisites and other supporting activities required. These are examined in the sections below.

Test Methods and Process Validation

It is important to consider test methods early on in the validation life cycle. Before you can begin to consider process validation, test methods should be understood and in place.

A test method is a process or an action used to verify that a product feature meets a predefined specification. Tests methods can be physical or analytical in nature. Test method validation should be completed in advance of process validation to allow the proper assessment of process and product outputs meaning it is often a pre-requisite to process validation.

Examples of test methods include simple visual inspection by microscope, measurement of a dimension with a callipers or measurement of a dimension using an automated optical inspection system. Some test methods will involve MSA (Measurement System Analysis) studies, for example, a measurement of a dimension by an operator using a microscope. In contrast, a test method to determine organic residuals would require an analytical test method validation.

The equipment must be qualified (installation qualification and operational qualification) before the method is validated. Remember – testing completed in contract laboratories or specialist services also require validation! Test methods are critical to the success and integrity of process validation as they assess the outputs. E.g. what are the dimensions, physical attributes or chemical properties of the product and how do they conform to specifications?

Stages of Process Validation

The three stages of process validation include:
- Process Design
- Process Validation
- Continued Process Monitoring

The commercial manufacturing process must be established during the process design phase. Some typical activities include:
- Definition of process inputs
- Effects of inputs
- Process outputs – CQAs (critical quality attributes)
- Establishing process windows
- DFMEA /PFMEA (design/process)

Design Control procedures should be developed to allow proper management of the process design stage. At the process design stage, the business must define the manufacturing process. This often involves liaising with vendors and Subject Matter Experts (SMEs). The process qualification stage looks at the validation of process design to confirm process is operating as intended and is capable of consistently producing product to meet quality requirements. Finally, stage 3, Continued Process Verification provides ongoing assurance through regular testing and verification to ensure the process is in control. Stage 3 is often referred to as In-Process Control or In-Process Testing (IPC/IPT). This data provides feedback to engineers allowing them to trend the performance of output data. This can identify deficient equipment, changes in wear tooling etc.

Fundamentals of Process Validation

The most important point when it comes to validation is that validation is neither exploratory nor investigative. Equally, it is not an engineering study. If you are ready to validate a system or process, all of the groundwork must be completed. This means critical parameters must be defined and documented, with technical rationale on why such parameters are critical etc. This body of work is typically done during a process development study or protocol. Process validation is confirming that a process is capable of consistently manufacturing product under anticipated conditions. Remember, validation should be representative of the commercial process, so any issues in process validation will be repeated in commercial manufacturing.

Consistency, a core principle of process validation, is typically demonstrated by producing three batches/runs for a Process Performance Qualification (PPQ). These batches should be representative of normal production i.e. the size of the batch should be typical of commercial volumes. The PQ study should be executed at nominal conditions, (often termed "anticipated conditions") essentially referring to a controlled environment. Controlled material and controlled parameters (CPPs) are required. Nominal settings should be selected for PQ.

Process Validation and Dominance Factors

The concept of dominance is a term used to describe the "influential" or "dominating" effect on a system or process. Typical examples include the injection moulding process, and packaging process. For example, an injection moulding process can be said to have *material* as a dominant factor. Batch-to-batch differences of resin or raw material may cause a change to outputs such as the dimensions of a product or component. If dominant factors cannot be identified or understood a "Designed Experiment" (DoE) technique can be used to properly determine them.

Dominance can be categorised into five sections: (1) setup dominance (2) time dominance (3) worker dominance (4) information dominance and (5) component dominance.

Setup Dominance

Setup Dominance - The Process or equipment relies principally on a procedure or process setup. Process should be stable once "set-up".

Examples include ovens and package sealers. With regard to the oven, the setup would generally be controlled by a recipe or program. This program would be selected by the operator through the Human Machine Interface (HMI). The setup with the correct version of the recipe that contains the desired temperatures, times and pressures is therefore a critical input to the process. With regard to the packaging machine (blister packaging), the correct setup for the tooling and program are critical inputs. If setup dominance is significant, it is best practice to have three separate set-ups/changeovers in the Performance Qualification (PQ).

Time Dominance

The Process or equipment is subject to changes over time (drift over time in temperature, solvent cleanliness, tool wear etc.) The process may need a schedule of process checks and adjustments to ensure process consistency. Examples include CNC Machinery (tool wear) or aqueous based cleaning systems. The tool may only be able to manufacture 1000 parts before defects or quality issues are encountered. If time dominance is significant, three time-points or cycles of expected variation should be made e.g. three points in the cycle (start, middle and end) or three points in a shift (start of shift, middle of shift and end of shift).

Worker Dominance

For worker dominance, the process requires operator experience and skill. Examples include manual or hand finishing. If dominance is significant, ensure there are a minimum of three operators involved in the manufacturing/ activity.

Information Dominance

With information dominance, the process or equipment requires the transmission and/or analysis of information. Examples include LIMS, MRP and ERP systems. A minimum of three information transmissions in the PQ should be completed.

Component Dominance

The process is influenced by the variability of the input materials and/or components. It requires robust inspection and sorting procedures as well as process adjustments. When component dominance is significant, ensure there are a minimum of three component/raw material batches in the PQ sampling plan. If component dominance is significant, this can be mitigated by including the material/component variation in "worst case" testing as part of the Operational Qualification Process (OQ-P)

Process Operational Qualification (OQ-P)

During the Operational Qualification-Process (OQ-P) study, worst-case process conditions are normally employed. This may be worst case temperatures, speeds, feeds etc. The OQ-P should challenge the manufacture/processing of product at the limits of the processing window. If no worst-case conditions exist, then an OQ may not be required and only a performance qualification is required. A family or matrix approach is often used where similar products are to be validated. A particular product size or product configuration may be selected to represent the worst-case product. Therefore, by qualifying the worst case, all other products within that family of products would be considered validated. However, this approach must be clearly documented and technical rationale provided in advance of any qualification activities. This can be addressed in a validation plan or within a protocol.

Protocol Approval Check list

The validation protocol is the means in which objective evidence is documented and gathered. The validation protocol is therefore a critical document. It should clearly set out the approach to the validation, detailing methods, tests and verifications to be completed and the acceptance criteria that applies to such tests and verifications. Remember, a validation document is a legal and regulatory document and can be subject to detailed scrutiny. Below are some suggested general checks to apply when writing validation protocols.

Author:
- SOP available - Protocol conforms to validation procedure.
 - Ensure item numbers and batch size are correct.
 -Test methods are correct.

SME Reviewer:
- Is the protocol number correct?
-Review content of protocol for accuracy and completeness.
- Protocol conforms to validation procedures.
- Procedure and evaluation table are appropriate and correct.

Engineering:
- Review content of protocol for accuracy and completeness.
- Specifications and operating parameters are correct.

QC / Laboratory:
- Review content of protocol.
 - Raw material specifications are in place.
- Finished product specifications are in place.

- Testing and sample size is correct.

Quality:
- Review content of protocol.
- Protocol conforms to SOPs.
- Evaluation and acceptance criteria are appropriate.

Process Performance Qualification

The purpose of the PPQ is to demonstrate the capability of the process to consistently manufacture product to pre-determined specifications under normal operating conditions and defined parameters.

Key principles of Process Performance Qualification

Validation is confirmation, so process validation is confirming that a process is capable of consistently manufacturing product under anticipated conditions.

- Lots should be produced consecutively (in sequence)
- Lots must meet the acceptance criteria set out in the protocol
- The lot size should be reflective of the intended lot size and also take into account normal variation
- If a family approach or matrix approach is used, the product selection must be clearly justified and documented
- Execute under anticipated conditions; essentially this refers to a controlled environment. Controlled material, controlled parameters (CPPs)
- Nominal settings should be selected for PPQ

Yield Data (aka Process Yield Data)

Process yield is a term used in manufacturing to represent the overall process performance. Yield is most often expressed as a percentage of goods/passing products. It reports the percentage of compliant units, that is units or products that meet the product acceptance criteria (e.g. CQAs). The remaining "bad" units are classified as defects or scrap. In some manufacturing processes, rework is possible or permitted.

Yield data often forms part of the acceptance criteria for a validation. The overall process yield for each batch should be calculated and compared to the starting process weights or units to determine loss due to processing as it is common to lose material during processing.

Continued Process Verification

Once the initial validation is completed it is important that the system or process remains within the validated state, meaning that the system remains in a state of controlling process systems that capture information and data about the performance of the process. The use of statistical trending techniques should be considered. Data analysis of process and product should also include trending of raw materials, components and finished product. The purpose of process monitoring is to ensure critical parameters remain within control limits. It also helps to identify increasing variability or instability within the process which can then be investigated. All processes must have an upper and lower limit. If a process parameter only has a one-sided limit, then provide rationale in the OQ protocol to justify why a one-sided parameter window is acceptable. This requirement is not applicable to parameters that are set points.

Revalidation (or Maintaining a Validated State)

Revalidation is sometimes required if the original validation is no longer valid or representative of the process. Some instances where revalidation must be considered include changes to the process that can affect the product quality or efficacy, a removal, or the addition of a processing step or transfer of the equipment to a different location. In many companies an impact assessment is conducted if there is a proposal to modify a manufacturing process. Some changes may not require any validation while others may require a verification run. When changes are proposed to the validated state of a process, the proposed changes must be fully understood in terms of the impact to product quality and the validated state. A risk assessment should be conducted to determine risks and appropriate mitigations.

Other Scenarios – Maintaining a Validation State

Line Addition-Product (New Product or Product Transfer)

This may be required if a new product has been introduced but uses the same process(es) for manufacturing. Typically this can apply if a new size has been introduced. For example, a new size of surgical blade.

Line Extension

A line extension commonly refers to a scenario where the product/process is different or considered outside the existing range or processing parameters. **Note:** The impact on the validated state for line additions and line extensions should be assessed formally and documented. A line extension may require a new validation or addendum to the existing validation, whereas a line addition to add a new product may be within the scope of existing validations.

Acceptance Criteria

The acceptance criteria contained in validation protocols are normally based on established product specifications. For example, a contact lens manufacturing company may produce a lens with a diameter of 18mm ±0.2mm. The product produced during a process validation must be inspected to record the diameters of lenses being manufactured. Disposition of product is based on the product specification and determines if the product feature measured receives a pass or fail. In addition to product specifications, it is common to have acceptance criteria such as yield, and OEE. The acceptance criteria for these conditions are normally driven by an internal company procedure or alternatively can be detailed in the validation plan or protocol study.

Validation Strategies

A Family Approach (a.k.a. Bracketing, Matrix Approach) to validation is often used where a variety of similar products are manufactured using the same equipment. For process validation, a product that is representative of the family or group of products may be selected. Alternatively, a 'worst case' product may be selected as it presents the greatest challenge to manufacture to product specifications.

Principles of Worst Case Selection

Worst Case is a particular condition, set of conditions, and/or set of process parameters, generally made up of processing limits. Worst case conditions present the greatest chance of process issues or the greatest chance of failures due to product quality. Worst case conditions are used at the OQ-P stage to provide the greatest level of challenge, however, this is outside of normal operating conditions.

Requalification

During the lifetime of a process or piece of equipment, the need to re-qualify may arise. Such need should be assessed according to a validation procedure. Generally, the same tools used in the original validation can be re-applied to identify the need to re-qualify and indicate what requirements must be included.

The first step must be a review of the existing qualification, as changes may not impact the validated state, or may only require a limited requalification. For example, moving a piece of equipment may only require requalification of the utilities such as compressed air or process water if the operation of the equipment is not impacted by the movement and re-siting. Some examples where re-qualification may be required include:

- Transferring a process from one plant to another plant
- Changes to the process settings which may impact the product quality
- Changes to the design of the product
- Changes to manufacturing aids (e.g. cleaning agents, jigs and fixtures)

Process Validation - Examples of Deficiencies

Identification of Critical Process Parameters

No procedure in place to define or identify Critical Process Parameters (CPP). Not documenting from where these CPPs would be obtained when writing process validation protocols. In a drying process for an oven, no rationale was documented as to why time and pressure were not considered CPPs. Temperature was only considered a CPP.

Reproducibility

For a concurrent validation, three separate reports were drawn up and each batch had been concurrently released. However, no summary report was generated to assess the reproducibility of the process.

Risk Assessment

No linkage between the validation protocol and the various controls in the manufacturing process that had been identified in the risk assessment as important for risk mitigation.

Case Studies on Process Validation

The following case studies review key considerations when it comes to process validation. For your benefit, we have focused on the critical elements which include:

1) General principles of process validation
2) Dominance factors
3) Parameters and settings
4) Other considerations

Case Study - CNC Grinder–Performance Qualification (PQ)

System Overview

A Computer Numerically Controlled (CNC) Grinder uses grinding wheels to machine complex geometries or modify surfaces. As with any process validation, it should factor in when to influence the process (dominant factors). The machine settings are also another factor which need to be considered for process validation. Machine settings should also have a tolerance or +/- value associated with them. However, with CNC machines, as they are numerically controlled, parameters such as spindle speed tend to be accurate. In the other case studies, we will see other parameters such as temperature and how tolerances are associated with settings.

General Principles of Process Validation applied to a CNC Grinder

As the spindle speed of the machine is "set-point" only and is known to be accurate, an Operational Qualification (OQ) is not required as there are no 'worst case' settings. All system settings are set-points and the only variable is 'wheel life'. As the wheel is used to manufacture/grind parts, over time the wheel's grinding performance decreases, therefore, the wheel has a life expectancy. Wheel life assessment should form part of the performance qualification. A performance qualification manufactures components at start, then middle and at the end of wheel life. First article inspection of (FAI) should be performed at the start of each run in order to confirm the "first off" is according to specification and that the machine setup is correct.

Dominance Factors

For a CNC grinder, appropriate dominant factors include time and setup. Time is a factor as the wheel life changes over time. Setup is also a factor as the tooling and fixturing must be setup accordingly and this can be a source of variation that must be challenged as part of the PQ.

Parameters and Settings

Coolant temperature, grinding feed rate, cutting dwell time and spindle speed all have the ability to affect performance and product quality, therefore they can be considered controlled parameters. Coolant temperature should have a tolerance specified in the validation protocol e.g. +/- 2 degrees.

Other Considerations

The incoming raw materials e.g. mild steel bar stock, should confirm to an acceptance criterion of Certificate of Analysis.

Case Study – Cleanline Operational Qualification and Performance Qualification (OQ-P/PQ)

System Overview

Cleanlines are used to clean fixtures, equipment parts and products within the medical device industry. Cleaning processes can be loosely divided into intermediate cleaning processes and final cleaning processes. A higher acceptance criterion is applied to final cleaning processes. Cleanlines are aqueous or solvent-based systems. Aqueous systems typically use deionised water and detergent, followed by DI rinsing and dryings steps. Solvent systems are also effective at removing grease and oils from parts and components.

General Principles of Process Validation Applied to a Solvent Based Cleanline

A cleaning process is made up of: 1) cleaning parameters such as temperature, time and ultrasonics 2) manufacturing agent, in this case solvent and 3) worst case conditions. The heating or cooling of liquids (although temperature controlled by a PLC-temperature probe) is subject to drift. This drift can vary on a given day, or due to the use of the equipment. Therefore, it is important to quality-check an operating range for the equipment and the cleaning process. This operating range is qualified by the execution of an OQ-P, Operational Qualification-Process. The OQ-P represents the limits at which products will be cleaned. The OQ-P aims to demonstrate that if the process settings are subject to drift in time or temperature, the cleaning process is still effective as it operates within the validated operating range. This operating range is also known as the process window. The following strategy should be applied for the Operational Qualification- Process and Performance Qualification of a cleaning system:

OQ Low, 1 lot
OQ High, 1 lot
PQ nominal, 3 lots
PQ nominal rework, 1 lot

Full loading conditions should be applied for all cleaning run/conditions. This means that the basket or "carrier" that holds the pieces while in the washer, is full to the normal, anticipated capacity. This is to create the conditions as they are intended during commercial manufacturing.

Dominance Factors

The dominant factors for a cleaning process are considered to be time based. As the system is used, the water/solvent in the tank gets more "soiled" as time moves on. It essentially gets dirtier; therefore, there may be a difference in the cleaning performance of the system at the start of the day (clean water with fresh detergent) as opposed to the end of the day, after a full shift of cycles (dirtier conditions).

Parameters and Settings

Parameters include rinse temperatures, times and ultrasonic frequencies. Some equipment may be fitted with in-line conductivity meters and/or pH probes which indicate the "cleanliness" of the solution. If a drying stage is incorporated into the system, this will also have temperature and time settings and tolerances. Note, that some basic systems may only have fixed ultrasonics which cannot be varied.

Other Considerations

An important consideration when conducting cleaning process validation for medical devices is the manufacturing agents (greases, oils etc.) that are used up-stream in intermediate steps or processes. The chosen cleaning detergent or cleaning solvent should be selected in a systematic way to ensure it is effective in removing the "soiled" material.

5.4 Packaging Validation

The content of this chapter provides an outline of six key stages of the packaging validation life cycle. In order to describe the packaging life cycle in a structured and manageable format, the process can be sub-divided into 6 stages. It must be noted that an individual company will have its own interpretation of the required stages with different terminology or specific requirements. However, the intent of any approach should broadly align.

Stage 1 - Design and Development of Packaging
Stage 2 – Material(s), Equipment and Process Technology
Stage 3 - Material Performance and Suitability Testing
Stage 4 - Stability Testing
Stage 5 - Packaging Performance Testing
Stage 6 – Packaging Validation

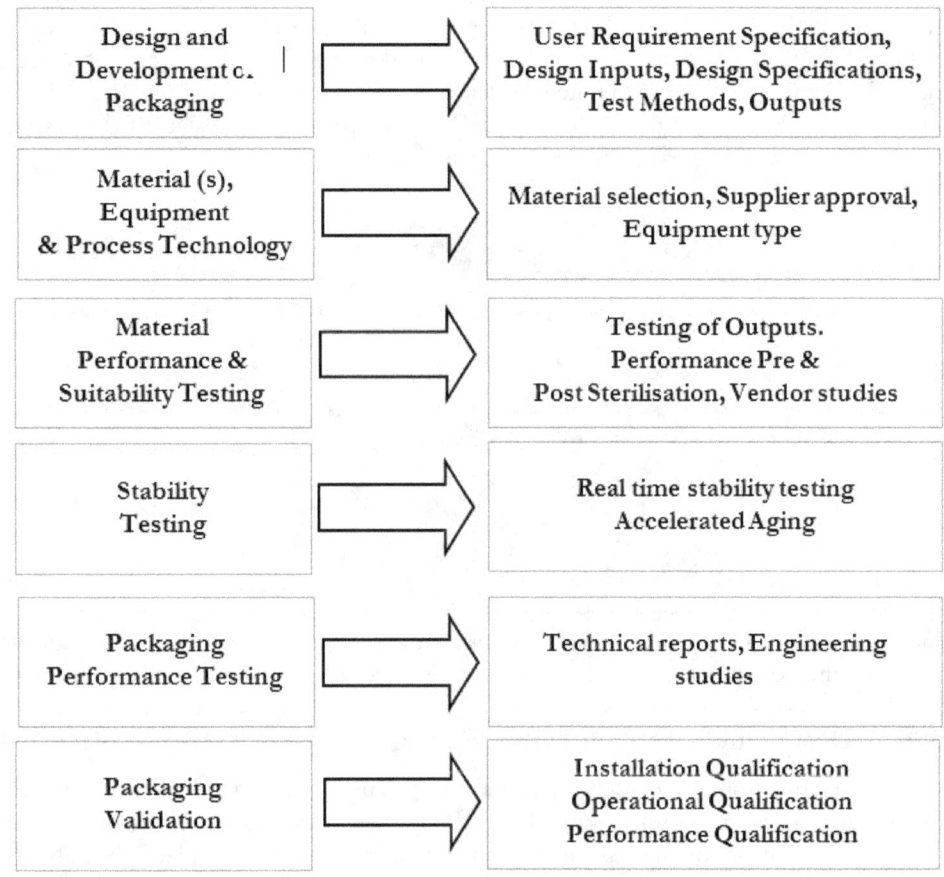

Blister packing is a process where pre-formed plastic packaging (blisters) manufactured using a cavity of a defined shape and size are used to pack a range of items including various personal tech-ware goods, foods, pharmaceuticals and medical devices. The primary component of a blister pack is a cavity or pocket which is made by forcing a material, usually a plastic into the cavity under vacuum and at deformation temperature. The resulting blister is often closed with cardboard, paperboard, foil or plastic depending on the product. The combination of the blister and lid or lidstock helps to protect products from the environment such as microbial contamination, humidity and foreign matter.

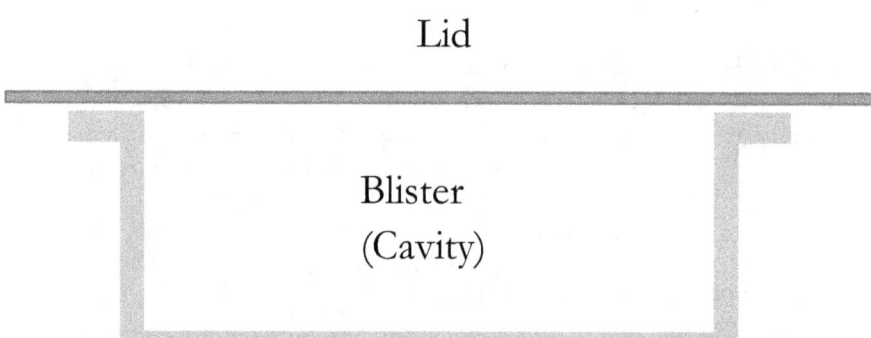

Figure: Simple representation of a blister cavity and lid or lidstock.

Most blisters made from plastics will provide good protection to the inner product. Often, it is the lid or lidstock that is the weaker of the two. Some lid materials are prone to tearing or puncturing and degradation over time. In addition, the area where the lid is bonded to the blister is a point of interest and must be inspected adequately to ensure a proper seal integrity is achieved. Not only is the movement of the product and force of the product a risk factor in damaging the package, but equally the design of the system should also account for handling and a degree of inappropriate handing as a safety factor.

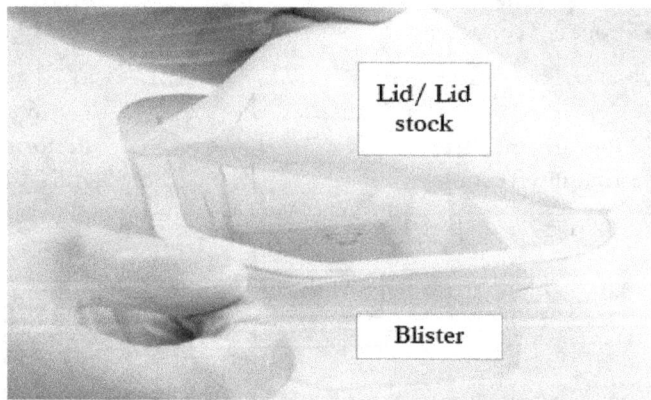

Some manufacturers offer breathable-type seals which reduce the risk of condensation forming on the inside of the pack due to temperature and humidity differences that can occur during shipping or storage.

Stage 1 - Design and Development of Packaging

The design of medical packaging is equally as important as the products they contain. They ensure products are kept clean, sterile (if applicable) and essentially make them safe and effective when used. They need to serve the requirements of the regulatory bodies and international standards but also provide ease-of-use to patients and users. Packaging systems can include lids (lidstock), pouches, bags, trays and blisters that are used to contain the drugs and medical devices.

General Requirements for Design and Development

Validated Test Methods: Test methods are used to verify the outputs of manufacturing processes. In the case of packaging, some examples of test methods include seal strength.

Robustness of Process: Tests selected for use must adequately address the robustness of the packaging system and process being tested. The rationale for the tests selected for use must be documented in the development protocols.

Design: The design of the blister and lid are properly scoped out, documented and controlled.

Sample Size: Test sample sizes must be based on statistical rationale. The rationale should be documented in the development protocols.

Key Requirements of Medical Packaging

The specific requirements on a given packaging system depend on the classification of medical device. Medical devices are classified based on the application of the devices and the level of risk associated with their use. In general, the level of risk is understood to increase with the (1) duration of use (2) level of invasiveness. Therefore, the packaging requirements for a particular product must be specified correctly and in keeping with the intended use.

High-density polyethylene or HDPE is a popular material that aids in the prevention of microbial contamination. However, the contents and environment must meet microbial standards if a sterile product is being manufactured. The effectiveness of HDPE is based on the amount of very fine filaments and their random orientation makes it an effective barrier.

During the product design phase, the packing configuration must be selected. This selection impacts the equipment and manufacturing technology required in order to deliver the designed barrier system. Most products in a modern manufacturing facility will require a medium to high-volume manufacturing. This is typically driven by market need. If high volumes are not anticipated, a high cycle time may be required to allow the bare minimum number of machines to deliver the throughput required. Storage and shelf life of medical products is also a key concern for the patient and user. In particular, HDPE lidstock can maintain sterility for up to five years. The process of sterilisation involves the controlled release of gases or steam that "penetrates" the packed product but can then quickly escape from the packaging and leave the product sterile and unaffected. Materials must be suitable for the type of sterilisation process.

Inputs

Packaging materials must be compatible with the chosen method of sterilisation, therefore, during the design and development stage, the type of sterilisation must be selected carefully. This is based on regulatory requirements and the capability of the product as well as primary and secondary packaging materials. The requirements with regard to acceptable foreign matter, visual defects, seal strength and integrity criteria must also be developed to form what will be acceptance specifications.

User Requirement Specification (URS) A URS is a requirements document that specifies the intended use of the equipment along with specific operating and process requirements unique to a particular product. The scope of a URS document can be sub-divided into three main sections: (1) installation requirements (2) operational requirements and (3) process requirements.

Installation Requirements: These relate to the type of facility and space available. The footprint and weight of the packaging machine may be a factor for some premises. The customer may also which to procure a packaging solution that is mobile.

Utilities: Typically, electrical and pneumatic supply options are specified in a URS and the machine manufacturer must confirm that the equipment can be successfully operated with the available utilities onsite.

Operational Requirements: Operational requirements may be specific to a particular product or family of products. Some typical examples include:

- Equipment must reach operating parameters from standby within 15 minutes of cold start-up
- It shall be possible to operate the equipment with one person
- Equipment shall be capable of processing a minimum of 20 blisters per minute

Process Requirements: Process requirements can also relate to a specific product or packaging configuration. For example, "the seal shall meet the minimum width specification of 6mm".

While other process requirements are more generic:
- No smudge marks, burn marks or tool marks shall be visible
- Seal areas must be free from creases or any other defects

Outputs

Outputs are essentially the packaging features or attributes that need to be validated and found to be within acceptance levels. For blister packaging, the typical requirements are listed below:

Integrity of sterile barrier
Tensile strength of seal and delamination
Seal width
Cosmetic requirements

Outputs must be considered from the very beginning of a packaging project prior to packaging validation. The choice of materials, equipment and technology can all impact the outcome of the final packaged product. Choosing the wrong materials may cause long delays when issues are discovered or encountered. Choosing the wrong technology, equipment or process may not provide the required capability or necessary quality, especially for regulated products (e.g. medical devices and sterile products).

Sterile Packaging

Packaging materials must be compatible with the chosen method of sterilisation. Sterilisation methods include ethylene oxide, electron-beam, gamma, electron-beam, steam (under controlled conditions) to name but a few. Packaging must provide a high microbial protection and breathability along with acceptable levels of tear resistance and durability.

For sterile medical devices, regulation requirements per EN ISO 13485:2013 include:

"Devices delivered in a sterile state must be designed, manufactured and packed in a non-reusable pack and/or according to appropriate procedures to ensure that they are sterile when placed on the market and remain sterile, under the storage and transport conditions laid down, until the protective packaging is damaged or opened."

"Devices delivered in a sterile state must have been manufactured and sterilised by an appropriate, validated method. Devices intended to be sterilised must be manufactured in appropriately controlled (e. g. environmental) conditions."

"Packaging systems for non-sterile devices must keep the product without deterioration at the level of cleanliness stipulated and, if the devices are to be sterilised prior to use, minimize the risk of microbial contamination; the packaging system must be suitable taking account of the method of sterilisation indicated by the manufacturer."

"The packaging and/or label of the device must distinguish between identical or similar products sold in both sterile and non-sterile condition".

Stage 2 – Material(s), Equipment and Process technology

Supplier Requirements

Above all, materials must be suitable for the intended use and classification of medical device. Most vendors will operate to a quality management standard such as ISO 9000 or ISO 13485. When dealing with vendors and external suppliers of packaging materials, all medical device manufactures should adopt a vendor approval procedure to specify the supplier requirements.

Materials

Polyvinyl Chloride (PVC)

PVC is a low cost material and suits blister packaging due to its suitability to thermoforming. However, it offers limited barrier protection against moisture and oxygen ingress. PVC blisters provide good protection for physical pharmaceutical solid dose tablets and caplets. PVC sheet thickness is typically between 200μm to 300μm depending on the cavity size and shape. PVC does not provide the highest protection with regards to water vapour ingress. This can be improved by laminating processes using PVDC. To meet suitability for use requirements, PVC formulations need to meet standards such as the US Pharmacopoeia 661, FDA 21 CFR and local regulatory requirements.

PVDC

Polyvinylidene chloride or PVDC is often combined with PVC film by using a lamination technique in order to gain better moisture and oxygen barrier performance. PVDC coated blister films are the most common and prevailing barrier films used for pharmaceutical blister packs.

Cyclic Olefin Copolymers (COC)

Cyclic olefin copolymers (COC) or polymers (COP) can provide moisture barriers to blister packs, typically in multi-layered combinations with polypropylene and polyethylene. Cyclic olefin copolymers have good thermoforming properties even in deep cavities, leading some to use COC in blister packaging as a thermoforming enhancer, particularly in combination with polypropylene or polyethylene.

Lidstock

As previously mentioned, high-density polyethylene or HDPE is a preferred barrier material as it provides excellent protection against water and oxygen ingress. It also can be manufactured to provide good tensile strength offering protection to the product.

An alternative to HDPE lidstock is foil based lidstock. Foil based lidstock can be designed to be heat sealed to polymer blisters such as polypropylene and it suitable for steam sterilisation at high temperatures (over 100°C. Foil based lidstocks also achieve good tensile strength protection for the packaged product.

Equipment and Process Technology

This section describes two approaches to blister packaging of medical devices. The first approach adopted by some manufacturers (especially at initial launch where volumes are relatively low) is for the project or packaging engineer to select a simple manual process. Such a process may consist of the manufacturer receiving pre-formed blisters from a vendor or supplier. The manufacturing only then needs to worry about placing the contents in the blister and sealing it with a lid. The preformed blisters are loaded into each position, the product is placed in the cavity, a lid is placed on the top side of the blister and the sealing process can begin. A more advanced method of blister packaging involves more automated equipment that completes both the forming of the blister and sealing of the lid-to-blister.

Blister sealing is typically completed through pressure heat transfer over a short period of time. Controlled parameters include:

- Seal Temperature
- Seal Time
- Seal Pressure

The seal settings for the above parameters must be determined during process development prior to the commencement of any process validation (OQ and PQ).

Stage 3 - Material Performance and Suitability Testing

The material performance and suitability is demonstrated through testing. This often involves the development of technical reports or the execution of engineering studies to gather the evidence and appropriate rationale to support suitability for use. Prior to any functional or physical testing, test methods need to be validated in advance to ensure they are fit for purpose.

Test Methods

Test methods are used to measure both variable and attribute data that is generated as part of a validation. Variable outputs refer to data that is parametric in nature or continuous. Non-variable data, also known as attribute data is non parametric (such as pass/fail visual inspection)

Variable Outputs

Test method validation must address the following parameters for test methods with variable output:
Accuracy
Precision
Range
Resolution
Attribute Outputs

Test method validation addresses the following parameters for test methods with attribute output:
 Effectiveness
 Probability of false alarms
 Probability of misses

Prior to any test method validation, the test method itself (SOP/ procedure) should be available in draft form. The test equipment along with any software should also be qualified and fit for purpose.

Seal Width Measurement

Seal width measurement is a process output that helps to determine the seal integrity post blister sealing. The variable data can be used to monitor the seal quality of blisters and identify any changes in the process that might affect the barrier system.

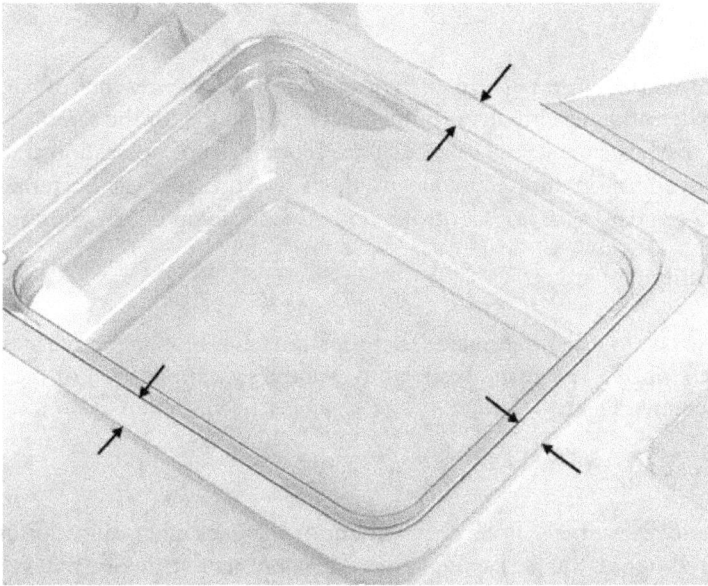

Figure: Seal width

Tensile Testing

Seal strength and seal integrity are critical outputs of the blister sealing process. The worst case sterilisation condition for a material is dependent on the material and the method of sterilisation. However, sealing and subsequent sterilisation at upper and lower worst case conditions should be completed. Tensile testing should be completed in accordance with ASTM F88 or another recognised standard.

Dye Penetration Testing

This test is designed to evaluate the integrity of a sterile barrier system. A blue dye is syringed into a sealed blister with the lidstock intact. The dye should then be allowed come into contact with seal for a defined period of time. Dye penetration testing should be completed in accordance with ASTM F 1929.

Packaging Inspection (Cosmetic)

Packaging inspection should be completed pre and post sterilisation in order to ensure product being sterilised is not damaged or compromised prior to sterilisation. Some typical cosmetic checks are listed below. Packaging inspection requirements should be detailed in an approved specification.

- The entire package must be free of foreign material
- The maximum permitted number of inclusions should be specified along with the max size of the inclusion in mm2
- No smudge marks or burns
- No tool marks
- No voids or bubbles
- No pinholes or tears

Note: For dye penetration, there should be no evidence of penetration across the complete width of the seal.

Bioburden Testing

Bioburden testing will be performed on devices pre-sterilisation to determine the levels of viable organisms that are naturally present on the product or introduced artificially.

Bacterial Endotoxin

Bacterial endotoxin testing is performed in order to test for the presence of bacterial endotoxins of the product. An endotoxin detection test involves testing the liquid sample (or the sample extract) with Limulus Amebocyte Lysate (LAL). LAL is an aqueous extract of the blood cells of horseshoe crabs. LAL forms a clot or changes in colour, depending on the technique used, in the presence of bacterial endotoxin. The test sample is compared to a standard series of Control Standard Endotoxin (CSE) dilutions.

Biocompatibility

Testing is necessary in order to evaluate that the material is biocompatible and appropriate to the intended use of the material/ finished product. Testing ensures the safe application of the device if it is in contact with the body or in used invasively.

Additional Testing

The below testing is normally not a requirement of the packaging validation itself, however, testing may need to be completed during the packaging system development stage or process development in order to ensure the design requirements are met:

Puncture resistance testing is used to measure the toughness of a material punctured via a standard method to determine the relative ability to experience a puncture failure.

Abrasion resistance is a test to determine failure due to rubbing of the product or part of the package against the primary sterile barrier and the potential to create pinholes or other failures.

Stage 4 - Stability Testing

The purpose of Stability Testing for packaging is to verify that the packaging materials meet requirements over time. Examples of requirements include sterility, functionality, safety, efficacy and visual appearance. Stability testing is also known as "ageing." Stability testing is required for all packaging materials, including blisters, films (lids), cartons and so on. The project team in conjunction with the packaging engineer determines the appropriate testing and time-points and conditions required during testing to verify the packaging system remains fit for purpose over the shelf-life of the product.
In addition to stability testing over real time, stability testing is also conducted in an accelerated manner. This is referred to as "accelerated" testing or "accelerated ageing".

Testing intervals should be determined based on the shelf life specification of the product in question. The first time point is always t=0 when packaging has just been completed. Sterilised and non-sterilised product should be tested at t=0. Testing must also examine the stability of labels to ensure they remain intact, the material does not crease and the artwork remains legible.

Stage 5 - Packaging Performance Testing

Performance testing the packaging system challenges the acceptability of the entire package system. Performance testing evaluates the interaction between the packaging system and the product in response to the stresses (events) imposed by the production processes and limits, sterilisation processes, storage and transportation environment. The testing is intended to demonstrate that the SBS and protective packaging are adequate to protect the product while maintaining sterility to the point of use. The worst case product configuration is used for this testing.

The following should be considered for inclusion in engineering trials or technical reports:
- Product or representative product is necessary for performance testing.

- Testing should be completed on worst-case product. What is the worst-case product? How was the worst-case product determined?
- What is the worst-case sterilisation process?
- Is the labelling and final packaging reflective of the process/product going forward?

Stage 6- Packaging Validation

Medical Packaging Process Validation

The ultimate aim of process validation for a given packaging system is to demonstrate the manufacturing packaging process is fit for purpose and robust enough to meet the acceptance criteria as set out in product specifications.

Post Equipment Qualification (IQ- Installation Qualification/ OQ- Operational Qualification: Process validation which consists of Process Operational Qualification (OQ) and Performance Qualification (PQ) is required to be completed.

Operational Qualification: Operational Qualification challenges the worst-case process sealing settings to ensure that worst case settings, product seal strengths and the other outputs meet specifications. OQ is typically completed for both high and low worst case conditions e.g. high temperature, high pressure and high seal times versus low temperature, low pressure or low time.

Performance Qualification: Performance Qualification provides a high degree of assurance that the sealing process will consistently produce a packaging system that meets predetermined specifications under normal operating conditions. Any product used in the PQ should be representative of the commercial process going forward.
A minimum of three lots is normally required for the PQ testing for an initial validation. Lots should be based on statistical rational and be reflective of commercial sizes.

Influencing Factors

Process validation aims to prove the consistency of a process under normal operating (anticipated) conditions. Every manufacturing process (however stable) is subject to influencing factors within day-to-day anticipated variation. These influencing factors can be categorised as (1) Worker, (2) Setup, and (3) Material.

For packaging equipment, these factors can influence the performance and consistency of a process and therefore need to be challenged during performance qualification.

A worker or operator can be a source of variation if the process is manual or not fully automated. Different people may have varying levels of expertise, experience and concentration. Even if an operator follows the instructions accurately and complies with standard operating procedures, there will likely be differences in how different operators handle raw materials and the product.

For packaging equipment that requires manual placement of raw materials, multiple operators should be used during the execution of validation builds.

Packaging systems are made up of more than one raw material or component. Whether components are manufactured in-house or supplied externally, they can be subject to variation, even if within the acceptance criteria. Therefore, for packaging validation, a minimum of three distinct lots of each material or component should be used. If multiple tooling is available, set-up can also be a source of variation. In these cases, a minimum number of setups should be completed as part of the validation. If the equipment is intended to run on different days or different shift patterns, the process may be subject to drift. Therefore, the validation should take account of this to ensure normal variation is captured.

Statistical Methods

The packaging process validation OQ and PQ must be performed using sample sizes and numbers of runs and batch sizes that are based on statistical rationale. A risk based approach should be taken when executing packaging validation.

Sample Size Determination and Sampling Rationale

The number of samples produced during the validation must meet predetermined acceptance criteria that are statistically relevant. This information should be documented in the pre-approved protocol.

Normality

For variable data (continuous data) a test for normality should be completed initially as this determines what type of statistical tool should be used. The normality of variable is verified by completing a normality test using a statistical package such as Minitab© or SPC© If the test returns a P-value ≥ 0.01, the data is considered normal. For non-normal data, accepted statistical methods for data transformation or non-parametric analysis should be used.

General Principles of Blister Packaging Validation

- The minimum and maximum critical process parameter set points at which product meeting all critical quality attributes can be manufactured should be defined during process development, ***prior*** to process validation.

- The calibrated range of all critical instruments must be greater than the process range (operating window).

- OQ-P testing will be executed by manufacturing worst case product at minimum process range settings (OQ-P min) and maximum process range settings (OQ-P max) for each critical process parameter. Devices manufactured at worst case minimum OQ-P min and OQ-P max settings must meet acceptance criteria for all critical quality attributes.

- Performance Qualification (PQ) should be manufactured at nominal process settings. Process capability (Cpk) or Process Performance (Ppk) critical quality attributes for each PQ batch.

Variable Print Packaging

This section examines some of the requirements for variable print which is either printed or laser etched onto the surfaces of labels, blisters or cartons.

What is Variable Print?

Variable print is when characters or other shapes or designs printed or laser etched on packaging materials need to change from lot-to-lot or product to product. Examples of variable print include lot/batch numbers, expiry date, date of manufacturing and other unique or variable information required to provide traceability and proper identification of the product.

What Is Fixed Print?

Fixed print is when characters or shapes do not change between lots or batches. The information (characters or shapes) is pre-printed by the component supplier per approved artwork.

Packaging Definitions

Wicking: is the process in which through capillary action, moisture moves from the inside to the surface. During a dye penetration test, wicking can result in a false fail if the dye is exposed to the seal for too long.

Primary Packaging: The labelled inner container in which product is placed and sealed. This generally is the blister pack itself.

Secondary Package: The outer container into which one or more inner containers or "primary packages" are inserted into to form the complete finished product.

Ink Jet Print: applied by a printer that discharges liquid ink, one drop at a time, onto the component.

Laser Etching: "print" is applied by a laser that etches the characters or shapes into the material.

Print and Apply: Labelling method that prints variable print information on a label, then applies the label to a package.

Further Reading

ASTM D1922: Test Method for Propagation Tear Resistance of Plastic Film and Thin Sheeting by Pendulum Method

ASTM D1938: Test Method for Tear-Propagation Resistance (Trouser Tear) of Plastic Film and Thin Sheeting by a Single-Tear Method

ASTM D1242: Resistance of Plastic Materials to Abrasions

ASTM D3420: Pendulum Impact Resistance of Plastic Films

ASTM F1306: Slow Rate Penetration Resistance of Flexible Barrier Films and Laminates

ASTM D1709: Standard Test Method for Impact Resistance of Plastic Film by the Free Falling Dart Method

5.5 Cleaning Validation

Introduction

Cleaning and disinfection must take place according to defined procedures and programmes, with on-going environmental monitoring to ensure compliance to the microbiological limits and to detect the development of resistant strains of organisms. The effectiveness of all disinfectants must be validated with reference micro-organisms and local isolates. Hard surfaces of equipment, premises and materials that are decontaminated can be selected on a risk-based approach. The choice of disinfectants must be adequate to maintain good results on the viable environmental monitoring trend analysis.

What Is Cleaning?

Cleaning can be defined as the process of removing potential contaminants from process equipment and maintaining the condition of equipment so that it can be safely used for subsequent product manufacture. It is complicated by many different chemicals used to produce medicinal drug products and other chemical agents used in the manufacturing process or in the cleaning process.

Why Clean Equipment or Products?

Facilities can be multi-product facilities, i.e. the same equipment is shared over different products. However, dedicated facilities also require cleaning evaluation and strategies. Cleaning minimises the transfer (or carry-over) of one product into another product, to an acceptable level, by means of product residue.

Some product residues are considered so toxic/potent if carried-over to another product that they are required to be manufactured in a dedicated facility (e.g. penicillin).

Equipment: Clean-in-place, often abbreviated to CIP, allows equipment cleaning to occur with minimal disassembly of equipment. CIP programs allow different products using similar or different materials to be manufactured on the same equipment.

Products: Supplies of products and medical devices used by patients or healthcare professionals must be clean and free of contamination.

Regulatory Requirements: It is the aim of every manufacturer to provide safe and effective products for use by patients and end users such as doctors and nurses. Companies are granted licenses to supply markets with products based on regulatory compliance and product safety. Cleaning compliance is a key part of achieving a state of compliance and more importantly, supplying safe products.

Verification and Validation

Verification: Verification means confirmation by examination and provision of objective evidence that specified requirements have been fulfilled[1]. When it comes to cleaning, if the cleaning procedures have not been fully validated, the effectiveness of the cleaning procedure should be verified at the completion of cleaning. This is "verification".

Validation: Validation means confirmation by examination and provision of objective evidence that the particular requirements for a specific intended use can be consistently fulfilled[1].

Definitions
Clean Hold Time (CHT): The total time the parts or equipment are held clean post-cleaning.

Cleaning Agent: The chemical agent or solution used as an aid in the cleaning process.

Cleaning Process Parameters: The parameters that are critical in the cleaning process. Subsequent cleaning process monitoring may or may not utilise these parameters.

Critical Process Parameter (CPP): A control parameter that has a direct relationship to the quality, safety, effectiveness or performance of the intermediate or final product.

Dirty Hold Time (DHT): The total time the parts (or equipment) are held dirty prior to cleaning.

Maximum Allowable Carry Over (MACO): The amount of allowed product residue carry-over from lot-to-lot, batch-to-batch, etc. This limit is based on the lowest of:

(1) Limited based on toxicity,

(2) Limit based on smallest therapeutic dose, and

(3) Worst case dose methodology

Residue: Substances left on surfaces of equipment after cleaning that may pose a risk for subsequent use. Example: residues that may require cleaning include: products, excipients, raw materials/intermediates, non-volatile solvent, non-intrinsic cleaning agents such as detergents, etc.

Worst Case Conditions: Conditions considered to pose the greatest chance of process or product failure. The highest or lowest value of a given control parameter or set of parameters.

Visual Inspection: With regard to cleaning, visual inspection should be completed by appropriately trained and experienced personnel on completion of equipment/process clean-down.

Surfaces should be visibly clean and free of visible residue. Hard-to-clean places should be examined in particular.

cGMP: Current Good Manufacturing Practices

Concurrent Validation: Validation activities occurring at the same time as one another or concurrent to a product launch.

Prospective Validation: This is when validation is done in advance of commercial manufacturing.

Protocol: An approved document that contains the tests and verifications to be conducted during the validation. Validation protocols include test methods and test conditions, acceptance criteria and parameters required.

---

[1] 21CFR820.3

FAT (Factory Acceptance Test): Typically classified as an engineering activity, the purpose of the FAT is to verify that the equipment or system meets the requirements of the URS.

Deviation: An event which results in failure with respect to the acceptance criteria in the protocol.

Process window: The selected operating range of machine settings/parameters that will produce product to meet all quality and product specifications.

Installation Qualification: Establishing through documented evidence that all functionality of the process equipment meets the manufacturer's specification and company requirements.

Equipment Qualification: Providing confidence through documented evidence that the equipment is suitable for the intended use and is capable of consistently operating within set limits and tolerances.

Operational Qualification: Providing confidence through documented evidence that the product can be manufactured to specifications within set limits and tolerances.

(MVP) Master Validation Plan: A governing document which sets out the validation approach and provides details of deliverables. An MVP should be written as soon as possible and it should align and reflect with the "current" validation strategy.

Precision Cleaning Systems: A precision cleaning system is a piece of equipment that can remove soil or dirt from parts or components. Most precision cleaning systems are made up of several stages, e.g. clean-rinse-dry. The simplest cleaning systems consist of one stage, e.g. an ultrasonic bath containing heated water.

Clean-in-Place (CIP): CIP is a cleaning method used to clean the inner surfaces of piping, vessels and process equipment without the need for disassembly.

PIC/s: The Pharmaceutical Inspection Convention and Pharmaceutical Inspection Co-Operation Scheme (referred to as PIC/S) are two international bodies between countries and pharmaceutical inspection authorities that co-operative on subjects relating to the field of GMP.

Skid: This is essentially a modular process that can be plugged into a process onsite with little construction or integration required. Skids are used as part of clean-in-place solutions within the food and beverage industry as well as pharmaceutical industries.

Regulatory Requirements

Key regulatory and international publications are included below:

➢ FDA – Food and Drug Administration – Guide to Inspections of Validation of Cleaning Processes

➢ EU GMP – European Commission – EudraLex Volume 4: EU Guidelines to Good Manufacturing Practice, Medicinal Products for Human and Veterinary Use, and Annex 15 (section 10 "Cleaning Validation")

➢ ICH Q7 – International Council on Harmonisation - Good Manufacturing Practice

➢ Guide for Active Pharmaceutical Ingredients (section 12.7 "Cleaning Validation")

➢ ICH Q9 – International Council on Harmonisation – Quality Risk Management

> PIC/S PI 006-3 – Pharmaceutical Inspection Co-Operation Scheme – Recommendations on Validation Master Plan, Installation and Operational Qualification, Non-Sterile Process Validation, Cleaning Validation (section 7 "Cleaning Validation")

> WHO TRS 937 – World Health Organisation - Specifications for Pharmaceutical Preparations; Annex 4: Supplementary guidelines on Good Manufacturing Practices: Validation; Appendix 3: Cleaning Validation

FDA – Food and Drug Administration - Guide to Inspections of Validation of Cleaning Processes

Introduction

Cleaning validation programmes are important requirements for both bulk pharmaceutical processing and biotechnology. As with validation of other processes, there may be more than one way to validate a cleaning process. Once the manufacturer can establish inspection consistency and repeatable outcomes that ensure predetermined acceptable criteria are met, a cleaning procedure can be deemed effective. This is a driven process which should support claims of consistent outcomes.

It isn't solely the FDA that has an expectation that cleaning procedures (processes) be validated; PIC/s, ICH, EudraLex and WHO guidance and requirements also specify the need to validate cleaning procedures.

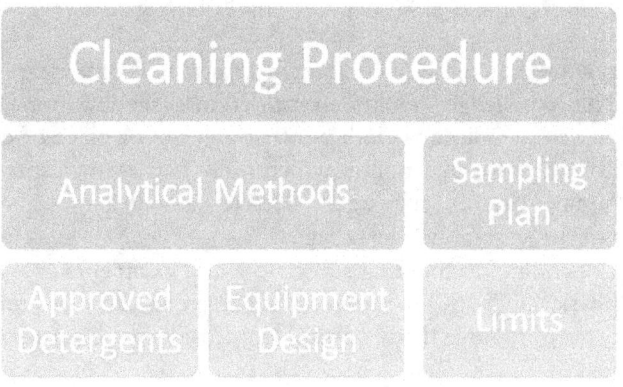

Figure: Components of a Cleaning Programme

Historically, the FDA was mostly concerned about the contamination of non-penicillin drug products with penicillin or the cross-contamination of drug products with potent hormones or steroids. One event which increased FDA awareness of the potential for cross-contamination due to inadequate procedures was the 1988 recall of a finished drug product, Cholestyramine resin USP.

In this instance, the bulk pharmaceutical used to produce the product had become contaminated with low levels of both intermediates and degradants.

The cross-contamination in this case was attributed to the reuse of recovered solvents. The recovered solvents had been contaminated because of a lack of control over the reuse of solvent drums. Drums that had been used to store recovered solvents from a pesticide production process were later used to store recovered solvents used for the resin manufacturing process. Some shipments of this pesticide-contaminated bulk pharmaceutical were supplied to a second facility at a different location for finishing. This resulted in the contamination of the bags used in that facility's fluid bed dryers with pesticide contamination.

Upon inspection by auditors, the following requirements are expected in order to demonstrate a robust and suitably validated cleaning programme (procedure):

➤ Written procedures (SOPs) detailing the cleaning processes used.
➤ Register or list of dedicated equipment. Fluid bed dryer bags are another example of equipment that is difficult to clean and are often dedicated to a specific product.

➤ A written procedure on how cleaning processes are validated.

➤ Written validation protocols detailing the sampling procedures and analytical methods to be used including the sensitivity of those methods and acceptance criteria.

➤ A validation report which is approved in advance of commercial manufacturing. The data generated during the validation should demonstrate that residues have been reduced to an "acceptable level".

Evaluation of Cleaning Validation:

The main focus of an auditor in respect of cleaning validation is to evaluate the evidence that aims to demonstrate the effectiveness of the approach and processes used to clean equipment.

The following questions are relevant when evaluating the cleaning process:

➤ At what point does a piece of equipment /system become clean?
  o This knowledge should be captured in cycle development and development of the cleaning process. Studies may indicate that a vessel or piece of equipment requires three rinses with hot water-for-injection at which point it meets acceptance criteria. However, an additional number or rinses may be included to provide a level of confidence in the cleaning process.
➤ Does it have to be scrubbed by hand?
  o Depending on the drug substances and excipients or other chemicals, residues may tend to physically "stick" to surfaces or behave as tarry or gummy which may require mechanical force to remove them, or a solvent rinse may be sufficient for removal.

When the cleaning process is used only between batches of the same product, a company may only meet the criteria of "visibly clean" for the equipment. This can often be referred to as a batch-to-batch clean. Such between-batch cleaning processes do not require validation. Change-over from one product to a different product of different materials requires a more comprehensive clean, potentially requiring multiple cleans or rinses.

EU GMP – European Commission – EudraLex Volume 4: EU Guidelines to Good Manufacturing Practice, Medicinal Products for Human and Veterinary Use, and Annex 15 (section 10 "Cleaning Validation")

Section 10 of Annex 15 provides a number of bullet points with regard to cleaning validation:

"Cleaning validation should be performed in order to confirm the effectiveness of any cleaning procedure for all

product contact equipment. Simulating agents may be used with appropriate scientific justification. Where similar types of equipment are grouped together, a justification of the specific equipment selected for cleaning validation is expected.

A visual check for cleanliness is an important part of the acceptance criteria for cleaning validation. It is not generally acceptable for this criterion alone to be used. Repeated cleaning and retesting until acceptable residue results are obtained is not considered an acceptable approach.

It is recognised that a cleaning validation programme may take some time to complete and validation with verification after each batch may be required for some products, e.g. investigational medicinal products. There should be sufficient data from the verification to support a conclusion that the equipment is clean and available for further use.

Validation should consider the level of automation in the cleaning process. Where an automatic process is used, the specified normal operating range of the utilities and equipment should be validated.

For all cleaning processes an assessment should be performed to determine the variable factors which influence cleaning effectiveness and performance, e.g. operators, the level of detail in procedures such as rinsing times etc. If variable factors have been identified, the worst case situations should be used as the basis for cleaning validation studies.

Limits for the carryover of product residues should be based on a toxicological evaluation. The justification for the selected limits should be documented in a risk assessment which includes all the supporting references. Limits should be established for the removal of any cleaning agents used. Acceptance criteria should consider the potential cumulative effect of multiple items of equipment in the process equipment train.

The risk presented by microbial and endotoxin contamination should be considered during the development of cleaning validation protocols.

The influence of the time between manufacture and cleaning and the time between cleaning and use should be taken into account to define dirty and clean hold times for the cleaning process.

Where campaign manufacture is carried out, the impact on the ease of cleaning at the end of the campaign should be considered and the maximum length of a campaign (in time and/or number of batches) should be the basis for cleaning validation exercises.

Where a worst case product approach is used as a cleaning validation model, a scientific rationale should be provided for the selection of the worst case product and the impact of new products to the site assessed. Criteria for determining the worst case may include solubility, cleanability, toxicity and potency.

Cleaning validation protocols should specify or reference the locations to be sampled, the rationale for the selection of these locations and define the acceptance criteria.

Sampling should be carried out by swabbing and/or rinsing or by other means depending on the production equipment. The sampling materials and method should not influence the result. Recovery should be shown to be possible from all product contact materials sampled in the equipment with all the sampling methods used.

The cleaning procedure should be performed an appropriate number of times based on a risk assessment and meet the acceptance criteria in order to prove that the cleaning method is validated.

Where a cleaning process is ineffective or is not appropriate for some equipment, dedicated equipment or other appropriate measures should be used for each product as indicated in chapters 3 and 5 of EudraLex, Volume 4, Part

I.

Where manual cleaning of equipment is performed, it is especially important that the effectiveness of the manual process should be confirmed at a justified frequency."

Ref: EU GMP V4, Annex 15, 2017

ICH Q7 – International Council on Harmonisation - Good Manufacturing Practice:

"Cleaning procedures should normally be validated. In general, cleaning validation should be directed to situations or process steps where contamination or carryover of materials poses the greatest risk to API quality. For example, in early production it may be unnecessary to validate equipment cleaning procedures where residues are removed by subsequent purification steps. (12.70)

Validation of cleaning procedures should reflect actual equipment usage patterns. If various APIs or intermediates are manufactured in the same equipment and the equipment is cleaned by the same process, a representative intermediate or API can be selected for cleaning validation. This selection should be based on the solubility and difficulty of cleaning and the calculation of residue limits based on potency, toxicity, and stability. (12.71)

The cleaning validation protocol should describe the equipment to be cleaned, procedures, materials, acceptable cleaning levels, parameters to be monitored and controlled, and analytical methods. The protocol should also indicate the type of samples to be obtained and how they are collected and labelled. (12.72)

Sampling should include swabbing, rinsing, or alternative methods (e.g., direct extraction), as appropriate, to detect both insoluble and soluble residues. The sampling methods used should be capable of quantitatively measuring levels of residues remaining on the equipment surfaces after cleaning. Swab sampling may be impractical when product contact surfaces are not easily accessible due to equipment design and/or process limitations (e.g., inner surfaces of hoses, transfer pipes, reactor tanks with small ports or handling toxic materials, and small intricate equipment such as micronisers and microfluidisers). (12.73)

Validated analytical methods having sensitivity to detect residues or contaminants should be used. The detection limit for each analytical method should be sufficiently sensitive to detect the established acceptable level of the residue or contaminant. The method's attainable recovery level should be established. Residue limits should be practical, achievable, verifiable, and based on the most deleterious residue. Limits can be established based on the minimum known pharmacological, toxicological, or physiological activity of the API or its most deleterious component. (12.74)

Equipment cleaning/sanitation studies should address microbiological and endotoxin contamination for those processes where there is a need to reduce total microbiological count or endotoxins in the API, or other processes where such contamination could be of concern (e.g., non-sterile APIs used to manufacture sterile products). (12.75)

Cleaning procedures should be monitored at appropriate intervals after validation to ensure that these procedures are effective when used during routine production. Equipment cleanliness can be monitored by analytical testing and visual examination, where feasible. Visual inspection can allow detection of gross contamination concentrated in small areas that could otherwise go undetected by sampling and/or analysis. (12.76)"

PIC/S PI 006-3 – Pharmaceutical Inspection Co-Operation Scheme – Recommendations on Validation Master Plan, Installation and Operational Qualification, Non-Sterile Process Validation, Cleaning Validation

(section 7 "Cleaning Validation")

PIC/s provides several pages of recommendations on cleaning validation. It clearly outlines the principles and purpose of conducting cleaning validation:

"Pharmaceutical products and active pharmaceutical ingredients (APIs) can be contaminated by other pharmaceutical products or APIs, by cleaning agents, by micro-organisms or by other material (e.g. air-borne particles, dust, lubricants, raw materials, intermediates, auxiliaries). In many cases, the same equipment may be used for processing different products. To avoid contamination of the following pharmaceutical product, adequate cleaning procedures are essential.

Cleaning procedures must strictly follow carefully established and validated methods of execution. This applies equally to the manufacture of pharmaceutical products and active pharmaceutical ingredients (APIs). In any case, manufacturing processes have to be designed and carried out in a way that contamination is reduced to an acceptable level.

Cleaning Vvalidation is documented evidence that an approved cleaning procedure will provide equipment which is suitable for processing of pharmaceutical products or active pharmaceutical ingredients (APIs).

The objective of cleaning validation is confirmation of a reliable cleaning procedure so that the analytical monitoring may be omitted or reduced to a minimum in the routine phase."

Validation Standards

➢ ASTM E2281-03 "Standard Practice for Process and Measurement Capability Indices".

➢ ASTM E2500-07 "Standard Guide for Specification, Design, and Verification of Pharmaceutical and Biopharmaceutical Manufacturing Systems and Equipment".

➢ ASTM E2709-09 "Standard Practice for Demonstrating Capability to Comply with a Lot Acceptance Procedure".

Stages of Validation

This section provides a background on general principals in respect of validation. Process validation can be divided into three stages which are:

➢ Process design
➢ Process validation
➢ Continued process monitoring

Similarly, cleaning validation can be divided into the same three stages; cleaning process design, cleaning process validation and continued process monitoring of cleaning processes. Prior to the commercial manufacture and distribution of drug products and medicines to consumers, a manufacturer must gain a high degree of assurance in the performance of the manufacturing process such that it will consistently produce APIs and drug products meeting those attributes relating to identity, strength, quality, purity and potency. A high degree of assurance and demonstrated consistency of a process must be evidence-based.

In order to best prepare a process that is consistent and produces safe and effective products, the following points must be understood and implemented:

> ➢ Understand the sources of variation
> ➢ Detect the presence and degree of variation
> ➢ Understand the impact of variation on the process and ultimately on product attributes
> ➢ Control the variation in a manner commensurate with the risk it represents to the process and product

Stage 1 — Process Design

The goal of this stage is to design a process suitable for routine commercial manufacturing that can consistently deliver a product that meets its quality attributes. The activity as this point is focused on defining the commercial manufacturing process that will be used going forward and that the records and associated documents support the shaping of the commercial process.

Designing an efficient process with an effective process control approach is dependent on the process knowledge and understanding of personnel. This is achieved through pilot studies and testing according to sound scientific methods and principles, including good documentation practices.

ICH Q10 Pharmaceutical Quality System also highlights the importance of process knowledge and the extent of its capability. It is also a regulatory requirement to establish a strategy for process control. Strategies for process control can be designed to:

> ➢ Reduce input variation
> ➢ Adjust for input variation during the manufacturing process
> ➢ Take into account a combination of reduction and adjustment

Having a method of process control reduces variability and gives a higher assurance of consistent product quality.

Stage 2 — Process Qualification

During the process qualification (PQ) stage of process validation, the process design is evaluated to determine if it is capable of reproducible commercial manufacture.

This stage has two elements:

(1) Design of the facility and qualification of the equipment and utilities

(2) Process performance qualification (PPQ)

Stage 3 — Continued Process Verification

The goal of stage three, continued process validation, is to provide assurance that the process remains in a state of control (the validated state) during commercial manufacture.

Systems for detecting unplanned departures from the process as designed are essential to accomplish this goal. Adherence to the cGMP requirements, specifically, the collection and evaluation of information and data about the performance of the process, will allow detection of undesired process variability.

Evaluating the performance of the process identifies problems and determines whether action must be taken to correct, anticipate and/or prevent problems so that the process remains in control (§ 211.180(e)).

An ongoing programme to collect and analyse product and process data that relate to product quality must be established (§ 211.180(e)). The data collected should include relevant process trends and quality of incoming materials or components, in-process material and finished products. The data should be statistically trended and reviewed by trained personnel. The information collected should verify that the quality attributes are being appropriately controlled throughout the process.

The FDA recommends that a statistician or person with adequate training in statistical process control techniques develop the data collection plan and statistical methods and procedures used in measuring and evaluating process stability and process capability. Procedures should describe how trending and calculations are to be performed and should guard against overreaction to individual events as well as against failure to detect unintended process variability. Production data should be collected to evaluate process stability and capability. The quality unit should review this information.

If properly carried out, these efforts can identify variability in the process and/or signal potential process improvements. Good process design and development should anticipate significant sources of variability and establish appropriate detection, control and/or mitigation strategies as well as appropriate alert and action limits.

Key FDA Recommendations:

- The use of quantitative, statistical methods whenever appropriate and feasible. Scrutiny of intra-batch as well as inter-batch variation is part of a comprehensive continued process verification programme under § 211.180(e).

- Continued monitoring and sampling of process parameters and quality attributes at the level established during the process qualification stage until sufficient data are available to generate significant variability estimates.

- Process variability should be periodically assessed and monitoring adjusted accordingly. Variation can also be detected by the timely assessment of defect complaints, out-of-specification findings, process deviation reports, process yield variations, batch records, incoming raw material records and adverse event reports.

- Production line operators and quality unit staff should be encouraged to provide feedback on process performance.

- The quality unit meets periodically with production staff to evaluate data, discuss possible trends or undesirable process variation and coordinate any correction or follow-up actions by production.

- Maintenance of the facility, utilities and equipment is another important aspect of ensuring that a process remains in control. Once established, qualification status must be maintained through routine monitoring, maintenance, and calibration procedures and schedules.

Reference: FDA Process Validation: General Principles and Practices January 2011 — Current Good Manufacturing Practices (CGMP) Revision 1

Some references that may be useful in understanding the general principles of process validation (which are also relevant to cleaning validation) include the following:

- ➢ ASTM E2281-03 "Standard Practice for Process and Measurement Capability Indices,"

- ➢ ASTM E2500-07 "Standard Guide for Specification, Design, and Verification of Pharmaceutical and Biopharmaceutical Manufacturing Systems and Equipment,"

> ➤ ASTM E2709-09 "Standard Practice for Demonstrating Capability to Comply with a Lot Acceptance Procedure."

Note: *This is not a complete list of all useful references on this topic.*

Validation: General Principles and Practices

Validation has a number of common requirements that apply to equipment qualification, process validation or cleaning validation.

- ➤ Documented evidence (records/reports)
- ➤ Demonstration of a high degree of assurance (data-driven studies and data analysis)
- ➤ Consistency (traditionally three PQ runs)
- ➤ Predetermined quality attributes (product specifications)

Cleaning Validation

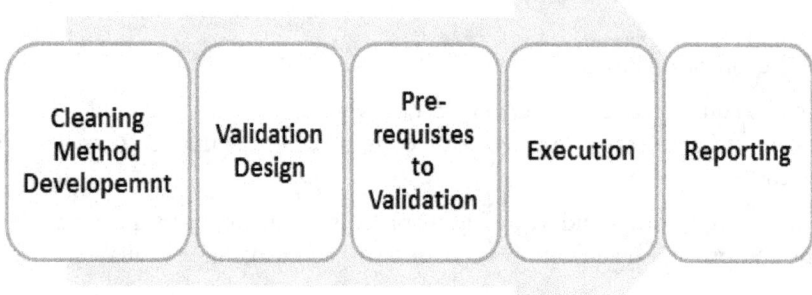

Figure: Life Cycle of Cleaning Validation

The purpose of cleaning validation is to provide objective evidence that methods and procedures are capable of removing product residues, contaminants, cleaning agents used, by-products, solvents and degradants to below a predetermined level.

All contamination is referred to as soiling. A simple method of examining sources of contamination is reviewing the input materials (product ingredients, manufacturing agents etc.) of a process. Risk assessment must be applied to decide on the extent of cleaning validation studies. Cleaning validation for a CIP system design involves the intersection of two similar or different products. Take a simple example: a pharmaceutical company manufactures two types of paracetamol caplets (tablets).

Product A contains the active ingredient paracetamol, preservatives and other excipients.

Product B is also a paracetamol product but it contains an additional ingredient, caffeine. Therefore, product B is branded differently and marketed with a more discerning customer in mind.

Where multiple products are manufactured on the same equipment or machinery, the process is often referred to as non-dedicated.

As with the above example, if the same equipment is used to produce product A and Product B, an intersection of products occurs.

Product A- Cleaning must be effective enough to remove residues to acceptable levels.

Product B- When manufacturing commences, the residue levels must not contaminate the product.

Residue is any substance or trace of substance left on equipment or surfaces after cleaning. It is near possible to remove all residues from surfaces, so a residue limit should be medically safe and at a level that does not cause product quality issues or concerns.

Visibly Clean

Within any cGMP environment, the requirement to maintain a clean and suitable manufacturing area is key to compliance and ensuring product quality and customer safety. Visual inspection of the cleaning process must be done before swabbing. Inspection should confirm the equipment is visually clean and dry and no adverse odours are present.

Upon completion of visual inspection, swabbs should then only be taken if required by procedure. For areas that cannot be accessed for visual inspection or swabbing, a rinse sample can be taken in place of a swab.

Sometimes it is not possible to obtain a swab or rinse sample, therefore visual inspection may be the only method used to verify cleaning effectiveness.

In any validation, an important theme is to challenge the consistency of a process. Therefore, samples must be representative to ensure a proper picture is painted. Sampling sites should be taken from "hard-to-clean" areas as well as "easy-to-clean" areas to ensure that samples are representative of the equipment.

Soils and Their Behaviour

Soils are a source of contamination to products and therefore can present a risk to patients or users. Soils can be introduced by unplanned and unintended events, but they are likely a part of the process, or as a result of a manufacturing agent that has been used within a manufacturing process. Examples include coolant of cutting fluid used in a machining process. The fluid is required to achieve a good surface finish and reduce tool wear. The presence of this soil of parts can potentially be:

➢ Dried on during subsequent process step
➢ Compacted
➢ Dried on during dirty hold time
➢ Baked on during an oven process

Detergents

Cleaning agents and solvents must be sourced and approved according to a supplier qualification procedure. The following information should be considered for inclusion in a supplier qualification file.

➢ Certification to a quality management system such as FDA, ISO
➢ Supplier questionnaire
➢ Product specification
➢ Material safety data sheets

➢ Change notification policy
➢ Expiration dating (format and controls)
➢ Onsite audit
➢ Statements of suitability
➢ Sample certificate of analysis

Acceptance criteria for cleaning agent residues should be based on the lowest **LD50** of each chemical in an agent's formulation.

The standard LD50 cleaning agent's classification applies to all agents which contain chemical components whose LD50 is greater than 100 mg/kg. For low LD50 cleaning agents, the classification applies to all agents which contain chemical components whose LD50 is less than or equal to 100 mg/kg.

Validation Strategies

In this section we examine validation strategies. There are two main approaches for consideration:

(1) Direct approach

The direct approach consists of validating the cleaning procedure for all pieces of equipment and for all the products made.

(2) Matrix approach (aka grouping approach, family approach, bracketing)

A family or matrix approach can also be used where similar products can be grouped together with a representative.

A matrix can be formed and justified by defining a set of parameters and characteristics so that limited numbers of parameters or quality attributes are representative of the group. It should also focus on the worst-case parameters and quality attributes. All of the information that provides a rationale for implementing a matrix approach should be documented with a risk assessment created.

The following criteria may be considered to define the worst-case product with regard to cleaning validation:

➢ PDE
➢ Solubility
➢ Cleanability

With regard to medical devices, a particular product size of product configuration may be selected to represent the worst-case product. Therefore, by qualifying the worst case, all other products within the family of products would be considered validated. With regard to pharmaceutical products, e.g. solid dose manufacturing of pain killers, products of a similar chemistry/content can be grouped together.

Summary
Grouping/matrix approach can be done by:

– product (soil)

– equipment

— worst case

Advantages of grouping/matrix approaches:

– Potential to simplify the amount of validation work
– Fewer validation runs

Establishing Residue Limits and Acceptance Criteria

Typical target residues on product within precision cleaning systems include:

➢ Organic residuals
➢ Particulate
➢ Bioburden
➢ Endotoxin

Typical target residues for CIP systems include:

➢ Drug active
➢ Cleaning agent
➢ Bioburden
➢ Endotoxin
➢ Degradation products or by-products

How Are Acceptance Levels Defined?

Several considerations need to be accounted for when establishing safe and effective residue levels.

– Consider the potential effects of target residue on subsequent products or raw materials

– Pharmacology of residue

– Toxicity of residue

– Stability issues

As per European guidance, (Reference Human Drug CGMP Notes, 9:2, 2Q 2001) equipment does not have to be as clean as the best possible method of residue detection or quantification, as absolute cleanliness is required or feasible. However, it should be as clean as can reasonably be achieved – "to a residue limit that is medically safe and that causes no product quality concerns....."[2]

Historical Context of Limits

Pre-1993 Industry Acceptance Limits:

➢ 1/10th of therapeutic dose

➢ 1/50th of max therapeutic dose

---

[2] Human Drug CGMP Notes, 9:2, 2Q 2001)

➢ Less than smallest therapeutic dose

➢ 3ppm (arsenic)

➢ 30ppm for cleaning agents

➢ Detection limit of method

This approach was inconsistent from company to company, arbitrary and not based on risk.

1993 Eli Lilly Article:

Gary Fourman and Dr. Mike Mullin in "Pharmaceutical Technology" April 1993 proposed:

➢ $1/1000^{th}$ of a dose in max daily dose

➢ 10ppm of product in another product (next product)

➢ No residue visible

Uses of the Term "Limit"

L0 = Daily amount allowed per patient (µg or mg)

(L zero)

This has been called the Acceptable Daily Intake (ADI), Acceptable Daily Exposure (ADE), Permitted Daily Exposure (PDE). The limit is based on safety and toxicity information. Typical values used for L0 include: 0.001 of minimum daily dose of active ingredient based on toxicity information.

L=1 Concentration in next product: PIC/s guidance suggests a maximum of 10ppm.

L=2 Absolute amount in manufacturing vessel train (mg): [MAC – maximum allowable carryover] – L2

This limit uses the absolute amount in manufacturing vessels. It is calculated by multiplying L1, limit, times the batch size of subsequent product to be manufactured.

L3=Amount per surface area (µg/cm²): This is calculated by dividing L2 by shared surface area of the equipment train (the sum of surfaces).

L4a = Amount per swab (µg): The amount per swab depends on the limit per surface area; (L3) – swabbed area.

Calculate:

L4a = L3 X Swabbed Area

L4b Conc. in swab extraction solution:

Concentration in swab extract depends on:

o   Limit per surface area (L3)
o   Swabbed area

o   Amount of solvent present for extraction

Calculate:

L3 X Swabbed Area/ solvent extraction amount.

L4c = Conc. in "rinse" water (µg/g): L4b change based on volume for extraction

If sample is extracted into 10g of solvent:

100 µg / 10 g = 10 µg/g

NOTE: L=0, Daily amount allowed is also known as:

Acceptable Daily intake (ADI)

Acceptable Daily Exposure (ADE)

Permitted Daily Exposure (PDE)

Safe Daily Intake (SDI)

Values for L0 can be a minimum daily dose of active 0.001 or a value based on toxicity data (MSDS sheets etc.)

PDA Technical Report No. 29

The PDA technical report proposes the following limits:

- ➤ 1/10th -1/100th for topicals
- ➤ 1/100th – 1/1,000th for oral products
- ➤ 1/1,000th – 1/10,000th injections & opthalmics
- ➤ 1/10,000th – 1/100,000th for research or investigational products

Calculation of MACO (Maximum Allowable Carryover)

There are two steps in calculating MACO residue levels.

Firstly, it is necessary to **calculate the MACO from one batch to the next batch.**

The second step is to calculate the **allowable "drug" or "residue" of each unique product contact surface, for each piece of equipment.**

This then provides a **calculation based on the overall equipment train** (aka the equipment line).

The **MACO** is based on three calculations, which are:

- • **MACO** based on toxicity
- • **MACO** based on the smallest therapeutic drug dose

- **MACO** based on smallest solution batch size

$$NOEL = \frac{LD50 x NHW}{2000}$$

$$ADI = \frac{NOEL}{SF}$$

$$MAC0 = \frac{ADIxSSBS}{LNDD}$$

NOEL = No observed limit effect

LD50 = Lethal dose of drug

NHW = Nominal human weight

2000 = Is a constant factor for calculating NOEL

ADI = Allowable daily intake

SF = Safety factor e.g. 1000

SSBS = Smallest solution batch size

MACO based on smallest therapeutic drug dose (STDD):

$$Product\ Carry\ Over = \frac{STDD}{SF}$$
$$MAC0 = Product\ Carry\ Over\ x \frac{SSBS}{LNDD}$$

MACO based on smallest solution batch size (SSBS):

$$Worst\ Case\ Number\ of\ Doses = \frac{SSBS}{LNDD}$$

$$MACO = LNDDxWorst\ Case\ Number\ of\ \frac{Doses}{SF}$$

Using the above calculations, **the MACO for the equipment train can be determined.**

The MACO for each individual piece of **equipment of surface can then be calculated.**

MACO for Each Piece of Equipment

To calculate the MACO (allowable residue for each piece of equipment) you will need to have the following information available:

- **MACO** (per calculations above).

- **Surface area** of each piece of equipment and the total of the equipment train.

Coverage Testing

Process equipment often contains critical surfaces that need to be cleaned according to validated procedures. Examples include mixing vessels and freeze-dryer chambers. The removal of any residual contamination from these surfaces needs to be demonstrated. This is typically done with an easily detectable tracer such a riboflavin (e.g. for simple visual detection).

It is important to make the distinction between (1) coverage testing of equipment and (2) coverage testing of components that are subject to cleaning. However, the principle is the same in both instances. For equipment, coverage testing should be performed as part of equipment qualification for all process-contacting surfaces. Coverage testing verifies that all process-contacting surfaces are wetted by cleaning liquids and identifies any potential blind spots or hard-to-clean locations on the equipment. Locations on equipment that are not adequately cleaned can be identified through riboflavin fluorescence testing using UV light inspection.

Regulatory requirements do not specify a requirement for spray coverage testing. However, as per US Code of Federal Regulations and EudraLex Volume 4 Part II, equipment should be of appropriate design to facilitate cleaning. The Pharmaceutical Inspection Convention and Pharmaceutical Inspection Cooperation Scheme (PIC/S) specify "critical areas (i.e. those hardest to clean) should be identified, particularly in large systems that employ semi-automatic or fully-automatic clean-in-place (CIP) systems."

A 2004 FDA warning letter included two separate mentions of inadequate spray ball coverage: "Your firm failed to establish and follow written procedures to ensure the cleaning and maintenance of equipment used in the manufacture, processing, packing or holding of a drug product 21 CFR 211.67(b) and 600.11(b).

For example, cleaning validation for the clean-in-place (CIP) process vessel which is utilised in the aseptic formulation of trivalent bulk influenza vaccine, did not include an assessment of the spray ball coverage for the vessel. The spray ball is used for cleaning product contact equipment.... In addition, the cleaning validation did not include an assessment of the spray ball coverage for the tanks."

Ref: FDA Warning letter issued to Chiron corporation, December 9, 2004)

<u>Examples of Riboflavin Solution Strengths</u>

250mg/1 litre of water (1:4 ratio), generally suitable for spiking components.

2g per 1 litre (1:5),

200mg/L (0.2g/L) solution of riboflavin, suitable for testing interior surfaces of tanks/vessels.

1g of riboflavin in 10 Litres of water 100ppm solution, suitable for coverage testing of equipment surfaces.

Cleaning Validation Protocol

The validation protocol is a formal document that is preapproved prior to its use of execution. It defines the prerequisites, methods, specific requirements, activities and acceptance criteria for the cleaning validation at hand.

The protocol should address the following:

- ➢ Scope and Objectives
- ➢ Approval by cross-functional team
- ➢ Responsibilities
- ➢ Signature and training log
- ➢ Equipment/area to be cleaned under study
- ➢ Critical cleaning parameters
- ➢ Sampling methods and sample plan
- ➢ Maximum hold times
- ➢ Acceptance criteria
- ➢ Number of cycles

The dirty hold time (DHT) aka dirty equipment hold time (DEHT) is the time lapse between the end of manufacture and the start of cleaning. The purpose of this time control is to limit the difficulty of the cleaning before residues or remaining products are allowed to dry out.

The clean hold time (CHT) aka clean equipment holding time (CEHT) is the maximum time equipment can sit idle before a re-clean is required prior to its use. The main purpose of this time control is to limit microbial contamination. The drying time (DT) is another time control that aims to limit microbial growth within vessels or equipment. It is important to dry equipment immediately after it has been cleaned.

PIC/S Guidance on Limits

The Pharmaceutical Inspection Convention and Pharmaceutical Inspection Co-operation Scheme (jointly referred to as PIC/S) are two international instruments between countries and pharmaceutical inspection authorities which provide together an active and constructive co-operation in the field of GMP.[3] The most important point to remember when it comes to limits is that residues meet predefined criteria, the most stringent criteria as listed below:

---

[3] http://www.picscheme.org/

(a) No more than 0.1% of the normal therapeutic dose of any product should appear in the maximum daily dose of the following (next) product,

(b) No more than 10 (parts per million, ppm) of any product will appear in another product, (this value is not always the default)

(c) No quantity of residue should be visible on the equipment after cleaning procedures are completed. Spiking studies should determine the concentration at which most active ingredients are visible. [4]

The method of determining residue limits of active ingredients is based on an approach developed by Fourman and Mullen (1993) and is referenced in PIC/s guidance amongst other publications.

Test Methods

It is important to consider test methods and test method validation early on in the validation life cycle. A test method is a process or an action used to verify that a product feature or particular requirement meets a predefined specification. Test methods can be physical or analytical in nature. Test method validation should be completed in advance of cleaning as the test method is used to verify the outputs of such cleaning validations.

<u>ICH, Q7, Validation of Analytical Methods</u>

"Analytical methods should be validated unless the method employed is included in the relevant pharmacopeia or other recognised standard reference. The suitability of all testing methods used should nonetheless be verified under actual conditions of use and documented. (12.80)

Methods should be validated to include consideration of characteristics included within the ICH guidance on validation of analytical methods. The degree of analytical validation performed should reflect the purpose of the analysis and the stage of the API production process. (12.81)

Appropriate qualification of analytical equipment should be considered before initiating validation of analytical methods. (12.82)

Complete records should be maintained of any modification of a validated analytical method. Such records should include the reason for the modification and appropriate data to verify that the modification produces results that are as accurate and reliable as the established method. (12.83)

(Reference: Q7 Good Manufacturing Practice Guidance for Active Pharmaceutical Ingredients Guidance for Industry September 2016.)

Test method validation must address the following parameters in order to demonstrate suitability to the intended use of the method and ensure it is capable of achieving consistent performance.

<u>Definitions</u>

Specificity: The ability to assess unequivocally the analyte in the presence of components, which may be expected to be present. These can include impurities, degradants etc. (ICH Q2)

Linearity: The ability of an analytical procedure (within a known range) to obtain test results that are directly proportional to the concentration of analyte in the sample. (ICH Q2)

---

4 http://www.picscheme.org/publication.php?id=4

Range: The interval between the upper and lower concentrations (amounts) of analyte in the sample (including these concentrations) for which it has been demonstrated that the analytical procedure has a suitable level of precision, accuracy and linearity. (ICH Q2)

Accuracy: Expression of closeness of agreement between the value which is accepted either as a conventional true value or an accepted reference value and the value found. (ICH Q2)

Robustness: A measure of the capability of an analytical procedure to remain unaffected by small, but deliberate variations in method parameters and which provides an indication of its reliability during normal usage (ICH Q2)

Precision: Expression of the closeness of agreement of an analytical procedure (degree of scatter) between a series of measurements obtained from multiple sampling of the same homogeneous sample under the prescribed conditions. Precision may be considered at three levels: repeatability, intermediate precision and reproducibility. Repeatability expresses the precision under the same operating conditions over a short interval of time. Repeatability is also termed intra-assay precision. Intermediate precision expresses within-laboratories variations: different days, different analysts, different equipment, etc. Reproducibility expresses the precision between laboratories (collaborative studies, usually applied to standardisation of methodology). ICH Q2)

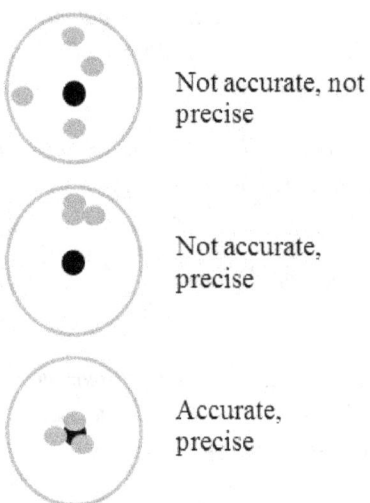

Figure: Illustration of accurate versus precise.

Detection Limit (LOD): The lowest amount of analyte in a sample that can be detected but not necessarily quantitated as an exact value for an individual analyte procedure (ICH Q2)

Quantitation Limit (LOQ): The lowest amount of analyte in a sample which can be quantitatively determined with suitable precision and accuracy for an analytical procedure. The quantitation limit is a parameter of quantitative assays for low levels of compounds in sample matrices and is used particularly for the determination of impurities and degradation products (ICH Q2).

Total organic carbon (TOC) analysis is a fast and effective analytical test method used for cleaning verification and validation in pharmaceutical manufacturing. It is used to test for residues of previously manufactured products (actives and excipients), cleaning detergents, chemicals, solvents, degradants and microbial contaminants. Detergent selection is a critical step in the development a cleaning programme. The purpose of the detergent is to remove residues; however, detergent levels remaining post-cleaning are undesired. If detergents remain post-cleaning they can affect the result of analytical tests.

Cleaning Process Design

For clean-in-place (CIP), the key elements to be considered in the design stage include:

– Equipment to be cleaned
– Soils to be removed
– Cleaning methods
– Cleaning agents
– Cleaning mechanisms
– Cleaning parameters
– Residue limits

CIP recipes should include the following information and parameters at a minimum:

> ➢ type of water (DI, USP purified, WFI) for pre-rinse,
> ➢ wash, and post-rinse
> ➢ volumes, times, flow rates, and pressures
> ➢ pump speeds
> ➢ process air blow times
> ➢ cleaning agent identification, concentration
> ➢ fill volumes to achieve circuit volume/flow
> ➢ alarm set points for parameters being monitored
> ➢ temperature and conductivity (rinses and cleaning solutions)

With regard to precision cleaning of medical devices, the design inputs are similar to clean in place. However, the focus of the cleaning is on the product that is processed, not the equipment itself.
– Product material and product type to be cleaned
– Soils to be removed (e.g. greases, oils)
– Cleaning methods (solvent or aqueous-based systems)
– Cleaning agents (detergents, nitric acid)
– Cleaning mechanisms (ultrasonic, heat, agitation)
– Cleaning parameters (temperatures, times, ultrasonic frequency)
 – Residue limits (acceptance criteria)

Equipment Considerations

Firstly, for precision cleaning systems, the choice of equipment must be based on the intended purpose of the equipment, for example, will it be used for intermediate cleaning or for final cleaning? As you may expect, there are more stringent acceptance criteria for cleanliness when it comes to final cleaning precision equipment.

For CIP systems, the intended purpose of the equipment is a key design consideration. Materials of construction should be in keeping with the process, maintenance and cleaning regimes associated with it. Material certification should be provided by the supplier or vendor to ensure the correct grade of materials is used from approved suppliers.

In summary, the design should take into account:

> ➢ Difficult-to-clean locations
> ➢ What legacy systems are in place (hence knowledge)?
> ➢ Materials of construction cleaning agents to be used

> ➤ Cleaning parameters
> ➤ Individual cleaning versus cleaning as an equipment train

Cleaning Agent Approval

(a) Precision Cleaning Equipment

For aqueous-based systems, detergents are the preferred cleaning agents used. Detergents are water-soluble cleaning agents that "stick" or cause soils and dirt to "bind" together.

Detergents are typically diluted with de-ionised water or suitably clean water. Dosing can range from 5%-10%, though the dose depends on the type of soils present and the equipment. Ultrasonics, temperature and agitation provide for a quicker cleaning cycle.

(b) Clean-in-Place (CIP)

Cleaning agents used with CIP should address the following points:

> ➤ Effectiveness with regard to API/material
> ➤ Rinsability
> ➤ Supplier certification and compliance
> ➤ Stability
> ➤ Toxicity
> ➤ Supply and global availability
> ➤ EHS compliance

Critical Cleaning Parameters

The key parameters for cleaning can be remembered by using the acronym TACCT.

Time

Action

Cleaning chemistry

Concentration

Temperature

Mixing/flow

Water quality

Rinsing

Other cleaning parameters include flow rate, consideration of turbulence, water quality and rate of rinsing.

Critical cleaning parameters should be challenged at lower end (least stringent) of the operating window conditions during validation.

In cleaning development and design studies, parameters can be pushed to the point of failure of below the anticipated operating ranges. By completing more in-depth testing in the design phase, process knowledge is gained. It also can allow the cleaning validation to be more focused and avoid a protracted validation and may cut down on the number of cycles required if the evidence is there to risk assess a family or matrix validation approach. Example: Temperature set-point is expected to be 80 ± 5°C. – Perform cleaning development studies at set-point of 75°C (this covers the expected range and accounts for calibration tolerances).

Cleaning Pipes

The efficiency of a cleaning process is influenced by the type of flow within the system; the two types of flow are laminar flow or turbulent flow.

Laminar flow is when fluid particles move in parallel layers, at a constant velocity.

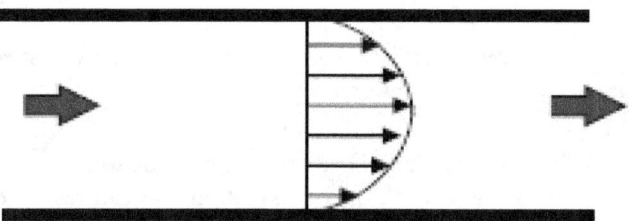

Turbulent flow is when the movement of fluid particles varies in velocity and direction.

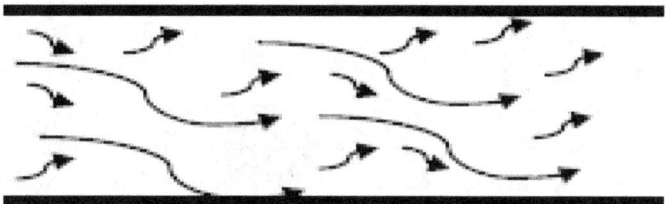

The Reynolds number of a system determines if the flow is turbulent or laminar. A Reynolds number (Re) greater than 4000 is described as turbulent flow.

$$Re = \frac{316Q}{dK}$$

Q– volumetric flow, (gal/min)
d= internal diameter (inches)
k= viscosity (cP)

Dead Legs

A dead leg is the world of piping terminology refers to an area of piping where there is insufficient flow or a tendency for water build-up or stagnation. The formal definition of a dead leg states that pipelines for the transmission of purified water for manufacturing or final rinse should not have an unused portion greater in length than 6 diameters (6D rule) of the unused portion of pipe measured from the axis of the pipe in use.[5]

<u>Connections and Tie-Ins</u>

Precision cleaning systems and CIP systems both have the necessity to be connected to utility supplies such as process and de-ionised water. Precision cleanliness will require tie-in of water supply and drainage on a continuous or defined frequency. Welding is the preferred permanent method of connecting pipes. Non-permanent connections are also used which allow the disconnection and swap-out of piping, vessels and equipment. Orbital welding is a common method of welding when joining piping assemblies and vessels. *(see next section).*

<u>Welding</u>

Stainless steel process piping can be orbitally welded. Quality inspection is typically done real-time by designated quality inspectors using a fibrescope.

Welding should meet necessary standards such as the visual weld criteria as detailed in the materials joining part of the ASME BPE-2000 standard. Discolouration of the weld can be evident as a result of the high degree of heat. The discolouration is a result of the oxidation and can result reduce the corrosion resistance of the weld. In general, welds should not exhibit cracks, crevices or other surface deformities or visual defects. In the event of a weld failure, this can lead to system contamination and would result in the system not being in compliance with 21 CFR 211(a).

Figure : Acceptable weld penetration

Figure : Mismatch or misaligned weld

Figure : Outer Diameter concavity

---

[5] FDA guidance

Figure : Inner diameter concavity, aka suckback

Figure : insufficient penetration

Valves

Clamp-type connections can also be used for non-permanent connections. With regard to the use of valves, electromechanical valves that can be PLC controlled are preferred to manual valves. The level of automation depends on the complexity of the system. For example, a precision clean line used to clean metallic hip implants may have five or six clean and rinse tanks, all fitted with inlets and inlet valves. Having automated control is essential to running a complex line safely and efficiently

Materials of Construction

When it comes to materials of construction, the same selection criteria can be applied to precision cleaning systems and CIP equipment trains. Above all, materials and their surfaces should be non-reactive, non-corrosive and non-porous. Stainless steel of a high grade is often the preferred material of construction. Examples of grades used include 304, 316 and 316L. For surfaces that are product contacting, material certificates are required to provide evidence that the materials and their constituents are of the correct make-up and suitable grade.

Stainless Steels

Stainless steels (SS) are crystallised solutions of at least 11% of corrosion reducing elements like chromium and nickel in iron. Generally, they are iron based with 12 to 30% chromium, 0 to 22% nickel and minor or no amounts of carbon, columbium, copper, manganese, molybdenum, nitrogen, phosphorus, selenium, silicon, sulfur, tantalum, and titanium.

Casting grades generally are designed with more sulphur to facilitate welding and have more ferrite (a less corrosion resistant phase) to prevent the formation of micro-cracks on cooling.

Preferred stainless steels for use in the life sciences are manufactured by VIM – vacuum-induction-melt followed by a secondary VAR – vacuum-arc-re-melt process with sulphur add-back and dispersion in order to minimise inclusions (stringers) and control the amount of sulphur used.

The surface of stainless steel can also be contaminated with the electrolyte solution used in electro-polishing if it is used an excessive amount of times or if rinsing steps are not adequate.

This solution builds up iron and other contaminants that can be transferred to the part being electro-polished if the conditions and chemistry are not carefully controlled.

To prevent these problems from occurring, all electrolyte solutions must be removed from the surface by using a chemically pure water rinse until the conductivity of the rinse from the stainless steel is equal to the conductivity of the water being supplied for rinsing.

Pressure Testing

Piping or system integration can be required for:

> Precision cleaning systems where the utilities need to be "tied" in to the system

> Installation of a pharmaceutical process within a facility e.g. a skid[6] plug in.

After installation, (and before passivation if required) piping systems are pressure tested by filling the system with clean air to 150% of the design pressure or 150psi, whichever is the greatest value. The pressure is then monitored over a 4-hour period to see if there is any drop in pressure.

Passivation

Passivation can be described as the active chemical process used to obtain a uniform chromium oxide layer on stainless steel (SS) surfaces. The chromium oxide layer or film forms a protective coating that gives corrosion resistant properties.

The protective layer naturally forms from the reaction of oxygen in air with the chromium on the metal surface but this naturally forming layer can be non-uniform or patchy due to impurities and surface chemistry defects.

When stainless steel is worked such as in welding, machining, mechanical polishing etc., its uniformity of the naturally forming protective layer can be damaged and oxides of other compounds forming the stainless steel composition can occur. Corrosion can begin at these non-uniform sites and, because stainless steel contains over 60% iron, the corrosion can proliferate from the surface through the body of the metal if no opportunity for protective layer reformation is given.

There are three passivation processes that are used to enhance the corrosion resistance of stainless steel:

> Treatment with oxidising acids
> Treatment with chelants
> Treatment by electro-polishing

Layer formation is a dynamic chemical process where the chromium atoms are combined with oxygen (and hydroxyl ions in aqueous environments) to form a complex surface layer that prevents attack on corrosion-prone atoms such as iron. Nickel and molybdenum may play a role in formation of the passive film but the mechanism has not been proven.

Passivation Process

---

6 see definition in introductory pages

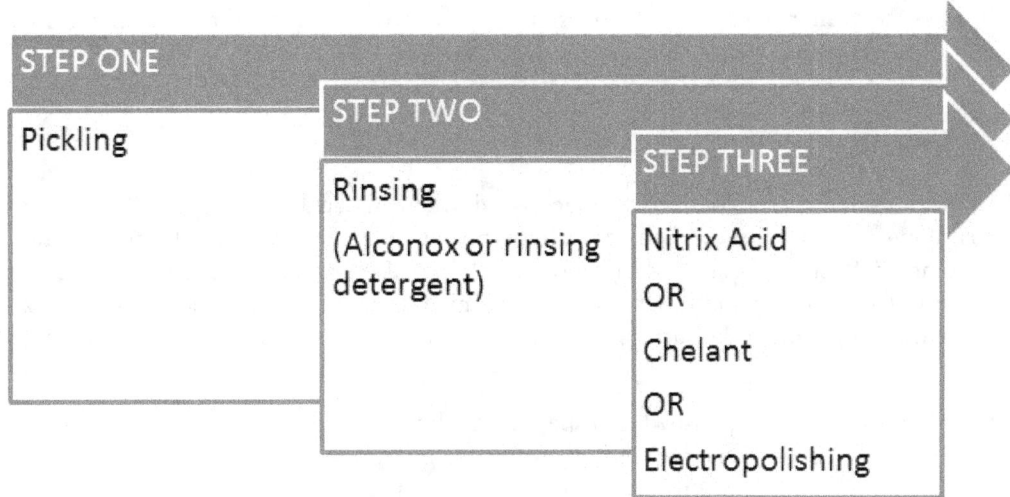

Figure: Passivation - three step process.

Passivation must start with a surface free of any oxide scale (including heat tints and oxide corrosion products) and organic contamination (machining lubricants, oils, coolant and grease).

The first step in a successful passivation process is pickling the metal at the mill. Pickling is a complete surface element and impurity removal process using a mixture of concentrated hydrofluoric and nitric acid.

The second passivation step occurs after fabrication and is the removal of organic contamination by washing of the surface with a basic trisodium phosphate (TSP) such as Alconox or other commercially available chelants-containing, free-rinsing detergent at elevated temperatures.

After the organics are removed, there are three commonly used processes used to complete passivation. The first uses a hot mineral acid solution (commonly nitric acid). The second method uses chelants with milder organic acids (citric acid) and sequestrants. The third is electro-polishing.

Mineral Acid:
This is a fast and affordable method of passivation; however, it comes with environmental and safety risks.

Nitric acid is the acid of choice because of its oxidising properties. The solution is usually heated to facilitate the oxidation reactions. However, the concentration has to be kept below 20% due to the metal surface removal that occurs at higher concentrations. Ph, temperature and conductivity of the acid are also monitored during the process.

Chelant:
Chelant processes are chemical in nature and the materials and their concentrations used can be adjusted to target particular contaminants or all likely residues on the metal.

The published data shows that chelant passivation can achieve chromium enrichment on the surface of stainless steel down to a depth of approximately 20 angstroms. This is a much higher enrichment and a greater depth of penetration than can be achieved by mineral acid passivation processes.

Chelant processes are proprietary but have the following points in common:

➢ Heating of solutions to assist the chemical kinetics of the processes. The passivation solution is usually heated.
➢ Mild organic acid to oxidise the surface iron to soluble ferric ions and insoluble ferrous ions.

> ➢ One or more chelants and sequestering agents to capture the ferric and ferrous ions and prevent their precipitation or depositing on the surface of metal.

Electro-polishing:

Electro-polishing uses a conductive, aqueous salt, reducing acid bath using sulfate and phosphate salts and acids along with a substantial direct current power source to remove 0.1 to 2.5 mils of surface metal preferentially from the peaks and high points.   It can also remove surface inclusions, free the surface from iron and nickel, carbon and other surface contaminants to a maximum depth of approximately angstroms. This removal will enrich the surface of the stainless steel in chromium and therefore a highly passive surface is developed.

The article to be electro-polished is suspended in the conductive liquid and connected to the anode of the power supply. A second electrode is also suspended in the conductive liquid and is connected to the cathode. In order to achieve an even metal removal and chromium enrichment, it is important to achieve constant electrical potential across the surface of the article. Electro-polishing is limited to improving the surface evenness by approximately 10 Ra.

Stainless Steel and Rouging

Rouging is a form of corrosion found in stainless steels. It can be due to iron contamination of the stainless steel surface due to welding of non-stainless steel for support columns, or other temporary means, which when welded off leaves a low chromium area.

There are three classes of rouging:

> ➢ Class I – the stainless steel surface and the Cr/Fe ratio of the metal surface beneath such deposits usually remain unaltered.
> ➢
> Class II - iron particles originating in-situ on unpassivated or improperly passivated stainless steel surfaces. By their formation, the Cr/Fe ratio of the metal surface is altered.
> ➢ Class III - iron oxide (or scale) which forms on surfaces in high temperature steam systems. The Cr/Fe ratio of the protective film is usually altered.

(Ref: ASME-BPE)

References
> ➢ EN 2516:1997 – Passivation of Corrosion Resisting Steels and Decontamination of Nickel Base Alloys
> ➢ ASTM A380 – Practice for Cleaning, Descaling and Passivating of Stainless Steel Parts, Equipment and Systems
> ➢ ASTM A967 – Specification for Chemical Passivation Treatments for Stainless Steel Parts
> ➢ ICH Q3D – International Council for Harmonisation – Guidance for Elemental Impurities

P&ID

Introduction

A piping and instrumentation diagram (P&ID) can be defined as:

1. A diagram which shows the interconnection of process equipment and the instrumentation used to control the process. In the process industry, a standard set of symbols is used to prepare drawings of processes. The instrument symbols are typically based on ISO 10628 International Society of Automation (ISA) Standard (S5.1).
2. The primary schematic drawing used for laying out a process control installation.

They usually contain the following information:

➢ Process piping, sizes and identification, including:
➢ Pipe classes or piping line numbers
➢ Flow directions
➢ Interconnections references
➢ Permanent start-up, flush and bypass lines
➢ tag identifiers),
➢ Valves and their identifications (e.g. isolation, shutoff, relief and safety valves)
➢ Control inputs and outputs (sensors and final elements, interlocks)

P&IDs are originally drawn up at the design stage from a combination of process flow sheet data, the mechanical process equipment design and the instrumentation engineering design. During the design stage, the diagram also provides the basis for the development of system control schemes, allowing for further safety and operational investigations, such as a hazard and operability study (HAZOP). To do this, it is critical to demonstrate the physical sequence of equipment and systems, as well as how these systems connect. The most important symbols in relation to cleaning validation include understanding the vessels, tie-ins, valves and instrumentation associated with the equipment under validation. A valve regulates, directs, or controls the flow of a fluid by opening, closing, or partially obstructing passageways in a piping system. This category includes orifices and other types of valves. The following pages provide a non-exhaustive list of some examples of common symbols.

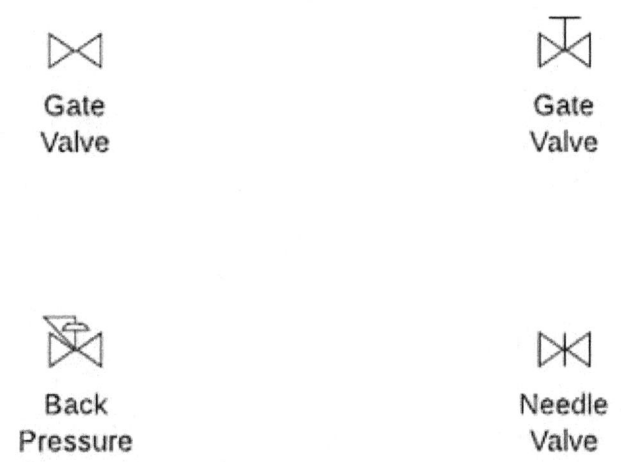

Gate
Valve

Gate
Valve

Back
Pressure

Needle
Valve

Above: Gate valve, back pressure valve and needle valve.

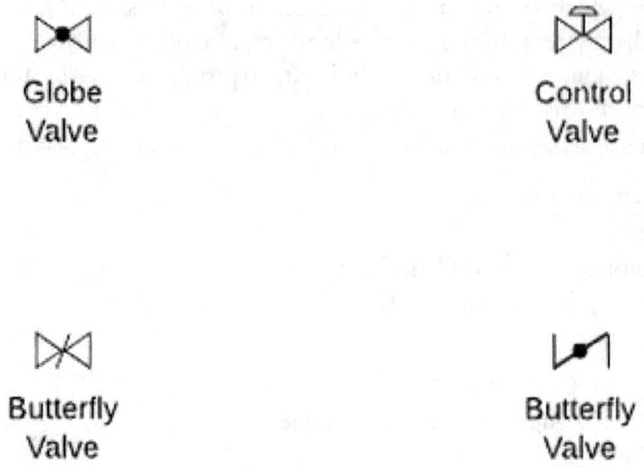

Globe
Valve

Control
Valve

Butterfly
Valve

Butterfly
Valve

Above: Globe valve, butterfly valve and control valve.

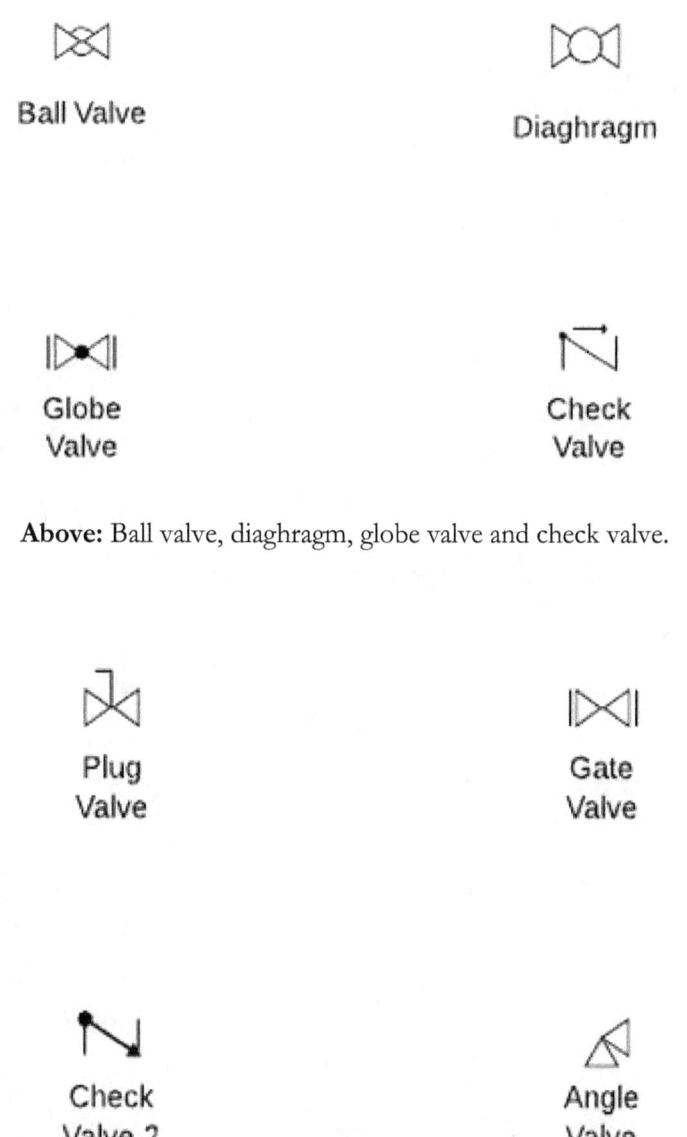

Ball Valve

Diaghragm

Globe Valve

Check Valve

Above: Ball valve, diaghragm, globe valve and check valve.

Plug Valve

Gate Valve

Check Valve 2

Angle Valve

Above: Plug valve, gate valve, check valve two and angle valve.

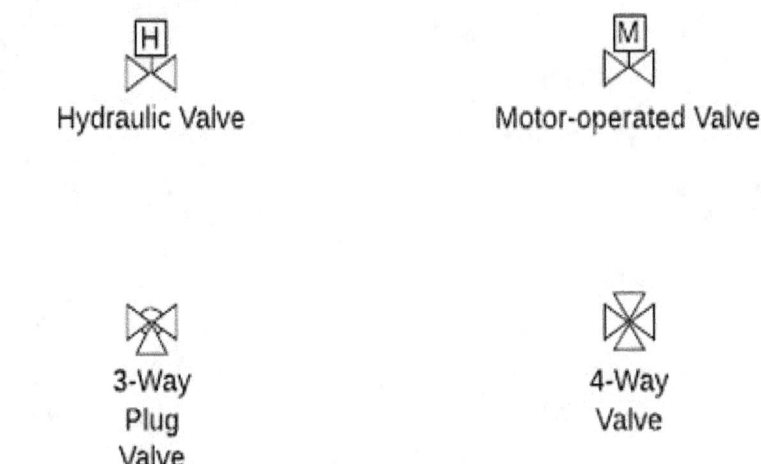

Above: Hydraulic valve, motor operated valve, three-way plug valve and four-way plug valve.

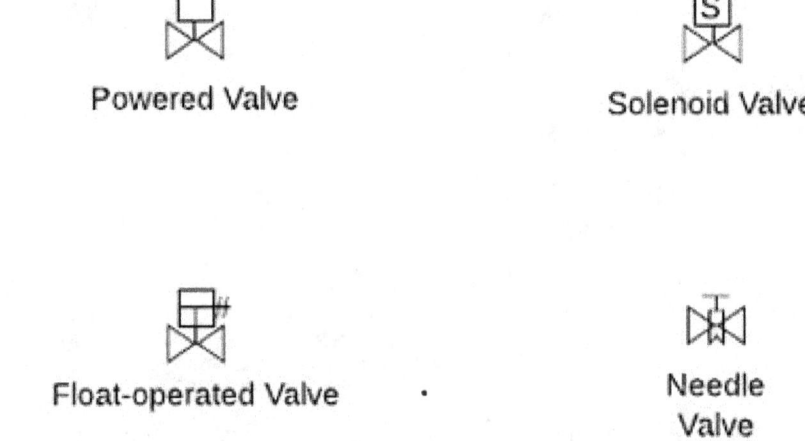

Above: Powered valve, solenoid valve, float-operated valve and needle valve.

Vessels

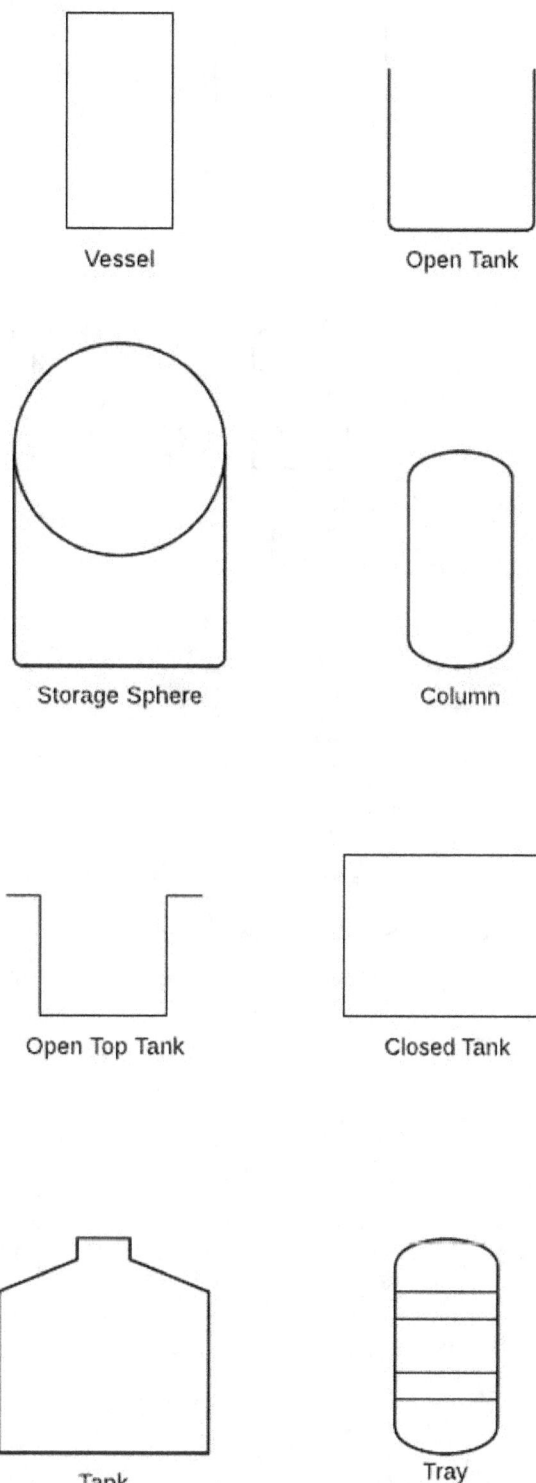

Vessel

Open Tank

Storage Sphere

Column

Open Top Tank

Closed Tank

Tank

Tray Column

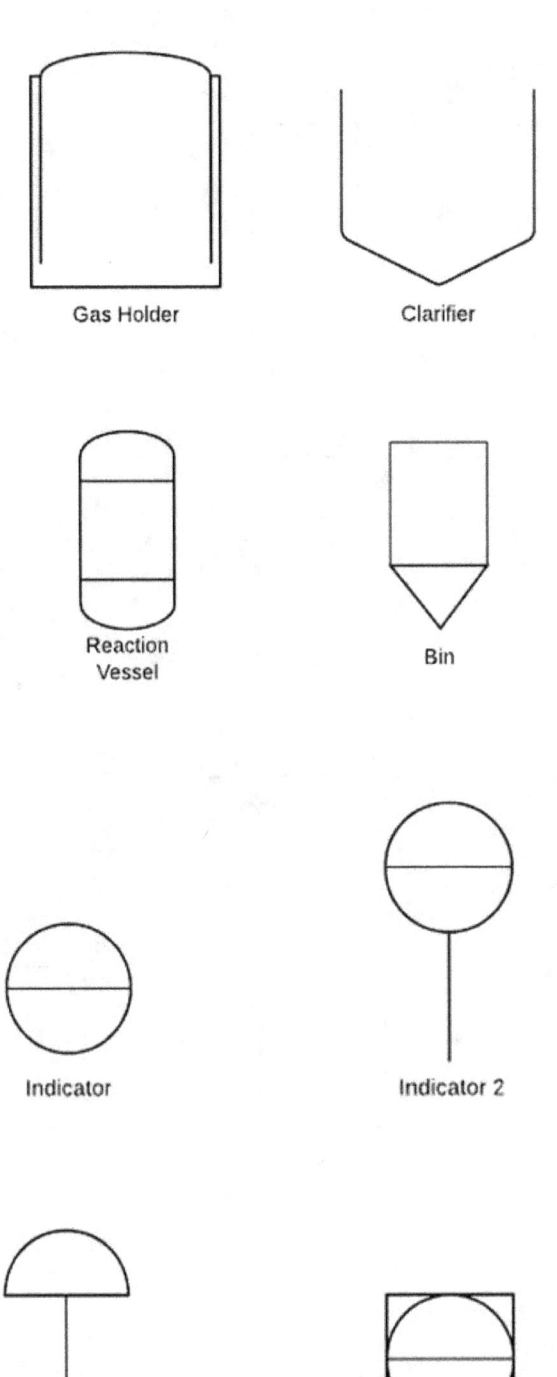

Gas Holder　　　　　Clarifier

Reaction
Vessel　　　　　Bin

Instruments

Indicator　　　　　Indicator 2

Indicator 5　　　　　Shared Indicator

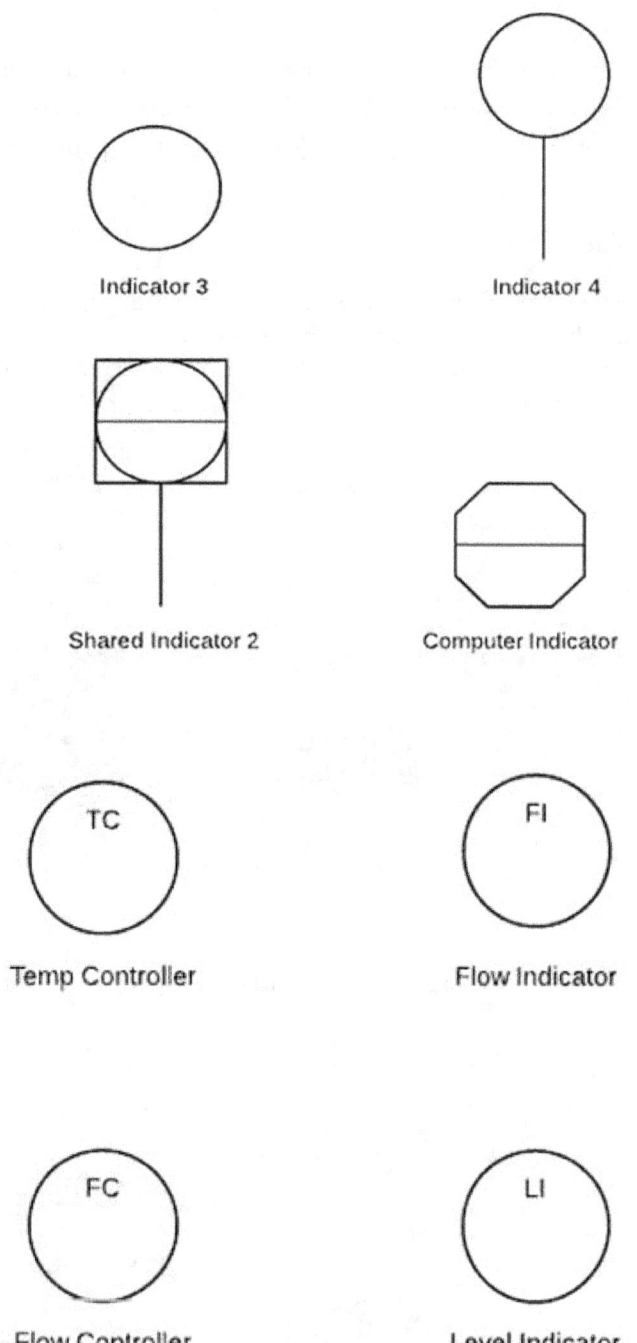

Indicator 3

Indicator 4

Shared Indicator 2

Computer Indicator

Temp Controller

Flow Indicator

Flow Controller

Level Indicator

Sampling

There are two main methods of sampling that are considered to be acceptable; direct surface sampling and indirect sampling (use of rinse solutions). A combination of the two methods is generally the most desirable, particularly in circumstances where accessibility of equipment parts can mitigate against direct surface sampling.

Direct Surface Sampling

(i)   The suitability of the material to be used for sampling and of the sampling medium should be determined. The ability to recover samples accurately may be affected by the choice of sampling material. It is important to ensure that the sampling medium and solvent are satisfactory and can be readily used. Ref: PIC/S PI 006-3.

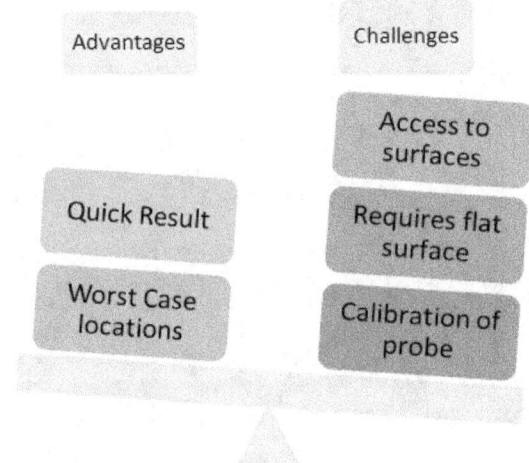

Figure: Direct surface sampling (Pros and Cons).

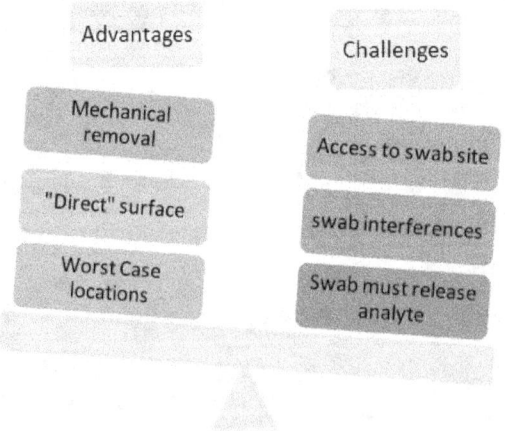

Figure: Swab sampling (Pros and Cons)

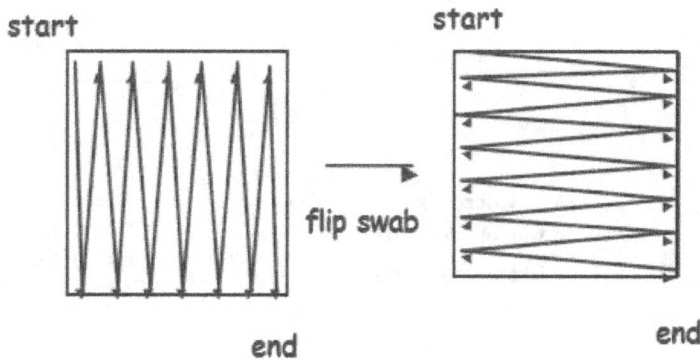

Figure: Method of swab sampling.

Rinse Samples

(i)   Rinse samples allow sampling of a large surface area. In addition, inaccessible areas of equipment that cannot be routinely disassembled can be evaluated. However, consideration should be given to the solubility of the contaminant.

(ii)   A direct measurement of the product residue or contaminant in the relevant solvent should be made when rinse samples are used to validate the cleaning process. Ref: PIC/S PI 006-3.

In rinse sampling a fluid (solvent) is used to rinse and make contact with all surfaces of the item. The sample is then tested quantitatively to remove the target residue.

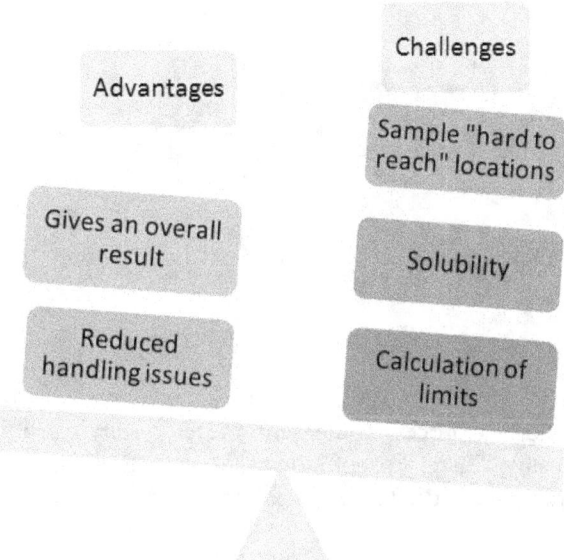

Figure: Rinse sampling (Pros and Cons)

Sources of Contaminants

Sampling aims to detect any residue drug content or solvents or other soiling left behind after the cleaning process. The visual inspection is also important in identifying any larger contamination of debris. Microbial sampling is also done to ensure no microorganisms are present in equipment or in areas of production.

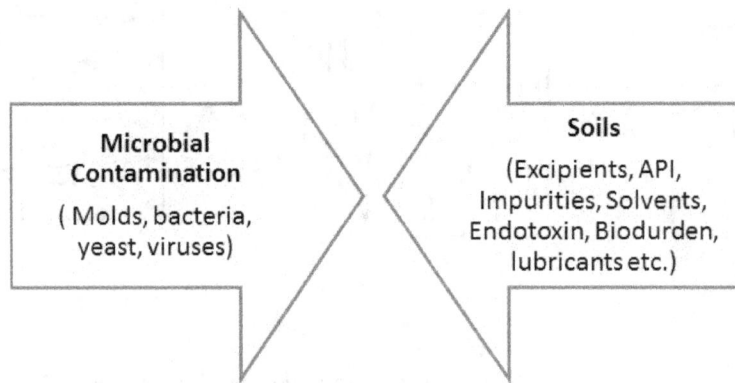

Figure: Sources of Contamination

Microbial Sampling

Microbial sampling of utilities such as water-for-injection, purified water, process air etc. is required to ensure no bacteria, moulds, fungi or yeasts are present which risk patient safety. Colonies can often be determined by visual inspection based on the attributes and appearance test plates/samples. If visual identification is not possible, the colony should be sent for gram stain analysis.

Visual checks involve assessing plates for:

➢ colour
➢ shape
➢ elevation
➢ size
➢ texture
➢ surface
➢ edge appearance

Utilities

The key utilities involved for cleaning include utilities such as water, compressed gases (air, nitrogen etc.) and the heating and cooling of process equipment. Water quality can impact the effectiveness of pre-rinsing, washing, and final rinsing. Therefore, both the water temperature and quality need to be tightly controlled and monitored. Gases are typically used in order to blowdown or blowout remaining fluids or they are used as a drying step.

The term "clean utilities" in the life science industry refers to utilities that have to fulfil regulatory requirements. The most common utility is water, which can be supplied in different pharmaceutical grades of purity. Purified water (PW or PUW), Highly Purified Water (HPW) and Water-for-Injection (WFI) are the most common. Water quality specifications can be found in the pharmacopeias, e.g. the US Pharmacopeia. Other clean utilities can also include clean compressed air, clean gasses (e.g. nitrogen, argon and oxygen), and clean steam.

<u>Key Definitions</u>

Alert limit: A value reached when the normal operating range of a critical parameter has been exceeded, indicating that corrective measures may need to be taken to prevent the action limit being reached.

At-Rest: A condition where the installation is complete with equipment installed and operating in a manner agreed upon by the customer and supplier, but with no personnel present.

Cleanroom: an area (or room or zone) with defined environmental control of particulate and microbial contamination, constructed and used in such a way as to reduce the introduction, generation and retention of contaminants within the area.

Containment: A process or device to contain product, dust or contaminants in one zone,

preventing it from escaping to another zone.

Contamination: The undesired introduction of impurities of a chemical or microbial nature, or of foreign matter, into or onto a starting material or intermediate, during production, sampling, packaging or repackaging, storage or transport.

Point extraction: Air extraction to remove dust with the extraction point located as close as possible to the source of the dust.

Pressure cascade: A process whereby air flows from one area, which is maintained at a higher pressure, to another area at a lower pressure.

Relative humidity: The ratio of the actual water vapour pressure of the air to the saturated water vapour pressure of the air at the same temperature expressed as a percentage. More simply put, it is the ratio of the mass of moisture in the air, relative to the mass at 100% moisture saturation, at a given temperature.

Turbulent flow: Turbulent flow, or non-unidirectional airflow, is air distribution that is introduced into the controlled space and then mixes with room air by means of induction. The process of identifying critical utilities can be done with the application of direct impact, indirect impact and no impact definitions. Risk assessments, CQAs and CPPs should also help identify critical utilities. When critical utilities are required as part of manufacturing and processing, the following points should be examined during the requirements and design stage:

- Materials of construction
- Internal surface finishes
- System sizing
- Flow rates, dead legs, drainage etc.

Compressed Air

Compressed air is used for valve actuation, instrument air and process air to name but a few applications. Only the point-of-use filtration and the gas quality instrumentation should be classified as level 1. When flow or pressure is a CPP, the measurement/monitoring should be performed by the system into which the gas is flowing. Additionally, the CQAs and CPPs should be routinely monitored through the calibrated monitoring system. For compressed air, the potential CPPs are listed below. For the physical system being evaluated, the use and the application of the compressed air will determine which (if not all) CPPs are needed to ensure the system produces product of the desired quality.

➢ Hydrocarbons
➢ Moisture
➢ Particulates
➢ Temperature

It is important that each point of use has appropriate sterile filters in place. If the filter is not placed directly at the point of use, control and counter measures should be implemented to address any risk of contamination downstream of the filter. Compressed air for bio-pharmaceutical use must be generated using oil free compressors with appropriate temperature controls in place.

Attribute	Clean Compressed Air (impacts product quality)	Sterile Compressed Air (impacts sterile product quality)
Oil content	*Not great than $0.1mg/m^3$	
Microbiological requirement	Meets requirements of the environmental zone served (e.g. Grade B,C etc.)	Sterile
Filtration requirement	Minimum $0.45\mu m$ membrane filter	$0.2\mu m$ membrane filter

*ISO 8573-1 Class 2

Water Systems

Water supply and the associated water Systems in biotechnology and pharmaceuticals are a vital component of the manufacturing process. They are used to clean equipment and vessels, to cool or heat processing pipes and systems, and in many circumstances certain grades of water are a component of the finished product (e.g. water-for-injection). Various grades of water service have a particular purpose.

Some common types include:

➢ Potable water
➢ Soft water
➢ Purified water
➢ Water-for-injection

Water used in process and in cleaning should be pure and free from microbial and chemical impurities. As the water gets easily contaminated by environmental conditions, diligence in the design is essential. Typically water systems are supplied on a continuous loop with recirculation.

CPPs typical for a water system include:

➢ Pressure
➢ pH
➢ Conductivity
➢ Level
➢ TOC
➢ Flow
➢ Temperature

> Resistivity

Water for Injection:

WFI is sterile and pyrogen-free water containing no less than 10 CFU/100ml (Colony Forming Units) with a sample size of between 100 and 300 ml and an endotoxin level < 0.25 EU/ml. The use of WFI is two-fold. Firstly, it can be used for critical processing steps such as washing and rinsing. It can also be used in injectable products. WFI is a key raw material for sterile intravenous and intradermal products. WFI is produced by a Multi-Column Distillation Plant (MCDP) and must meet the microbial requirements of regulated bodies.

Clean-in-Place (CIP) / Sterilise-in-Place (SIP) System

The cleaning of equipment, vessels and process piping is a critical activity. Any residue from a previous production batch needs to be removed in order to avoid cross-contamination. CIP and SIP skids are often utilised to allow efficient switchover between batches and/or products. Where possible, manual cleaning should be avoided unless essential due to the design of a system or particular location or configuration.

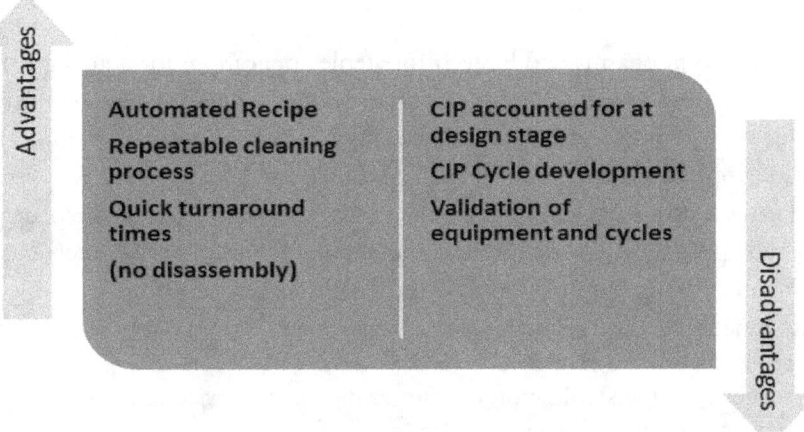

Figure: Advantages and disadvantages of CIP.

Clean Steam

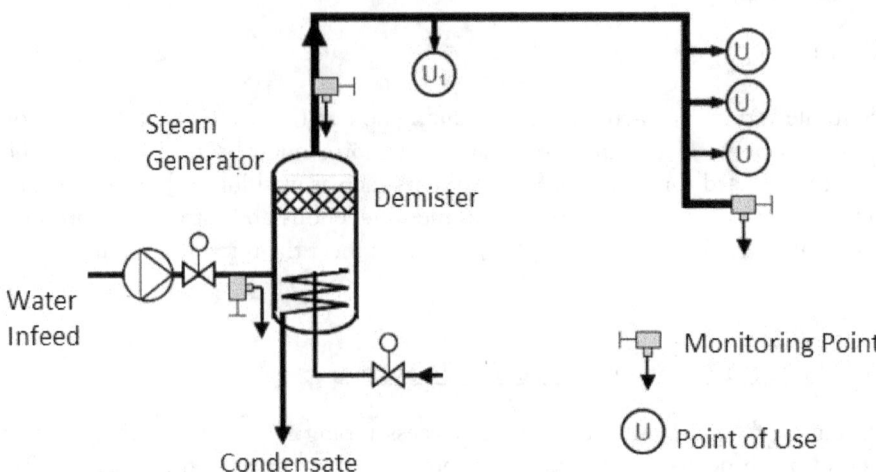

Figure: Simple Clean Steam Generation Piping and Instrumentation.

Pure steam is used in pharma and biotech for sterile application, for autoclave sterilisation etc. Distribution piping of clean steam is a critical aspect. Improper sizing of pipes may impact the production process and lead to a loss of time during sterilisation. Clean steam, also referred to as "pure steam", and gases used in manufacturing operations must be of a quality suitable for their intended purpose. The intended use of clean steam and gases must be understood in order to determine any risks to the patient or product. For example, gases that end up being part of the product must fulfil the regulatory requirements. Preventative maintenance and ongoing monitoring must be implemented for clean steam systems.

➤ Routine inspection and maintenance.
➤ Frequency of filter change
➤ Frequency of the sterilisation for the gas distribution system, if applicable
➤ Frequency for integrity testing of the sterile filter

Water systems for purified water, de-ionised water and Water-for-Injection (WFI) must provide a consistent and reproducible output. Where there is moisture, there is always a risk of microbial contamination. Therefore, the design of water systems should mitigate against such risks. Good engineering practices such as using circulation loops, no dead legs and polished-surface finishes all work to provide an effective and safe system. The design should also take into account ease of sampling at the point of use. The removal of endotoxins is a requirement for WFI. Ongoing sampling monitoring the quality of water is particularly important where water systems are concerned. Procedures should be in place to ensure that effective monitoring and testing is maintained. Action limits and acceptance criteria should be clearly documented in approved SOPs or the equivalent. Failure to meet limits or acceptance criteria should initiate an investigation. The potential CPPs are listed below for clean steam systems:

➤ Conductivity
➤ Flow
➤ Level
➤ Pressure
➤ Resistivity
➤ Temperature

OQ Testing

Operational qualification or OQ is a formal validation activity, and as such should be completed per an approved protocol. The purpose of OQ testing is to confirm the operational and functionality of the clean steam system. This should demonstrate that all critical aspects of a URS are fulfilled.

Key verifications include:

➢ Testing of temperatures and operating pressures
➢ Capacity testing (under load)
➢ Steam trap operation
➢ Verification of automated functions and alarms
➢ Check of automation systems, including PCS
➢ Correct function of valves and sampling points

PQ Testing

Due to the high operating temperatures and the associated lethality, clean steam systems are resistant to microbiological contamination.

Issues that arise can normally be attributed to equipment failures with the steam generator or contaminated water being supplied to the system. Bacterial endotoxin testing is used to monitor clean steam systems for both PQ purposes and throughout the life cycle of the equipment operation. Steam is condensed, sampled and tested. The condensate should meet WFI specifications with the exception of viable total aerobic count. Clean steam PQs are commonly completed using a three-phase approach to testing. The first phase ensures the system consistently operates within the required ranges and the steam provided meets the acceptance criteria. Typically, phase one bacterial endotoxin testing and physio-chemical testing is completed over a two-week period. For phase two, the same frequency and type of testing may be applied for an additional two weeks. After phase two testing, the system may be available for general use if allowed for within internal company procedures. Phase two testing at PQ should also provide a report with all results documented and reviewed.

Phase three of PQ is intended to demonstrate the effective and consistent operation of the system over a longer term (approx. 12 months). Sampling is typically performed weekly.

Further Reading on Clean Steam

- PIC/S PI009-3 – Pharmaceutical Inspection Co-Operation Scheme - Inspection of Utilities
- EN 285 – European Standard - Sterilisation, Steam Sterilisers, Large Sterilisers
- USP <1231> – United States Pharmacopeia - <1231> "Water for Pharmaceutical Purposes"
- USP– United States Pharmacopeia - Monograph "Pure Steam"

EN 285 – European Standard - Sterilisation, Steam Sterilisers, Large Sterilisers

Part 6
ENGINEER 4.0

6.1 Medical Software and "Apps"

Software applications also known as "apps" have become a mainstream phenomenon for mobile phones and tablets and other personal electronic devices. Apps have crossed over into the health and fitness sector that allow personal devices to track and report physical activity, sleep and provide health tips. Furthermore, medical devices also offer increased interaction with "apps" allowing real-time analysis of physiological data.

How to determine if a software app is a medical device?

In simple terms, any software app that provides a diagnostic function in order to determine disease or a medical condition is likely to fall under the definition of a medical device. Examples include:

- Apps that calculate medicine doses a patient is to take
- Apps that identify or inform you that you have a particular medical condition or disease
- Apps provide a risk-based assessment

CE marking of Devices

When an app developer applies a 'CE mark' they are claiming that the app is fit for the purpose (aka the intended purpose) and it is acceptably safe to use. The CE mark should be visible on the app when you are looking at it in the app store, the 'landing' page or on the app developer's product website or information page. The manufacturer has a duty to provide clear information that describes what the app can be used for and how to use it. Consumers should exercise caution when purchasing apps from unknown sources. Product not CE marked or products not assessed for safety pose a potential risk when used. Any medical device app that does not have a CE mark evident from literature, manufacturers information or interface should be reported to the relevant competent authority.

Personal data and apps

Many apps have the ability to capture and record physiological parameters such as heart rate, sleep patterns and activity. Many also support the input of additional data such as a persons weight and physical attributes.

It is very important that you have read the small print to understand what personal data you may have agreed to share with the developer by signing up to the app and how they might store or use your data or share your information with third parties. This includes information about you such as your name, address, date of birth and information about your health.

General Guidance on Use

Once you are sure the app is right for you and it is CE marked then you should follow the instructions carefully.

Be honest with the information you put into the app. If you enter wrong information about yourself, the app may not give you the right result. Ensure that you always update the app to the newest compatible version.

After using

If you are in doubt about the information that the app has given you or you are concerned about your health then you should consult a healthcare professional (a pharmacist, health visitor, practice nurse or GP) If you have any problems with the app not working as stated e.g.

• If the instructions aren't clear or the app is difficult to use

• If the app isn't giving you the results that you expected

• If you have concerns over the safety of the app or the information that it provides

European Requirements –Stand-alone Software

This guideline on the qualification and classification of stand-alone software was drafted by the European Commission after consultation of the competent authorities, commission services, industry and notified bodies

MEDDEV 2.1/6 rev. 1: Qualification and classification of stand-alone software
A stand-alone software must meet the following criteria in order to be classified as a medical device:

- o it has to be a computer program
- o the software has to have a different purpose than mere storage, archival, lossless compression, communication or simple search
- o the software has to be for the benefit of individual patients
- o the software has to have an intended purpose listed in Article 1(2)a of Directive 93/42/EEC.
  (In Germany, Article 1(2)a of Directive 93/42/EEC has been implemented as national law in Section 3 number 1 MPG.)

FDA Requirements

Food and Drug Administration FDA: "Mobile Medical Applications -Guidance for Industry and Food and Drug Administration Staff", February 2015

Introduction

Issued in 2015, the FDA provided non-binding guidance on "Mobile Medical Applications". The aim of the guidance is to inform industry on how the FDA intends to apply its regulatory authorities to select software applications intended for use on mobile platform/mobile apps. The guidance also states that although some mobile apps may meet the definition of a medical device, if they propose a lower risk to the public FDA intends to exercise enforcement discretion over these devices.

As mobile platforms become more user friendly, computationally powerful, and readily available, innovators have begun to develop mobile apps of increasing complexity to leverage the portability mobile platforms can offer. Some of these new mobile apps are specifically targeted to assisting individuals in their own health and wellness management. Other mobile apps are targeted to healthcare providers as tools to improve and facilitate the delivery of patient care.These software devices include products that feature one or more software components, parts, or accessories (such as electrocardiographic (ECG) systems used to monitor cardiac rhythms), as well as devices that are composed solely of software (such as laboratory information management systems).

On February 15, 2011, the FDA issued a regulation down classifying certain computer- or software-based devices intended to be used for the electronic transfer, storage, display, and/or format conversion of medical device data – called Medical Device Data Systems (MDDSs) – from Class III (high-risk) to Class I (low-risk).2

The FDA has previously clarified that when stand-alone software is used to analyze medical device data, it has traditionally been regulated as an accessory to a medical device3 or as medical device software.

As is the case with traditional medical devices, certain mobile medical apps can pose potential risks to public health. Moreover, certain mobile medical apps may pose risks that are unique to the characteristics of the platform on which the mobile medical app is run interpretation of radiological images on a mobile device could be adversely affected by the smaller screen size, lower contrast ratio, and uncontrolled ambient light of the mobile platform. FDA intends to take these risks into account in assessing the appropriate regulatory oversight for these products. This guidance clarifies and outlines the FDA's current thinking.

Definitions

Mobile Platform: For purposes of this guidance, "mobile platforms" are defined as commercial off-the-shelf (COTS) computing platforms, with or without wireless connectivity, that are handheld in nature. Examples of these mobile platforms include mobile computers such as smart phones, tablet computers, or other portable computers.

Mobile Application (Mobile App): For purposes of this guidance, a mobile application or "mobile app" is defined as a software application that can be executed (run) on a mobile platform (i.e., a handheld commercial off-the shelf computing platform, with or without wireless connectivity), or a web-based software application that is tailored to a mobile platform but is executed on a server.

Mobile Medical Application (Mobile Medical App): For purposes of this guidance, a "mobile medical app" is a mobile app that meets the definition of device in section 201(h) of the Federal Food, Drug, and Cosmetic Act (FD&C Act) - 7 - 4 ; and either is intended:

to be used as an accessory to a regulated medical device; or

to transform a mobile platform into a regulated medical device.

Mobile Medical App: the FDA defines a "mobile medical app manufacturer" is any person or entity that manufactures mobile medical apps in accordance with the definitions of manufacturer in 21 CFR Parts 803, 806, 807, and 820. They may include companies that initiates specifications, designs, labels, or creates a software system or application for a regulated medical device in whole or from multiple software components.

BfArM Germany

Guidance from the Federal Institute for Drugs and Medical Devices (BfArM), Germany

Differentiation between apps and medical or other devices as well as on the subsequent risk classification according to the MPG

BfArM as with other competent Authorities provides guidance on differentiation between:

(1) apps (in general: stand-alone software, not incorporated into a medical device, e.g. as control software) and
(2) medical devices.

In similar fashion, any this guidance is informative and non- binding. Qualification and classification needs to determine the intended purpose of the software and its classification must is the responsibility of the manufacturer.

The "intended purpose" is the use for which the medical device is intended according to the manufacturer's information, marketing, labelling and instructions for use (IFU). Thus, not only the explicitly described intended purpose is relevant e.g. for an authority decision on qualification as a medical device, but also the instructions for use and the promotional materials (e.g. website, information in App Store) regarding the specific   product.

BfArM guidance also points out that Stand alone software such as smartphone apps can indeed be classified as a medical device. However, the product must be intended by the manufacturer to be used for humans with a minimum of at least one of the criteria below fufilled:

- diagnosis, prevention, monitoring, treatment or alleviation of disease,
- diagnosis, monitoring, treatment, alleviation or compensation of injuries or handicaps,
- investigation, replacement or modification of the anatomy or of a physiological process,
- control of conception.

Essentially, the above criteria is a summary of the European regulations pertaining to medical devices. As opposed to mere provision of knowledge, e.g. in a paper or electronic book (no medical device), any type of interference with data or information by the stand-alone software is indicative of a classification as a medical device. Possible indicative terms in connection with the intended purpose of corresponding functions can be e.g.: alarm, analyse, calculate, detect, diagnose, interpret, convert, measure, control, monitor, amplify. Indicative functions for classification as a medical device can be among the following:

- decision support or decision-making software e.g. with regard to therapeutic measures
- calculation e.g. of dosing of medicines (as opposed to mere reproduction of a table from which users can deduce the dosage themselves)
- monitoring patients and collecting data e.g. by measurements if the results thereof have an influence on diagnosis or therapy.

Pure data storage, archiving, lossless compression, communication or simple search functions do not result in classification as a medical device. Like all other medical devices from own production, software applications from own production are medical devices and thus must fulfil the basic requirements of Council Directive 93/42/EEC.

Risk Classification

With the exception of in vitro diagnostic medical devices and active implantable medical devices, medical devices are allocated to risk classes that are mainly based on the potential damage that can be caused by an error/malfunction of the medical device. These risk classes range from Class I (low risk) and IIa and IIb to Class III (high risk). Class I products are additionally subdivided according to whether they require sterilisation (Is) or include a measuring function (Im) which is relevant for the further conformity assessment procedure. The classification is based on the rules laid down in Annex IX of Council Directive 93/42/EEC. The following rules are most suitable for the classification of stand-alone software.

Rule 9

"All active therapeutic devices intended to administer or exchange energy are in Class IIa unless their characteristics are such that they may administer or exchange energy to or from the human body in a potentially hazardous way, taking account of the nature, the density and site of application of the energy, in which case they are in Class IIb.
All active devices intended to control or monitor the performance of active therapeutic devices in Class IIb, or intended directly to influence the performance of such devices are in Class IIb."

Rule 10

"Active devices intended for diagnosis are in Class IIa,

- if they are intended to supply energy which will be absorbed by the human body, except for devices used to illuminate the patient's body, in the visible spectrum;
- if they are intended to image in vivo distribution of radiopharmaceuticals;
- if they are intended to allow direct diagnosis or monitoring of vital physiological processes, unless they are specifically intended for monitoring of vital physiological parameters, where the nature of variations is such that it could result in immediate danger to the patient, for instance variations in cardiac performance, respiration, activity of CNS in which case they are in Class IIb.

- Active devices intended to emit ionizing radiation and intended for diagnostic and therapeutic interventional radiology including devices which control or monitor such devices, or which directly influence their performance, are in Class IIb."

Rule 12

"All other active devices are in Class I."

Rule 14

"All devices used for contraception or the prevention of the transmission of sexually transmitted diseases are in Class IIb, ..."

The following definitions in accordance with Annex IX Section I No. 1 of Council Directive 93/42/EEC are to be observed:

- **Stand alone software**

  Stand alone software is considered to be an active medical device.

- **Active therapeutic device**

  "Any active medical device, whether used alone or in combination with other medical devices, to support, modify, replace or restore biological functions or structures with a view to treatment or alleviation of an illness, injury or handicap."

- **Active device for diagnosis**

  "Any active medical device, whether used alone or in combination with other medical devices, to supply information for detecting, diagnosing, monitoring or treating physiological conditions, states of health, illnesses or congenital deformities."

The afore-mentioned rules show that e.g. medical apps on smartphones and tablets will mostly be classified in risk Class I in accordance with Rule 12. If the medical devices are intended for diagnosis or monitoring of vital functions (e.g. cardiac functions), Classes IIa or IIb can also be considered.

Depending on the risk class there are different requirements for conducting a conformity assessment procedure as the prerequisite for affixing the CE marking and for correct marketing within the European Economic Area.

Thus, the manufacturer can perform a conformity assessment e.g. for Class I devices without involvement of a notified body; for all other risk classes (also in the case of Class I devices that require sterilisation or include a measuring function) it is mandatory to involve a notified body. If a stand alone software or app is placed on the market as a medical device it is subject to the same regulations as all other medical devices.

Examples for Qualification/differentiation

Decision supporting software

In general, software is usually considered a medical device when it is used for healthcare, if e.g. medical knowledge databases and algorithms are combined with patient-specific data and the software is intended to give healthcare professionals recommendations on diagnosis, prognosis, monitoring or treatment of an individual patient.

Software systems

If a software consists of several modules it is the manufacturer's responsibility whether he wants the modules as a whole to be qualified and classified or each module individually. If the entire system is qualified and if it consists both of software with and without the properties of a medical device, the system is subject to medical device legislation.

Telemedical software

In telemedicine the physician observes and assesses the patients' medical data using telecommunication technologies - e.g. via the internet. Patient and physician can be at different locations. Depending on the intended purpose, communication systems for telemedicine can either be non-medical devices (purely for transfer of data) or a combination of non-medical devices and medical devices (e.g. in order to support diagnoses).

Hospital information systems (HIS)

Hospital information systems that support patient management are generally not medical devices, especially if they have the following intended purpose:

- o   collection of data for patient admission
- o   administration of general patient data
- o   scheduling of appointments
- o   insurance and billing functions

However, hospital information systems can be combined with other modules that could be medical devices.

Picture Archiving and Communication System (PACS)

For example, if the manufacturer of the PACS software specifies in the intended purpose that the software is only meant for storage or archiving of pictures and not for diagnosing, this would indicate that it is not a medical device. However, if the manufacturer intends the PACS software for controlling a medical device or to have an influence on its use or to allow a direct diagnosis, this would support its classification as a medical device.

Stand alone software or apps that are **not** medical devices

Operating system software:
Operating system software (e.g. Windows, Linux) is neither a medical device nor is it an accessory to a medical device.

Software for general purposes

Software for general purposes is not a medical device even if it is used in connection with healthcare.

Software or apps as health or fitness products
When differentiating medical devices e.g. from health or fitness products, the decisive issue is whether they are intended for medical or non-medical purposes. This is defined by the manufacturer of the product. Software or apps merely intended for sporting activities, fitness, well-being or nutritional aims without a medical purpose claimed by the manufacturer are generally not medical devices

Medicines and Healthcare products Regulatory Agency, MHRA

The Medicines and Healthcare products Regulatory Agency (MHRA) is an agency of the Department of Health in the UK which is responsible for ensuring that medicines and medical devices work and are acceptably safe.

Formed in 2003 with the merger of the Medicines Control Agency (MCA) and the Medical Devices Agency (MDA). In April 2013, it merged with the National Institute for Biological Standards and Control (NIBSC) and as the MHRA. The MHRA has released guidance on Medical Device stand-alone software including apps

(Aug 2014). This guidance should be used in addition to MEDDEV 2.1/6.

Medicines & Healthcare products Regulatory Agency (MHRA): Guidance - Medical device stand-alone software including Apps, August 2014

Medical device stand-alone software including apps (including IVDMDs)

As well as medical device apps becoming a growth area in healthcare management in hospital and in the community settings, the role of apps used as part of fitness regimes and for social care situations is also expanding. However, in the UK and throughout Europe, standalone software and apps that meet the definition of a medical device are still required to be CE marked in line with the EU medical device directives in order to ensure they are regulated and acceptably safe to use and also perform in the way the manufacturer/developer intends them to. Health related apps and software that are not medical devices can be extremely useful but fall outside the scope of the MHRA.

But how do developers and users of this software decide whether apps qualify as medical devices and which are for health and fitness purposes?

When apps are not medical devices

Those apps that are not medical devices may be considered to be mHealth products. Work is ongoing at European level to determine a suitable legal framework.

The Concept of Mobile Health (mHealth)

Mobile health (hereafter "mHealth") covers *"medical and public health practice supported by*

mobile devices, such as mobile phones, patient monitoring devices, personal digital assistants

(PDAs), and other wireless devices"

It also includes applications (hereafter "apps") such as lifestyle and wellbeing apps2 that may connect to medical devices or sensors (e.g. bracelets or watches) as well as personal guidance systems, health information and medication reminders provided by sms and telemedicine provided wirelessly. mHealth is an emerging and rapidly developing field which has the potential to play a part in the transformation of healthcare and increase its quality and efficiency.
mHealth solutions cover various technological solutions, that among others measure vital signs such as heart rate, blood glucose level, blood pressure, body temperature and brain activities. Prominent examples of apps are communication, information and motivation tools, such as medication reminders or tools offering fitness and dietary recommendations. The expanding spread of smartphones as well as 3G and 4G networks has boosted the use of mobile apps offering healthcare services. The availability of satellite navigation technologies in mobile devices provides the possibility to improve the safety and autonomy of patients. Through sensors and mobile apps, mHealth allows the collection of considerable medical, physiological, lifestyle, daily activity and environmental data. This could serve as a basis for evidence-driven care practice and research activities, while facilitating patients' access to their health information anywhere and at any time.

mHealth could also support the delivery of high-quality healthcare, and enable more accurate diagnosis and treatment. It can support healthcare professionals in treating patients more efficiently as mobile apps can encourage adherence to a healthy lifestyle, resulting in more personalised medication and treatment.

It can contribute to the empowerment of patients as they could manage their health more actively, living more independent lives in their own home environment thanks to self- assessment or remote monitoring solutions and monitoring of environmental factors such as changes in air quality that might influence medical conditions. In this respect, mHealth is not intended to replace healthcare professionals who remain central to providing healthcare but rather is considered to be a supportive tool for the management and provision of healthcare.

6.2   21 CFR Part 11 -Electronic Records & Signatures

Introduction

This chapter covers the regulatory requirements of part 11 of Title 21 of the Code of Federal Regulations; Electronic Records; Electronic Signatures (21 CFR Part 11).

Part 11 of the FDA CFR is relevant to "records in electronic form that are created, modified, maintained, archived, retrieved, or transmitted under any records requirements set forth in Agency regulations." This first section of the book provides a background information and explanations of each section and requirement of the regulation. The second half of this eBook provides a clear and transferrable verification process for each requirement of 21 CFR Part 11, with suggested verification methods included.

As of 2007, several sections of the regulation have been identified as excessive and the FDA announced in guidance that it will exercise enforcement discretion on some parts of 21 CFR part 11. This has been welcomed by some manufactures but it has also causes a degree of confusion.

The requirements relating to access controls are the most fundamental requirements and are routinely enforced. The "predicate rules" that required organizations to keep records the first place are still in effect. If electronic records are illegible, inaccessible, or corrupted, manufacturers are still subject to those requirements.

If a regulated firm keeps "hard copies" of all required records, those paper documents can be considered the authoritative document for regulatory purposes. This then means that the computer system is not in scope for electronic records requirements, although subject to predicate rules which still require validation. If the "hard copy" is to be identified as the authoritative document, the "hard copy" must be a complete and accurate copy of the electronic source. The manufacturer must use the hard copy (rather than electronic versions stored in the system) of the records for regulated activities.

Definition of Records

The FDA has deemed the following records or signatures in electronic format subject to 21 CFR part 11:

"Records that are required to be maintained under predicate rule requirements and that are maintained in electronic format in place of paper format. On the other hand, records (and any associated signatures) that are not required to be retained under predicate rules, but that are nonetheless maintained in electronic format, are not part 11 records.

Records that are required to be maintained under predicate rules, that are maintained in electronic format in addition to paper format, and that are relied on to perform regulated activities. Records submitted to FDA, under predicate rules (even if such records are not specifically identified in Agency regulations) in electronic format (assuming the records have been identified in docket number 92S-0251 as the types of submissions the Agency accepts in electronic format). However, a record that is not itself submitted, but is used Contains Nonbinding Recommendations in generating a submission, is not a part 11 record unless it is otherwise required to be 205 maintained under a predicate rule and it is maintained in electronic format.

Electronic signatures that are intended to be the equivalent of handwritten signatures, initials, and other general signings required by predicate rules. Part 11 signatures include electronic signatures that are used, for example, to document the fact that certain events or actions occurred in accordance with the predicate rule (e.g. approved, reviewed, and verified)." The above definitions are taken from FDA guidance document entitled "FDA Guidance for Industry: 21 CFR Part 11 - Electronic Records and Electronic Signatures: Scope and Application, August 2003." This document also provides recommendations on documenting key decisions that may be taken in relation to 21 CFR Part 11 applicability and compliance.

Requirements and Specifications

The need for compliance to 21 CFR depends on type of technology and level of automation and computerisation involved in the manufacturing process or other actives that are GxP impacting. Does the system store electronic records? Does the system require a login? Is there an audit trial? If a complex system is to be procured, the requirements need to be communicated to the manufacturer as part of a User requirement specification and/or software requirement specification.

General Guidance on Requirement Specifications

While the Quality System regulation states that design input requirements must be documented, and that specified requirements must be verified, the regulation does not further clarify the distinction between the terms "requirement" and "specification." A requirement can be any need or expectation for a system or for its software. Requirements reflect the stated or implied needs of the customer, and may be market-based, contractual, or statutory, as well as an organization's internal requirements.

There can be many different kinds of requirements (e.g., design, functional, implementation, interface, performance, or physical requirements). Software requirements are typically derived from the system requirements for those aspects of system functionality that have been allocated to software. Software requirements are typically stated in functional terms and are defined, refined, and updated as a development project progresses. Success in accurately and completely documenting software requirements is a crucial factor in successful validation of the resulting software. Page 6 Guidance for Industry and FDA Staff General Principles of Software Validation A specification is defined as "a document that states requirements." (21 CFR 820.3(y)) It may refer to or include drawings, patterns, or other relevant documents and usually indicates the means and the criteria whereby conformity with the requirement can be checked. There are many different kinds of written specifications, e.g., system requirements specification, software requirements specification, software design specification, software test specification, software integration specification, etc. All of these documents establish "specified requirements" and are design outputs for which various forms of verification are necessary.

Validation of Computerised Systems

The requirement for computerised systems to be compliant to 21 CFR part 11, needs to be identified early on the project to ensure that the vendor or supplier of the systems or equipment can develop, build a system that meets the requirements of 21 CFR part 11. Computer system validation can be divided into 3 distinct phases which include: (1) Plan, (2) Design & Development, (3) verification and (4) Retirement. The requirement for computerised systems to be compliant to 21 CFR part 11, needs to be identified early on the project to ensure that the vendor or supplier of the systems or equipment can develop, build a system that meets the requirements of 21 CFR part 11.

Plan: This phase involves the planning of the validation effort required to implement the system and identification of key milestones and requirements. It requires supplier assessments, assessments of the regulatory and system risks, supplier, development of a validation approach and the identification of deliverables that will be generated, that will support the implementation and operation of the system.

Design & Development: This phase consists of the design, development and configuration of the hardware and software required to meet the system requirements. In case of custom software, design and developmental testing is important to ensure proper functionality prior to verification testing.

Verification: This phase confirms that requirements and specifications have been met. Testing is required to ensure the system operates as intended. Upon successful testing and verification, the system can be released for use. Once verification activities have begun any changes to the system must managed through change control. In case of successful completion of the verification activities (i.e. any deviation has been evaluated and addressed), the system is released for effective use. Operation This phase supports the need to maintain compliance and fitness for intended use after the system is accepted and released for use.

Retirement: This phase consists of the planning, executing and summarizing of the events required for system shutdown. It includes the appropriate handling of the supporting documents and the data contained within the system. While described here as a separate phase, a system's retirement can be handled as part of a new system implementation or as a separate project.

Best practice when it comes to Computer System validation is to adopt a life cycle approach for computer systems which requires the completed of activities in a systematic way from system conception to retirement. Life cycle activities could be scaled according to system impact on product quality, patient safety and data integrity, system complexity and novelty, supplier assessment and business risk.

Definitions

Computer System: A computer / automated system consisting of the hardware, software, and network components, together with the controlled functions (personnel, procedures, and equipment) and associated documentation.

Computer System Validation: A process that confirms by examination and provision of objective evidence that the computer system conforms to user needs and intended uses. Computer System validation is a process for achieving and maintaining compliance with GxP regulations and fitness for intended use by adoption of life cycle activities, deliverables, and controls.

GxP Regulated Computer Systems: Computer systems determined to have a potential impact on Product Quality, Patient Safety and Data Integrity; these systems are required to comply with the relevant GxP regulations.

Data Integrity: is the degree to which data is reliable and without error. Data must be accurate, attributable, contemporaneous, original, legible and available. A breach of data integrity occurs when any person manipulates or distorts data and submits the results of that data as valid.

Predicate rules: a predicate rule is any FDA regulation that requires companies to maintain certain records and submit information to the agency as part of compliance.

To gain a better understanding of the validation of computerized systems, consult the following publication- "FDA's guidance for industry and FDA staff General Principles of Software Validation." Industry guidance such as the GAMP 5 guide issued by ISPE is also a useful reference.

Electronic Records

When it comes to the regulated industries such as the medical device industry, every process and procedure must be documented. Documentation ensures that everyone is working in the same manner with the same procedures. However, documentation is more than just writing down procedures and processes. It is also concerned with how documents are controlled, how they are updated and how they are stored.

Electronic Document management systems

Electronic document management systems aka EDMS are now the norm and gold standard for most medium to large organisations. Many companies that provide medical device manufacturers with an EDMS can be customised to match the business processes particular to an organisation. With configurable or customisable software, validation and proper verification is important to ensure the system operates as intended. There are also regulatory requirements that stipulate the expectations and requirements of such system. For example, the application of electronic signatures and the presence of audit trials. FDA 21 CFR Part 11 details the requirements with regards to electronic records and electronic signatures. For medicinal products in Europe, GMP V4 Annex 11 specifies similar requirements.

Record Retention

Regard to the part 11 requirements for the protection of records to enable their accurate and ready retrieval throughout the records retention period (11.10 (c)) Persons must also comply with all applicable predicate rule requirements for record retention and availability such as (211.180(c) general requirements. The decision to follow 21 CFR part 11 should be justified and documented as part of a risk assessment and based on the value of the records over time.

FDA does not object to archiving of required records in electronic format to non-electronic media such as paper, or to a standard electronic file format (examples of such formats include, but are not limited to, PDF, XML, or SGML). Persons must still comply with all predicate rule requirements, and the records themselves and any copies of the required records should preserve their content and meaning. As long as predicate rule requirements are fully satisfied and the content and meaning of the records are preserved and archived, you can delete the electronic version of the records. In addition, paper and electronic record and signature components can co-exist as long as predicate rule requirements are met and the content and meaning of those records are preserved.

Electronic Signatures

Electronic signatures are computer-generated character strings that count as the legal equivalent of a handwritten signature. The regulations for the use of electronic signatures are set out in 21 CFR Part 11 of the FDA. Each electronic signature must be assigned uniquely to one person and must not be used by any other person. It must be possible to confirm to the authorities that an electronic signature represents the legal equivalent of a handwritten signature. Electronic signatures can be biometrically based or the system can be set up without biometric features.

Conventional Electronic Signatures

If electronic signatures are used that are not based on biometrics, they must be created so that persons executing signatures must identify themselves using at least two identifying components. This also applies in all cases in which a chip card replaces one of the two identification components. These identifying components, can, for example consist of a user identifier and a password. The identification components must be assigned uniquely and must only be used by the actual owner of the signature.
When owners of signatures want to use their electronic signatures, they must identify themselves by means of at least two identification components. The exception to this rule is when the owner executes several electronic signatures during one uninterrupted session. In this case, persons executing signatures need to identify themselves with both identification components only when applying the first signature. For the second and subsequent signatures, one unique identification component (password) is then adequate identification.

Audit Trail

Title 21 CFR details predicate rule requirements relating to documentation of, for example, date time, or the sequencing of events, as well as any requirements for ensuring that changes to records do obscure previous entries.

Making the decision on whether to apply audit trails, or other appropriate measures, or on the need to comply with predicate rule requirements should involve a justified and documented risk assessment. Any Risk assessment should determine the potential effect on product quality and safety and the integrity of the record.

Change Management

Validation programs are subject to change control. Each company or organisation should have a procedure detailing the change management process. Below is a suggested overview of a typical change control process. Any system, facility, document or process that has the potential to impact product quality and validated state is generally subject to following a change control process. Another term used in industry is Enterprise Change Control or Engineering Change Control. Essentially these terms are the same. The intent is to control and manage change consistently.

A change control can take the form of a document which drives the agenda and the specific requirement. Change control is also created with enterprise software such as Kintana, Documentum and SAP. While each company will have varying processes, some basics are common. These include the 3 stages of change control; pre-implementation, implementation and post implementation (if required). Below, 2 case studies are detailed where there is a change in manufacturing which requires a formal change control process to be applied.

Summary of 21 CFR Part 11.10

11.10 (a) Accuracy, reliability & consistent intended performance.

11.10 (b) Copies of records (Paper)- complete copies of records in both human readable and electronic form suitable for inspection, review, and copying.

11.10 (b) Copies of records (Electronic)- complete copies of records in both human readable and electronic form suitable for inspection, review, and copying.

11.10 (c) Protection of records- accurate and allow ready-retrieval throughout the records retention period.

11.10 (d) & (g) Authorised access - Limiting system access to authorized individuals in relation to accessing records, the operation or computer system input or output devices, altering of records, or performing the operation at hand.

11.10 (e) Computer systems (including hardware and software), controls, and attendant documentation maintained under this part shall be readily available for, and subject to, FDA inspection.

11.10 (f) Sequencing of steps - checks to enforce permitted sequencing of steps and events

11.10 (h) Input device authorisation - checks the validity of the source of data.

11.10 (i) Input device persons education, training, and experience

11.10 (j) Establishment of written policies - establishment of, and adherence to, written policies that hold individuals accountable and responsible for actions initiated under their electronic signatures, in order to deter record and signature falsification.

11.10 (k) Appropriate Controls - appropriate controls over systems documentation-(1) Adequate controls over the distribution of, access to, and use of documentation for system operation and maintenance. (2) Revision and change control procedures to maintain an audit trail that documents time-sequenced development and modification of systems documentation.

Electronic Records Verification Methods
(21 CFR 11.10 Electronic Records)

This section provides some simple verification methods for electronic records

11.10 (a) Accuracy, reliability & consistent intended performance
Verification Method: create test scripts to verify that all types of records generated and maintained by the system are accurate, consistent and contain the intended data. Repeat tests using different challenge conditions to cover any anticipated operating conditions. Ensure that test scripts contain the relevant acceptance criteria.

11.10 (b) Copies of Paper based records

Verification Method: Create paper copies for the record types under test and verification. Compare the paper hardcopies with electronic records that are stored and displayed in the computer system.

11.10 (b) Copies of electronic records

Verification Method: Create electronic copies (softcopies) and compare the copy to that contained in the computer system.

11.10 (c) Protection of records

Verification Method: Review the controls and procedures that ensure accuracy and ready retrieval throughout the records retention period

11.10 (d) & (g) Authorised access

Verification Method: Complete security verification for each account type e.g. operator manager etc. Verify that access is granted and denied via the login function.

11.10 (e) shall be readily available for, and subject to, FDA inspection

Verification method: maintain periodic inspections of the system

11.10 (f) Sequencing of steps

Verification Method: The purpose of this requirement is to ensure the correct sequencing of steps and events occurs. Create tests scripts to demonstrate that any series of steps are operating as intended e.g. approval or signature or login.

Note: This requirement relates to electronic records and is not intended to verify any Operational or functional sequences. These are typically covered in Equipment Validation.

11.10 (h) Input device authorisation

Verification method: The purpose of this step is to check the validity of the source of data. When data is networked or transferred form a device the user must verify that the device is a valid input device. This can be achieved by code review or physical testing.

11.10 (i) Input device person's education, training, and experience

Verification method: Ensure persons using the system are appropriately educated, trained and experienced. Attach evidence of same.

11.10 (j) Establishment of written policies

Verification method: verify the availability of written policies in relation to responsibilities for actions taken under their electronic signatures.

11.10 (k) Appropriate Controls - appropriate controls over systems documentation

Verification method: review access to SOPs to verify their distribution and use for operation of the system is controlled. Verify that chance control is practiced for system related documentation.

What is a Quality Risk Matrix?

A QRM is a risk assessment that identifies and manages the risk to patient safety, product quality and data integrity that relate to the systems processes.

Risk Scenarios or *potential causes* should be developed for each identified function or process step and then assessed for the impact on patient safety, product quality or data integrity. Risk mitigations and controls should then be introduced to address both medium and high levels of risk.

Quality Risk Matrix and 21 CFR Part 11

To ensure proper compliance to 21 CFR Part 11, a simple Risk assessment can help to identify any gaps or concerns. Below an example of a Quality Risk Matrix approach is shown. As this example is a top level initial assessment, the risk classification is estimated as Broadly acceptable, intolerable or as low as possible. This determination is based on the information available and application of relevant experience of dealing with the vendor in question and validation of automated systems.

As a project moves through the phases of planning and design and development, a more detailed risk assessment can be created. At this point a more detailed estimation or risk classification can be made. This can be done by applying 3 "assessments" in order to produce an estimation or overall Risk (Low, medium, high). The 3 assessments are: (1) Assess Likelihood, (2) Assess Detectability, (3) Assess Severity

Assess Likelihood

Determine the likelihood on an adverse event occurring. The risk likelihood should be determined according to a defined criteria or definition. Examples of such are shown below:

High: A standard system function or business process that has been customized by custom coding or by configuration of non-standard system parameters and/or options.

Medium: A standard system function or business process that has been significantly modified solely by configuration of standard system parameters and/or options.

Low: A standard system feature or process that has not been significantly modified by configuration or coding.

Assess Detectability

For each risk scenario a probability of detection should be determined using the following definitions or those set-down by internal company procedures.

High: errors in the output are checked by a standard system error check (i.e. integrity of data, format of data,) prior to completion of the function or process, or at the input to a downstream subsequent function.

Medium: Any errors in the output of the function will be checked by a standard system error check (i.e. integrity of data, format of data, data range) prior to completion of the function or process, or at the input to a subsequent function.

Low: Any errors in the output of the function are not checked by a standard system error check.

Assess Severity

In addition to identifying the likelihood of the risk and detectability, it is also necessary to identify effects or the severity on the business and the GxP status of the system

High: The process is used to create, update, or process data which may have direct impact on: (1) Product efficiency (2) integrity

Medium: The process is used to create, update or process data which has direct impact upon (1) Quantity, (2)Traceability

Low: The process is used to create, update or process data that may have a direct impact on activity that supports cGxP operations.

Quality Risk Matrix for 21 CFR Part 11

Note: Complete this matrix for the system or equipment that is been qualified or validated to meet the requirements of 21 CFR Part 11. Input from the vendor or manufacturer may be required along with examination of Quality management systems, procedures and design documentation.

Requirement	Vendor Software Complies? (Y/N)	SOP Controls required? (Y/N)	Implementation Notes	Risk Scenario	Risk Classification	Mitigation
11.10 Controls for closed systems	Y	Y	Closed system means an environment Closed system means an environment in which system access is controlled by persons who are responsible for the content of electronic records that are on the system.t in which system access is controlled by persons who are responsible for the content of electronic records that are on the system.	System uncontroll ed	Broadly Acceptable	System Validation

Requirement	Vendor Software Complies? (Y/N)	SOP Controls required? (Y/N)	Implementation Notes	Risk Scenario	Risk Classificatio n	Mitigation
11.10a System validation	Y	Y	The software was developed and tested following approved and controlled SDLC practices within a ISO 9001 certified Quality Management System.	Record Deletion	Broadly Acceptabl e	Windows controls/ folder access control.

Requirement	Vendor Software Complies? (Y/N)	SOP Controls required? (Y/N)	Implementation Notes	Risk Scenario	Risk Classification	Mitigation
11.10b Accurate and complete copies of records	Y	Y	Records can be viewed in electronic form within the software. Analysed data records can be printed by using the reporting function in our software.	Records can be viewed in electronic form within the software.	Broadly Acceptable	Access control

Requirement	Vendor Software Complies ? (Y/N)	SOP Controls required? (Y/N)	Implementation Notes	Risk Scenario	Risk Classification	Mitigation
11.10c Protection of records	Y	Y	Records created by the software are saved as files. It is up to the customer to create SOPs surrounding the proper storage and archiving of such files.	Improper storage Deletions and copying of files	Broadly Acceptable	Windows controls / folder access control. SOP to detail proper storage of files

Requirement	Vendor Software Complies? (Y/N)	SOP Controls required? (Y/N)	Implementation Notes	Risk Scenario	Risk Classification	Mitigation
11.10d	Y	Y	Management of user IDs, passwords and control of physical access is addressed through customer SOPs.	Unauthorized user	Broadly Acceptable	Windows controls / folder access control.

Requirement	Vendor Software Complies? (Y/N)	SOP Controls required? (Y/N)	Implementation Notes	Risk Scenario	Mitigation
11.10f	Y	Y	Customer SOPs ensure proper instrument calibration and usage.	Instrument Calibration Correct Operation of the Instrument	All equipment based instruments are subject to a calibration schedule Training shall be completed

Requirement	Vendor Software Complies? (Y/N)	SOP Controls required? (Y/N)	Implementation Notes	Risk Scenario	Risk Classification	Mitigation
11.10g	Y	Y	The software relies on Windows NT logon security for authorization. Management of user IDs, passwords and control of physical access is addressed through customer SOPs.	Unauthorised login	Broadly Acceptable	Windows controls/ folder access control.

Requirement	Vendor Software Complies? (Y/N)	SOP Controls required? (Y/N)	Implementation Notes	Risk Scenario	Risk Classification	Mitigation
11.10h	Y	Y	Customer SOPs ensure proper instrument calibration and usage.	Improper use	Broadly Acceptable	All equipment based instruments are subject to a calibration schedule

Requirement	Vendor Software Complies? (Y/N)	SOP Controls required? (Y/N)	Implementation Notes	Risk Scenario	Risk Classification	Mitigation
11.10i	Y	Y	Education and training of customer personnel is provided at start-up by qualified employees.	Use of equipment by untrained personnel	Broadly Acceptable	All users will be trained prior to use or operation

Requirement	Vendor Software Complies? (Y/N)	SOP Controls required? (Y/N)	Implementation Notes	Risk Scenario	Risk Classification	Mitigation
11.10j	Y	Y	Establishment of policies related to personnel accountability is the responsibility of the customer.	Unable to identify user	Broadly Acceptable	Windows controls/ folder access control.

Requirement	Vendor Software Complies? (Y/N)	SOP Controls required? (Y/N)	Implementation Notes	Risk Scenario	Risk Classification	Mitigation
11.10k	Y	Y	Document management is addressed through SOPs.		Broadly Acceptable	Software is controlled as part of vendor QMS. Software versions are verified and documented as part of the IOQ.

Summary of 21 CFR Part 11.50 Electronic Signatures (11.70, 11.200)

11.50 (a) (1) Signature content -printed name of the signer

11. 50 (a) (2) Signature content-the date and time when signature is executed

11. 50 (a) (3) Signature content- the meaning of the signature

11.50 (b) Printed content

11.50 (b) (1) Printed content

11.50 (b) (2) Printed content showing changes

11.50 (b) (3) Audit trail showing changes

11.70 Signature/record linking

11.200 (a) (1) (i) Biometrics requires two identification codes

11.200 (a) (1) (ii) Biometric Signing requirements

11.200 (b) Biometric identification – genuine owners

Electronic Signature Verification Method

(21 CFR Part 11.50 Signature manifestations)

This section provides suggested verification methods for signature verification

11.50 (a) (1) Signature content

Verification Method: View signed records. Electronics signatures should be verified for all functions where electronic signatures are used for different purposes. The print name of the person signing should be displayed in the record.

11. 50 (a) (2) Signature content

Verification Method: View a n electronically signed record to confirm the date and time of signed is included on the record.

11. 50 (a) (3) Signature content

Verification Method: View a signed record and confirm the meaning of the signature is shown on the record such as such as review, approval. responsibility, or authorship.

11.50 (b) (1) Printed content

Verification Method: Print a signed record and verify the record shows the name if signer, the date and time and the meaning of the signature.

11.50 (b) (2) Printed content showing changes

Verification Method: Complete a change to a signed record and verify that the change is rejected (not allowed) or the change is recorded in the audit trail with the original information still available.

11.50 (b) (3) Audit trail showing changes

Verification method: Complete a change to a signed record and verify that the change is rejected (not allowed) or the name, date & time are recorded in the audit trail.

11.70 Signature/record linking

Verification method: Using a log-in without signer authority attempt to delete, change, copy/paste an existing signature in the system.

Attempt to sign a record. It is not possible to remove or alter the signature associated with the record. It is not possible to link the existing signature to another record. It is not possible to sign a new record.

11.100 (a) Each electronic signature shall be unique to one individual and shall not be reused by, or reassigned to anyone else.

Verification method; verify unique signatures in printed format

11.200 (a) (1) (i) Signature identification components (continuous session)

Verification method: Using an appropriate User ID sign 3 records consecutively. The first signature required entry of 2 identification components (e.g. user name & password) – these may be included in initial log-on. The subsequent signatures each required re-entry of at least one of the identification components.

11.200 (a) (1) (ii) Signature identification components -signings not performed in the same period of controlled system access.

Verification methods: Verify by completing a signing for a record and then allowing a defined time lapse when logged in to the system. The system should request all electronic signature components to be entered when the signing is to be executed.

11.200 (b) Biometric identification

Verification method: Challenge the biometric element of the user identification e.g. user another person to attempt to represent another

Note: if biometrics is not used as part of the system, this requirement is not applicable.

Appendix I- 21 CFR Part 11.10 and 21 CFR Part 11.50

TITLE 21--FOOD AND DRUGS

CHAPTER I--FOOD AND DRUG ADMINISTRATION

DEPARTMENT OF HEALTH AND HUMAN SERVICES

PART 11 -- ELECTRONIC RECORDS; ELECTRONIC SIGNATURES

Subpart B--Electronic Records

Sec. 11.10 Controls for closed systems.

Persons who use closed systems to create, modify, maintain, or transmit electronic records shall employ procedures and controls designed to ensure the authenticity, integrity, and, when appropriate, the confidentiality of electronic records, and to ensure that the signer cannot readily repudiate the signed record as not genuine. Such procedures and controls shall include the following:

(a) Validation of systems to ensure accuracy, reliability, consistent intended performance, and the ability to discern invalid or altered records.

(b) The ability to generate accurate and complete copies of records in both human readable and electronic form suitable for inspection, review, and copying by the agency. Persons should contact the agency if there are any questions regarding the ability of the agency to perform such review and copying of the electronic records.

(c) Protection of records to enable their accurate and ready retrieval throughout the records retention period.

(d) Limiting system access to authorized individuals.

(e) Use of secure, computer-generated, time-stamped audit trails to independently record the date and time of operator entries and actions that create, modify, or delete electronic records. Record changes shall not obscure previously recorded information. Such audit trail documentation shall be retained for a period at least as long as that required for the subject electronic records and shall be available for agency review and copying.

(f) Use of operational system checks to enforce permitted sequencing of steps and events, as appropriate.

(g) Use of authority checks to ensure that only authorized individuals can use the system, electronically sign a record, access the operation or computer system input or output device, alter a record, or perform the operation at hand.

(h) Use of device (e.g., terminal) checks to determine, as appropriate, the validity of the source of data input or operational instruction.

(i) Determination that persons who develop, maintain, or use electronic record/electronic signature systems have the education, training, and experience to perform their assigned tasks.

(j) The establishment of, and adherence to, written policies that hold individuals accountable and responsible for actions initiated under their electronic signatures, in order to deter record and signature falsification.

(k) Use of appropriate controls over systems documentation including:

(1) Adequate controls over the distribution of, access to, and use of documentation for system operation and maintenance.

(2) Revision and change control procedures to maintain an audit trail that documents time-sequenced development and modification of systems documentation.

Reference (https://www.accessdata.fda.gov/scripts/cdrh/cfdocs/cfcfr/CFRSearch.cfm?fr=11.10)

Subpart B--Electronic Records

Sec. 11.50 Signature manifestations.

(a) Signed electronic records shall contain information associated with the signing that clearly indicates all of the following:

(1) The printed name of the signer;

(2) The date and time when the signature was executed; and

(3) The meaning (such as review, approval, responsibility, or authorship) associated with the signature.

(b) The items identified in paragraphs (a)(1), (a)(2), and (a)(3) of this section shall be subject to the same controls as for electronic records and shall be included as part of any human readable form of the electronic record (such as electronic display or printout).

(Reference) https://www.accessdata.fda.gov/scripts/cdrh/cfdocs/cfcfr/CFRSearch.cfm?fr=11.50

11.100 General requirements

(a) Each electronic signature shall be unique to one individual and shall not be reused by, or reassigned to, anyone else.

(b) Before an organization establishes, assigns, certifies, or otherwise sanctions an individual's electronic signature, or any element of such electronic signature, the organization shall verify the identity of the individual.

(c) Persons using electronic signatures shall, prior to or at the time of such use, certify to the agency that the electronic signatures in their system, used on or after August 20, 1997, are intended to be the legally binding equivalent of traditional handwritten signatures.

11.200 Electronic signature components and controls.

(a) Electronic signatures that are not based upon biometrics shall:

(1) Employ at least two distinct identification components such as an identification code and password.

(i) When an individual executes a series of signings during a single, continuous period of controlled system access, the first signing shall be executed using all electronic signature components; subsequent signings shall be executed using at least one electronic signature component that is only executable by, and designed to be used only by, the individual.

(ii) When an individual executes one or more signings not performed during a single, continuous period of controlled system access, each signing shall be executed using all of the electronic signature components.

(2) Be used only by their genuine owners; and

(3) Be administered and executed to ensure that attempted use of an individual's electronic signature by anyone other than its genuine owner requires collaboration of two or more individuals.

(b) Electronic signatures based upon biometrics shall be designed to ensure that they cannot be used by anyone other than their genuine owners.

Reference **https://www.accessdata.fda.gov/scripts/cdrh/cfdocs/cfcfr/CFRSearch.cfm?fr=11.200**

Recommended Reading

- General Principles of Software Validation; Final Guidance for Industry and FDA Staff (FDA, Center for Devices and Radiological Health, Center for Biologics Evaluation and Research, 2002) (http://www.fda.gov/cdrh/comp/guidance/938.html)

- Guidance for Industry, FDA Reviewers, and Compliance on Off-The-Shelf Software Use in Medical Devices (FDA, Center for Devices and Radiological Health, 1999) (http://www.fda.gov/cdrh/ode/guidance/585.html)

- Pharmaceutical cGMPs for the 21st Century: A Risk-Based Approach; A Science and Risk-Based Approach to Product Quality Regulation Incorporating an Integrated Quality Systems Approach (FDA, 2002) (http://www.fda.gov/oc/guidance/gmp.html)

- Code of Federal Regulations Title 21 (21 CFR), Food and Drugs. The Code of Federal Regulations, Title 21 includes parts such as Parts 210 and211. Part 11 (known as 21 CFR Part 11 is of particular importance for computer validation). This part deals with electronic records and electronic signatures.

- Annex 11 of the EU GMP Guideline, Annex 11 of the EU GMP guideline is divided into 19 points and covers topics ranging from requirements for configuration, operation and change control for computerized systems in a GMP Environment.

- FDA Guidance for Industry: 21 CFR Part 11 - Electronic Records and Electronic Signatures: Scope and Application, August 2003.

6.3 Lean Methodologies

Introduction

Lean is a globally recognised set of principles and practices that help build and continually improve businesses and organisations across different industries and sectors. Lean can be applied to small projects and large projects that are delivered over the short, medium or long-term. The origins of lean date back to over 50 years ago to the Toyota production system. This was an in-house methodology developed within the Toyota manufacturing company of Japan. From the early 1980s lean principles began to become more widely known and acknowledged and very quickly its popularity grew to impact different sectors in different countries. It is without question a proven way to ensure businesses are more effective and customer focused while maximising value and quality. Applying lean techniques can help companies reduce defects and deficiencies and most of all continually improve their systems and processes. It also works to eliminate waste of materials and wasting time and other resources.

What Is lean?

If you are new to lean in a manufacturing environment, you may question what is the essential meaning of the lean philosophy? At its core is a continual drive to do more with less. If Lean is correctly implemented, a company or organisation will (1) use less human effort to perform their work, (2) use less material and (3) manufacture less defects. Not only will lean achieve these goals, it also works to maintain and sustain the results over time.

Key Points of Lean:

- A focus on customer value
- Getting the whole team involved
- A philosophy of continuous improvement
- Reducing variation
- Eliminating waste
- Taking the long-term view
- Improving value
- Providing exactly what's needed at the right time based on customer demand and requirements maintaining flow and the right movement at the right time

A Roadmap to Lean

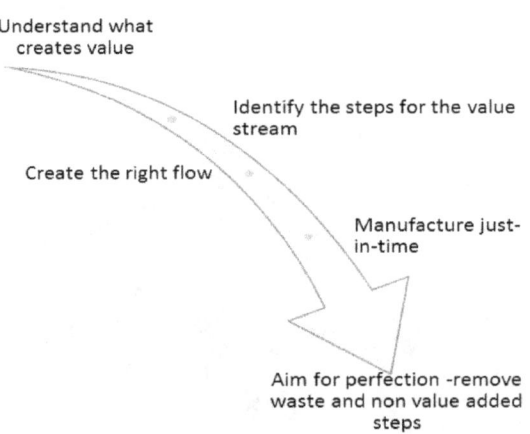

Understand what creates value

Identify the steps for the value stream

Create the right flow

Manufacture just-in-time

Aim for perfection -remove waste and non value added steps

(1) The true value of a product or service is must be based on the customer's perspective. This is an important

point to grasp. Value is not what an individual engineer or department deems it to be. It must be established from the customer. Therefore, the first step should be to clearly specify value as the customer sees it.

(2) In identifying all the steps within a value stream, this will allow the team to determine if steps are value adding or non-value adding.

(3) Make only those actions which create value flow.

(4) Only make what is pulled by the customer just-in-time.

(5) Perfection is the goal. Strive towards perfection by continually removing successive layers of waste.

The History of Lean

The early beginnings of lean can be dated right back to the 1900s. What was referred to as Time & Motion studies, pioneered by an American Engineer Frederick Winslow Taylor. It was one of the first business efficiency techniques (even this term wasn't yet defined). The idea of time and motion studies is to breakdown large tasks into small steps in the sequence they occur and the exact time taken for each "motion" or step to be completed. Henry Ford focused on reducing waste while developing the mass assembly manufacturing system for car production. This awareness of waste as a drain goes back to as early as 1915. Henry Ford himself, quoting from in My Life and Work (1922), provided a single-paragraph description that encompasses the entire concept of waste:

"I believe that the average farmer puts to a really useful purpose only about 5% of the energy he expends. Not only is everything done by hand, but seldom is a thought given to a logical arrangement. A farmer doing his chores will walk up and down a rickety ladder a dozen times. He will carry water for years instead of putting in a few lengths of pipe. His whole idea, when there is extra work to do, is to hire extra men. He thinks of putting money into improvements as an expense.... It is waste motion— waste effort— that makes farm prices high and profits low." Ford looked at the bigger picture and understood that poor arrangement of the workplace leads to waste and inefficiencies, such as the length of time to transport water manually.

Design for Manufacture (DFM) - a Ford Concept

This standardisation of parts was central to Ford's concept of mass production, and the manufacturing "tolerances", or upper and lower dimensional limits that ensured interchangeability of parts became widely applied across manufacturing. Decades later, the renowned Japanese quality guru, Genichi Taguchi, demonstrated that this "goal post" method of measuring was inadequate.

He showed that "loss" in capabilities did not begin only after exceeding these tolerances, but increased as described by the Taguchi Loss Function at any condition exceeding the nominal condition. This became an important part of W. Edwards Deming's quality movement of the 1980s, later helping to develop improved understanding of key areas of focus such as cycle time variation in improving manufacturing quality and efficiencies in aerospace and other industries. While Ford is renowned for his production line it is often not recognised how much effort he put into removing the fitters' work to make the production line possible. Until Ford, a car's components always had to be fitted or reshaped by a skilled engineer at the point of use, so that they would connect properly. By enforcing very strict specification and quality criteria on component manufacture, he eliminated this work almost entirely, reducing manufacturing effort by between 60-90%. However, Ford's mass production system failed to incorporate the notion of "pull production" and thus often suffered from over-production.

Dr. Deming's Management System

The contribution of Dr. Deming to Lean and modern manufacturing was most significant during the 1950s. With a strong background in statistics, Deming studied at New York University's graduate school of business.

Deming became acquainted with Walter A. Shewhart of the Bell Telephone Company. Shewhart was seen as a pioneer in the fields of statistical control of processes and the tools of the control charts. His influence led to Deming becoming interested in using statistical methods in the fields of manufacturing, industrial production and management. Furthermore, Shewhart's idea of common and special causes of variation resulted in the formation of Deming's theory of management. During the summer of 1950, Deming lectured hundreds of engineers, managers, and academics in statistical process control (SPC) and the concepts of quality. It is also believed that top management from key Japanese companies were also trained by Deming during this time. The implementation of Deming's systems led to improved quality and lower costs. This led to an increased demand for Japanese products across the globe.

Ford Motor Company was one of the first American corporations to seek help from Deming. Deming questioned the company's culture and the way its managers operated. To Ford's surprise, Deming did not talk about quality, but about management. He told Ford that management actions were responsible for 85 percent of all problems in developing better cars. By 1986, Ford had become the most profitable American auto company. In 1982, Deming's book Quality, Productivity, and Competitive Position was published by the MIT Center for Advanced Engineering and was renamed Out of the Crisis in 1986. Deming created a system of 14 key principles for management to follow for significantly improving the effectiveness of a business or organisation. Some principles are philosophical and cultural and some are more practical. The points were first presented in his book "Out of the Crisis."

However, the realisation of lean as a field of engineering is attributed to Taiichi Ohno who established the Toyota Production System. After some years of internal use and development Toyota began to see the powerful benefits of a lean culture and lean principles. It was opened up to the wider industrial community in 1973.

Taichi Ohno's two basic principles about lean & management thinking include:

(1)   The workplace needs to be transparent. Normally we are not alone in a company. Information needs to be readily available. Hoarding information by individuals is very harmful to the company.

(2)   The second message requires that you start with your own workplace first, before we can discuss others.

Understanding Flow

Flow and how it relates to a manufacturing environment is an essential part of lean. Flow is not just limited to the customer at the end of the supply chain. Issues within the manufacturing system may result in supply issues and delays to the customer. The term "flow" has a broader implication and needs to be applied across the different steps within a lean manufacturing process.

For product manufacturing, flow begins with the supply and introduction of materials and components to the production line or system. The right flow levels at the right times ensure each step of the process operates efficiently and therefore the end result will be greater overall efficiencies that benefit the customer and end user. Flow is when there is no queuing or delays between each value-added step. Not getting the flow levels correct can have a knock-on effect right across a value stream.

For instance, (1) if queues form between value added steps, this indicates a potential bottleneck or pinch point. This may therefore call for greater capacity, e.g. more machines, faster cycle times, more operators and so on. (2) If flow is not optimised correctly, there can be an impact on the levels of WIP, intermediate product and inventory levels. High levels of inventory is costly and ties up the cash flow of a company or organisation. In order to achieve the right balance, steps that actually create value need to occur in rapid sequence. Each step should be value added. The customer does not want to pay for non-value added activities and therefore they must be eliminated or continuously reduced.

Process Flow Illustrated

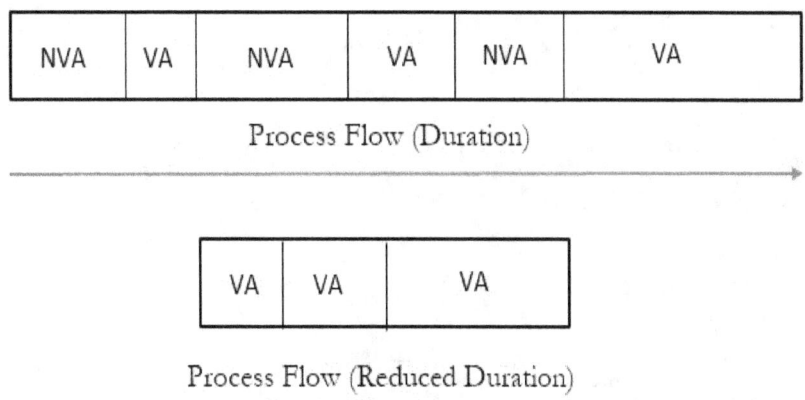

NVA= Non value added, VA= Value added

The first illustration shows a series of steps that are both non-value added and value added. The combination of both results in the total duration of the processing time. Removing NVA steps reduces the duration of the process flow making it more efficient.

The Goal of Lean?

The ultimate goal of lean manufacturing is to achieve a perfect process. The ideal process is simply one with the perfect performance where all sources and causes of waste are reduced to zero. As previously stated, the customer is the authority on value. They determine if the right combination of quality is provided at the right place, right time and at a cost effective price. Any steps in the process should be designed to add value to the customer; these steps can range from design activities, manufacturing, packaging and so on.

Within a lean manufacturing organisation the tenets and philosophy of lean must be practised on a regular basis. Not only this, it is the responsibility of every person to embrace lean culture and practices. Remember, lean fosters continual improvement of processes.

Lean & Six Sigma

Six Sigma helps companies in identifying and controlling variation in processes that most affect performance and process outcomes. Variation in its most severe form leads to defects and defects cost money. Controlling the variation can lead to more efficiently run processes. Six sigma practitioners such as Black Belts analyse root causes and implement corrective actions A black belt project can take from 4 to 6 months to complete, however, the cost saving and return to the company can be in the hundreds of thousands of pounds, dollars or euro in value.

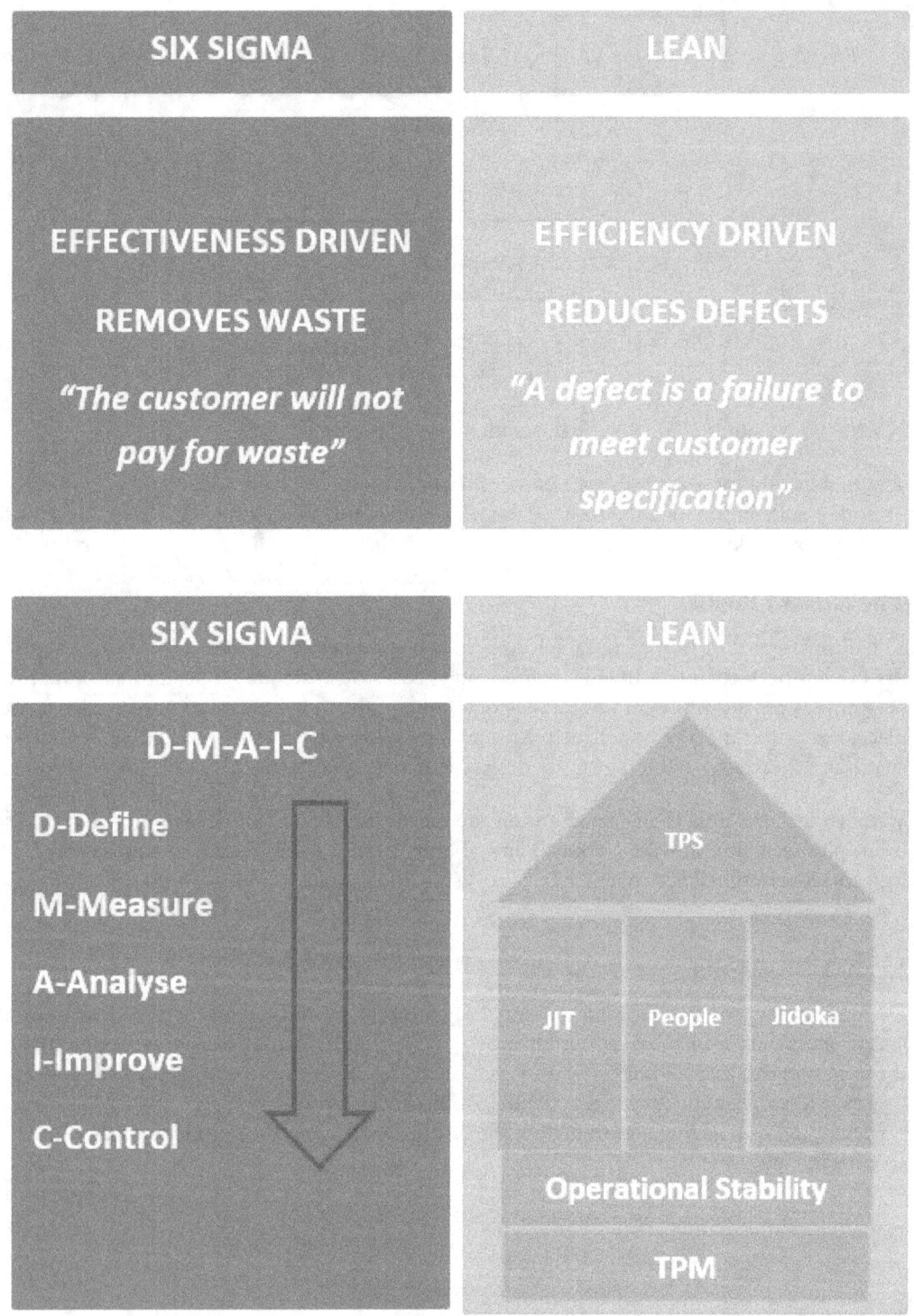

Figure: Six Sigma Vs Lean, a simplified representation of the focus areas of Six Sigma and Lean manufacturing.

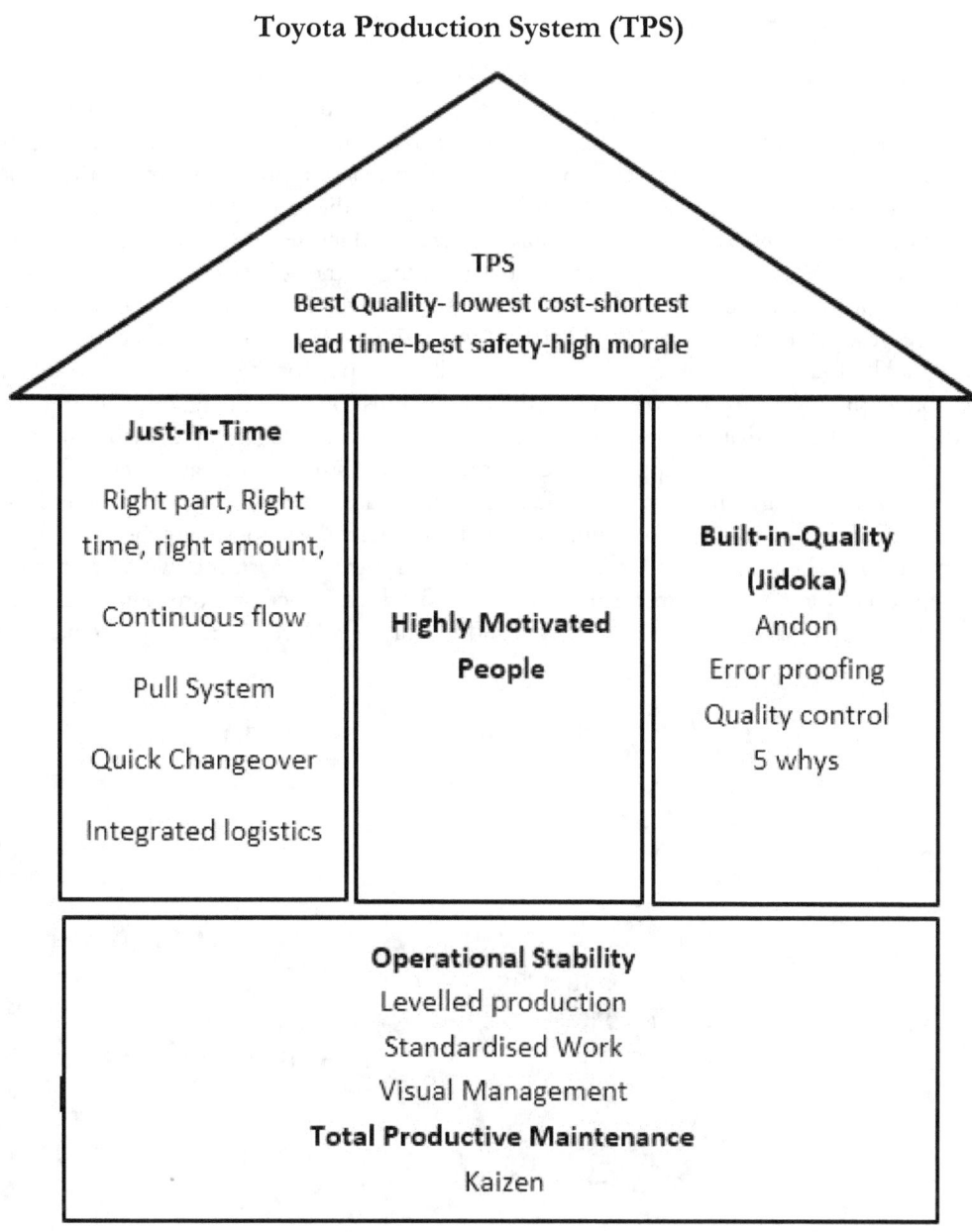

Figure: The Toyota Production System (TPS) represented in a house format illustrates the key elements of lean.

Value Stream Explained

Providing value to customers is a core aim of lean manufacturing. In the ideal world only value added activities are required during the manufacturing process. However, no system is likely to be perfect. When creating a value stream for a particular process or product you identify all the steps that occur to get the product or service to your customers along with key information on the steps and activities e.g. machine type, cycle times etc.

Value Stream Mapping

Value stream mapping is used to visualise and capture specific activities of a manufacturing process. It is often underestimated but listing the sequence of events or actions can raise a lot of questions within a group. Some may disagree with the sequence of steps, some may omit steps etc. However, these building blocks of the process must be noted down on paper. Above all, the best approach is to map the value stream in its current state to analyse how does it work "today", then later on the team can propose the ideal state. In simple terms, value is the worth placed upon something, either a product, service or something that a customer can express in terms of money. A key principle of lean is listening to the voice of the customer (VOC) and creating a clear picture of what value is from the customer's perspective. The customer is the one who can define the true value of a product or process and whether it is value added. However, there are 3 common principles that should be followed (1) the customer must be willing to pay for the activity (2) the activity must transform the product or service in some way and (3) the activity must be done the first time correctly. The above principles of value added activities apply to the whole process, consisting of all activities regarding people, processes systems and equipment. Applying the 3 principles consistently will allow non value added activities to be spotted quite easily. In contrast to value added activities, it is also important to understand what constitutes non value added activities. In manufacturing if an activity does not satisfy the above criteria we can determine the action to be non-value added. Simply put, this means the customer will not be willing to pay for it or the steps do not in any way transform or improve a product service or ultimately, it cannot be done correctly the first time and therefore is a waste of time and resources (which is paramount to increasing costs).

Examples of NVA and VA

Processing Orders	Waiting for Raw Materials
Completing Engineering Drawings	Testing
	Inspection
Manufacturing	Rework
Assembly	Moving WIP
Painting	Revising
Shipping	Tracking

Introducing the Lean Toolbox

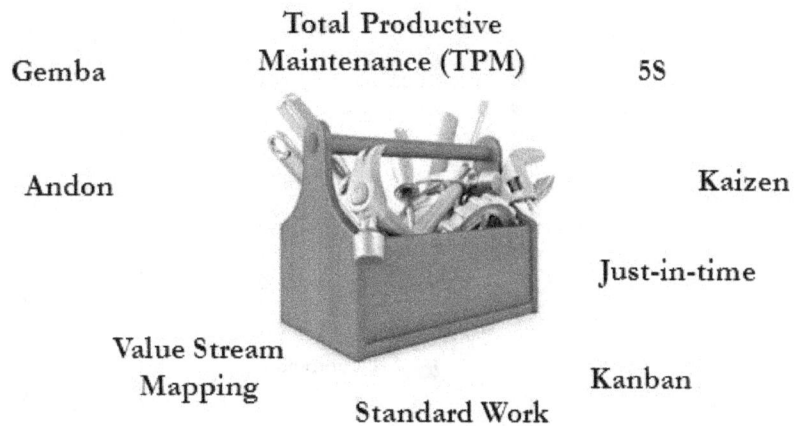

Gemba

Total Productive
Maintenance (TPM)

5S

Andon

Kaizen

Just-in-time

Value Stream
Mapping

Kanban

Standard Work

Theory of Constraints (TOC)

The Theory of Constraints (TOC) is a "thinking process" developed by Dr. Eliyahu M. Goldratt. The TOC can be used as a methodology to improve the operational running of a company or organisation. But what exactly does a "constraint" mean in manufacturing terms? A constraint is the most important limiting factor that prevents a goal being reached. The aim is to improve the constraint to a point where it can no longer limit the process.

The Theory of Constraints consists of the following:

- Five Focusing Steps (a methodology for identifying and eliminating constraints)
- The Thinking Processes (tools for analysing and resolving problems)
- Throughput Accounting (a method for measuring performance and guiding management decisions)

The Five Steps

(1) Identify the constraint
What is the constraint? What is the amount of work in the queue ahead of a process unit operation?
(2) Exploit the constraint
Once the constraint is identified, the process is improved or supported to meet the capacity without major investment or changes.
(3) Subordinate other processes to the constraint
After the constraining process is working at maximum capacity, the speeds of other subordinate processes are paced (often referred to as synchronised) to the speed or capacity of the constraint.
(4) Elevate the performance of the constraint
If the output of the overall system is not satisfactory, then further improvement should be made. A major change may be required to the constraint in question. Changes can involve capital improvement, reorganisation or other major expenditures of time or money.
(5) Repeat the process
After the major limiting constraint is broken, another part of the system or process chain becomes the new constraint. Repeat the cycle of improvement for the new constraint.

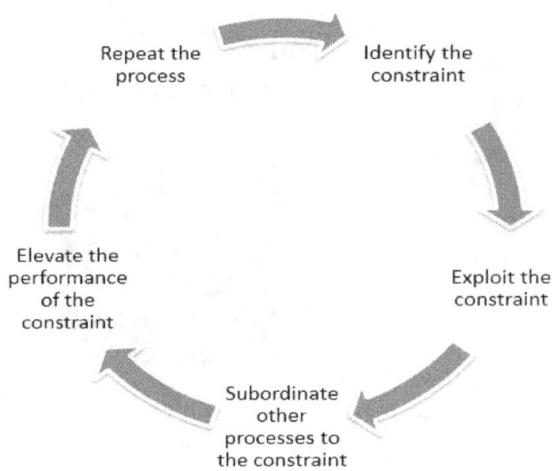

Figure: The five steps involved in the theory of constraints which help identify and eliminate constraints (also known as bottlenecks)

How They Compare

The diagram below shows the contrasting strengths and focal points of Six Sigma, lean and the theory of constraints.

SIX SIGMA	LEAN	THEORY OF CONSTRAINTS
EFFECTIVENESS DRIVEN REDUCES VARIATION	EFFICIENCY DRIVEN REDUCES WASTE	MANAGES CONSTRAINTS
1) Define 2) Measure 3) Analyse 4) Improve 5) Control	1) Identify value 2) Identify value stream 3) Flow 4) Pull 5) Perfections	1) Identify constraint 2) Exploit constraint 3) Subordinate processes 4) Elevate constraint 5) Repeat cycle
Problem Focused	Flow Focused	Constraint Focused

The 7 Wastes

There are 7 accepted categories of waste in lean thinking. However, to make it simple a certain *Mr. Tim Wood* will help us remember them easily (see below). Firstly, a definition of waste and why it is harmful is provided.

T   Transportation

I   Inventory

M   Movement

W   Waiting

O   Overproduction

O   Over-processing

D   Defects

Transportation Waste

Transportation Waste: Waste can be defined as the unnecessary movement of parts/elements between a process. It can be any process however, as it is lean manufacturing, it can also focus on the manufacturing cycle.

Examples of Transportation Waste

Movement of materials
Having to go get approval signatures
Storing files
Retrieving files

Impact

Increases production times
Consumes resources
Increases work in progress (WIP)
Potential damage to products

Inventory Waste

Inventory waste includes any material or work in progress (WIP) or finished goods which are not having value added to them. If they are sitting on a shelf, they are tying up money.

Examples of Inventory Waste

WIP stored in a warehouse
Files waiting to be worked on
E-mails waiting to be read
Impact

Adds cost – storage and resource to manage
Hides shortages and defects
Can become damaged
H&S risk

Movement Waste

Movement waste arises when movement within a process is unnecessary.

Examples of Movement Waste
Searching for equipment or files

Forgetting the right tools or equipment
Going to the wrong area

Impact
Increases production times
Interrupts the work flow
Adds time which equals cost

Waiting Waste

Waiting waste occurs when people or parts need to wait for a process to be completed.
Examples of Waiting Waste
Waiting for WIP to offline pack
Waiting for the PC to start up/re-boot
Waiting for suppliers to provide info
Waiting for approvals
Waiting while equipment is being turned around
Waiting to know which job to do

Impact
Stopping and re-starting of the process
Causes bottlenecks which need to be cleared
Increases lead times for customers
Creates WIP

Overproduction Waste

Overproductions is a result of producing too soon, too fast or in larger amounts than required by customer demand.

Examples of Over Production Waste
Manufacturing according to a sales forecast rather than a real order
Generating reports that are not used
Overproduction causes other wastes (waiting, conveyance, motion, inventory, defects) – the worst waste of all.

Impact

Costs money
Uses resources/materials ahead of plan

Over Processing Waste

Over processing is a symptom of processing above and beyond the standard required by the customer.

Examples of Over Processing Waste
Completing different forms with the same information
Multiple document checking

Impact
Requires resources
Increases processing time
Costs the business

Defects Waste

Defects waste occurs when a product does not meet the quality standards and requirements of the customer.

Examples of Defects Waste

Damaged product
Rejected product
Re-work

Impact

Can impact schedules and lead times
Consumes resources
Reduces customer satisfaction if defects reach the user
Defects adds cost

Standardisation

Standardisation or standard work helps deliver lean manufacturing. The more diverse the manufacturing and the larger the operation, the greater the importance of standard work and the greater the opportunity to see the benefits.

Standard can be broken down into 4 key areas:
(1) Takt Time
(2) Line Balance
(3) Standard in process stock
(4) Standard operations

Standard Work Versus Standard Operations

Standard Work: delivers customer satisfaction throughout a stable, repeatable process. This ensures consistent quality, cost & delivery in a safe environment.
Standard Operations: is centred on human movements, outlining efficient, safe working methods that eliminate waste, whilst ensuring proper use of equipment and tooling.

Benefits of standard work:

- Documenting work practices creates a baseline for continuous improvement
- Reduced costs
- Improved quality and reduced the levels of defects
- Can be used in training
- Makes work practices and processes more consistent and repeatable

Takt Time

The word "takt" is German for "beat". Takt time is a fundamental element of any production system where the output volume of a process needs to ensure on time delivery of products to customers.

Standard in Process Stock

Standard in process stock is the minimum number of pieces of in process stock required to run each process on demand:

Standard in process stock can be calculated using the following equation:

$$\text{SIPS} = \frac{\text{Process Cycle Time}}{\text{Takt Time}}$$

Overall Equipment Effectiveness

Overall Equipment Effectiveness typically shortened to OEE, is the percentage of time the equipment is used to produce products that are in specification at maximum optimised speed. OEE is a "best practice" approach used to monitor and improve the effectiveness of manufacturing processes. OEE is simple, practical and helps us analyse the problem and make improvements. It also helps measure our progress rate towards improvement.

$$\textbf{OEE} = \text{Availability (\%) X Efficiency (\%) X Yield (\%)}$$

Availability refers to the operating time the machine or system is available (the operating time). Down time caused by mechanical or other issues affects the availability %.

Efficiency (performance) is the operating time which is impacted by any speed losses within the process.

Yield is the throughput that meets specifications, yield is impacted by defective product which can be referred to as yield loss.

It should be cautioned that to use OEE to compare different processes, or different machines does not provide a like for like comparison and should be avoided. OEE is also not a useful executive Key Performance Indicator, KPI. Rather, it is a measure of improvement.

Future-State Questions

1. What is the takt time?
2. Will you manufacture for a finished-goods supermarket from which the customer pulls, or directly to shipping?
3. Where can you use continuous flow processing
4. Where will you need to use supermarket pull systems to control production of upstream processes?
5. At what single point in the production chain will you schedule production?
6. How will you level the production mix at the pacemaker process?
7. What increment of work will you consistently release and take away at the pacemaker process?
8. What process improvements will be necessary for the value stream to flow as your future-state design specifies?

Poka Yoke

Poka Yoke is a technique for avoiding simple human error in the workplace. Simply put, it aims eliminate mistakes and is often referred to as mistake-proofing or fail-safe work methods. Poka Yoke is a system designed to prevent inadvertent errors made by workers performing a process. The word "Poka-Yoke" is Japanese for mistake-proofing or mistake avoidance. It involves the design of products, work practices, fixtures and jigs etc. that prevent the mistakes or errors that result in defects. A secondary aim of Poka Yoke is to make any defect easy to recognise with minimum time, skill and expertise. It is accepted as a simple and inexpensive way of preventing defects from being made or identifying a defect so that it is not passed to the next operation, downstream process and ultimately, the consumer.

The benefits of Poka Yoke are extensive. The specific benefits depend on the nature of the work and also where the focus is placed when executing a Poka-Yoke programme. If it focuses on cost reduction, the metric might be a decrease in set up times or processing times. If the focus is on quality, a redesign of jigs or fixture may be required. Here are 12 benefits of Poka Yoke:

- Reduces set-up issues
- Improves product quality
- Improves yield
- Reduces rework
- Reduces manufacturing cost
- Decreases set-up time
- Decreases set-up complexity
- Improves housekeeping
- Removes dependence on high skill levels or experience
- Increases manufacturing flexibility
- Improves work attitudes
- Reduces manufacturing costs

Key Principles of Poka Yoke

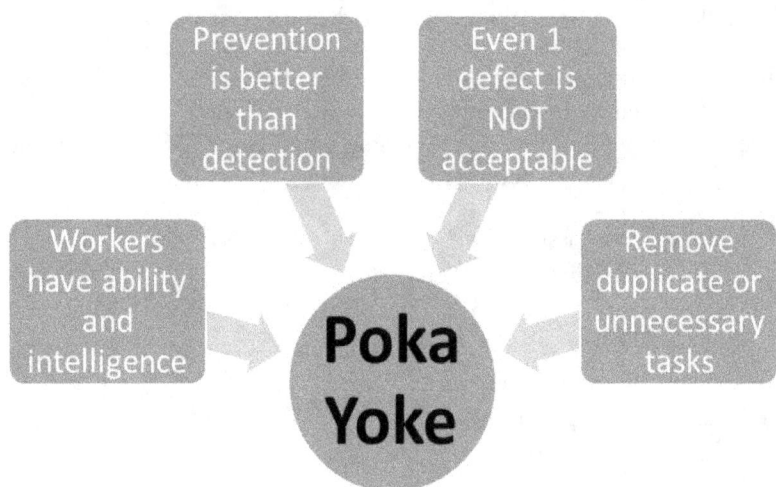

Workers Have Ability and Intelligence

For the application of any methodologies such as Poka Yoke, the right approach and attitude largely determines the success. The persons responsible for the rollout of such methodologies must fully support the programme and truly believe in the benefits. Typically, Top management are responsible for the rollout of programmes such as Poka-Yoke and Lean etc. Human intelligence can quickly see through any new initiatives that are not understood or supported by management.

There should also be an acknowledgement of worker's ability and aptitude. No matter what the role of the person, be it factory operator, technician, electrician or engineer, trust and confidence must be given to all stakeholders. It is a dangerous trait to overlook the input and contribution of junior staff.

Prevention Is Better Than Detection

Preventing defects saves money. Detecting defects costs money. Therefore, the preference is to prevent defects before they happen. This is where mistakes can be eliminated by Poka-Yoke techniques. If you eliminate mistakes or errors, this in turn works to eliminate defects. While inspection and detection systems will always have some use, the absence of defects will reduce the dependency of detection.

Even 1 Defect Is NOT Acceptable

Having zero defects is the golden rule. If defects are accepted as part of a culture, then complacency follows.

Remove Duplicate or Unnecessary Tasks

It may seem like a trivial observation but allowing duplicate tasks that are unnecessary (not required) can lead to mistakes causing defects. Why you may ask? Apart from the cost of completing extra manufacturing steps that are not required, extra steps or tasks generally result in extra handling. Extra handling may involve transfer to and from work stations, labelling, paperwork completion and stand down times. All of these activities can be a source of error generation.

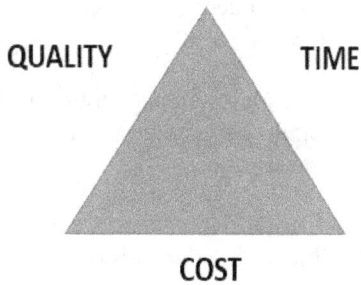

Figure: Adapted representation of the project management triple constraint

The above figure shows the 3 elements of the triple constraint theory often used in project management. This theory is also applicable to manufacturing and mistake proofing.

Cost

Cost is one of the most important factors to today's consumer. When faced with an abundance of choice, the sale price needs to be in keeping with other manufacturers. Even if the product is of superior quality and function, cost needs to be understood and managed effectively.

With regards to Poka Yoke and mistake proofing, there is an opportunity to reduce costs during manufacturing. If errors and mistakes are prevented, this reduces defects. Not having to rework product or dump defective products helps to save money and frees up cash within a factory. If you can reduce manufacturing costs by eliminating defects and wastage, this creates an opportunity to make a higher profit margin on the market to retail price.

Quality

Some defects may result in dissatisfaction for the customer. More serious defects may prevent the customer using the product at all or in a safe manner. It is in everyone's interest to manufacture a quality product that meets the user's requirements. Product quality is often a byword for safety which should always be priority to a manufacturer. It should be understood that changes in the level of product quality may impact upon both costs and time. It can be said that this is a balancing act, however, there should be a minimum standard of quality that is met every time.

Time

"Making mistakes increases the amount of work."

The time constraint also has implications on cost and quality. Take a manual process such as hand finishing wood. If the job is rushed or time is squeezed, the quality of the finish may not be up to standard.

If the hand finishing takes too long, it impacts upon the overhead costs associated with labour and the cost to deliver the product rises. It may be prudent to reduce the time required to complete a task or job without compromising the quality of the work.

The time to complete each manufacturing step contributes to the final cost of the product. This creates a lead time for each product type. Lead times become important in scheduling the right products and right volumes in order to meet the market requirements.

In recent years, with many consumer products, people expect next day delivery. This means the manufacturer most also be responsive and flexible to the demands of the market and wholesalers.

Customer Driven Companies

Every manufacturer wants to meet the expectations of its customers in terms of quality and other factors such as delivery times and costs. A popular standard which is used by manufactured worldwide is ISO 9000 Quality Management. The ISO 9000 is made up of the following standards:

- ISO 9001:2015 - sets out the requirements of a quality management system
- ISO 9000:2015 - covers the basic concepts and language
- ISO 9004:2009 - focuses on how to make a quality management system more efficient and effective
- ISO 19011:2011 - sets out guidance on internal and external audits of quality management systems.
- (Ref: http://www.iso.org/iso/iso_9000)

A key requirement of ISO 9000 is customer focus, the requirements of clause 5.2 deals with meeting customer requirements, and also managing the feedback from customers on an on-going basis.

Customer-Driven Practices and Quality Policies

A quality policy is a concise statement that sets out a company's commitment to the customer and the commitment to delivering quality products and services. Often a quality policy will be displayed in the reception area of a company or is available to download as a document on their website. The quality policy must be relevant to the business operations. While many themes and traits are common across different sectors, the quality policy for a service company would likely differ slightly to a company that manufactures physical products.

Example of a Quality Policy

"We practice continual improvement to achieve customer delight by providing customer-centric, cost-effective, timely and qualitative software solutions. We are committed to meeting the regulatory requirements of medical device manufacturing, and meeting our customer expectations"

Can errors be eliminated?

In the journey to achieve zero mistakes and zero defects, one school of thought promotes the belief that human error can be reduced to a minimum or indeed eliminated. There are several factors that help reduce the amount of mistakes people make. Training and experience are key parts, along with the proper systems and resources e.g. tooling, instruction, work area setup to name but a few.

Are errors inevitable?

An opposing view of errors is that people always make mistakes, no matter how small or low in occurrence. Even if we accept mistakes as a part of life, we still tend to blame the people who make them. The risk of adopting this philosophy is that defects can be missed during manufacturing which can result in defective products being sold commercially.

Sources of Errors

Poka-yoke tries to prevent or eliminate human errors. There are several common types of human errors some of which include:

Communication errors: Often mistakes occur due to a lack of communication or due to someone misinterpreting instructions. *Mitigation:* ensure any critical communications are written or available to review if there is doubt.

Rookie errors: Sometimes we make mistakes through lack of experience. For example, a new worker does not know the operation or is just barely familiar with it. *Mitigation:* Skill building and work standardisation.

Compliance Errors: If no consequences are perceived, sometimes we can overlook steps or processes. *Mitigation:* Foster a sense of personal responsibility and the impact of small defects on the customer.

Forgetting: Humans are prone to forgetting steps or tasks especially if they are repetitive and they are working on the same processes for long periods of time. *Mitigation:* Provide checklists to operators and workers in order to formalise the process. Paperwork documenting critical steps will alert the operator if they forget a step.

Procedure related errors: If instructions or standard operating procedures are inadequate it may lead to errors. Mitigation: Ensure existing work instructions are accurate and reflect the proper and necessary actions for safety, quality and prevention of mistakes.

There are various types of defects. The table below lists some common defects along with some suggested sources:

1. Omitted processing steps: Time pressures Carelessness Lack of training	6. Processing wrong workpiece: Workflow layout errors Incorrect labelling or identification
2. Processing errors: Skill deficiency Wrong specification	7. Misoperation: Poor equipment / tooling Operator error
3. Errors setting up work pieces: Lack of training / experience Lack of correct tools	8. Processing errors: Operator error Wrong machine
4. Missing parts: Oversight Workflow issues	9. Equipment setup errors: Lack of training Wrong settings Wrong program
5. Wrong parts: Parts not identified	10. Tools and jigs: Grease of coolant not applied Wrong jigs or tooling

Jigs and Fixtures

The terms "jig" and "fixture" are commonly used in the manufacturing industry, particularly in CNC machining and fabrication. Many machining processes require jigs and fixtures in order to achieve consistent and accurate results.

Jig

A jig is used to guide the item or component that has to be machined while a fixture holds in place of "fixes" the component to be machined or processed.

Fixture

A fixture is used to hold the component or part during the machining process. Its purpose is not to guide the part towards the machining tool. Fixtures are secured with the table surface of the mills in most of the cases. Fixtures reduce the need for other tools and facilitate more accurate machining and processes.

Common Poka-Yoke Tools

By using Poka-yoke tools, we are trying to eliminate human errors. These errors are usually oversights due to poor judgement or concentration. Poka-yoke helps prevent defects resulting from human error or mistakes. Human factors or human errors can lead to quality defects. Poka Yoke helps people including factory operators, fabricators, assembly personal and engineers to reduce defects due to errors. Some common examples of Poka Yoke tools include:

1. Checklists

Checklists are a practical and efficient way of detecting errors before they impact a process or product. Some of the best checklists are designed to be completed in a relatively short period of time, however, this can be influenced by the complexity of the task at hand. The principle still stands that designing a checklist that is to the point and easily completed will deliver the best benefits. Checklists should focus on factors that if overlooked in error can lead to defects.

2. Error detection - visual and audio alarms

A lot of automated pieces of equipment have in-built controls that will alarm visually and audibly when a process begins to drift or go out of control. For example, a parts washer may have temperature alarms to indicate if the temperature drops or rises below the process settings. Alarms and warning systems therefore can prevent mistakes and defects before they materialise. They can also help detect when errors do occur and help to ensure the customer gets a quality product.

3. Guide pins

Guide pins are a proven way to help force the proper set-up and assembly of parts. Typically guide pins of different sizes are used to orient and position components in the desired manner. This prevents misoperations such as drilling or machining in the wrong position. Guide pins also help to ensure components are assembled in the correct way.

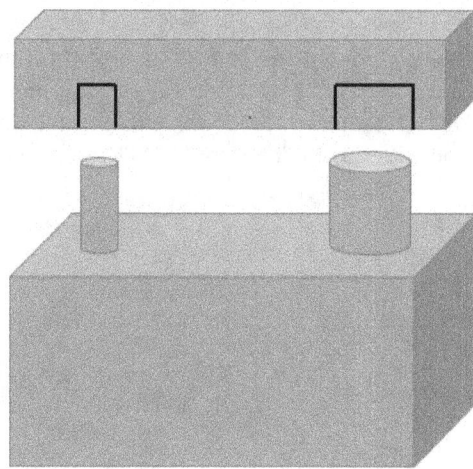

Figure: Example of guide pins in order to force the right assembly

4. Jigs

As previously described, a jig is used to guide the item or component that has to be machined while a fixture holds in place of "fixes" the component to be machined or processed. Jigs are very common and if quality upfront engineering is given to jig design, it can eliminate a lot of errors and defects during manufacturing, especially at high volumes.

Kaizen

Kaizen is the Japanese word for improvement though the term is often meant to infer not only improvement but "continuous improvement". Although the event can be technical in nature, the key components of a Kaizen event involve copper-fastening employee involvement along with the right attitude and a culture that supports improvement and lean principles. In practical terms, Kaizen can be described as a highly focused "assault" on process to problem in order to realise a rapid improvement. The event itself can take anywhere from 3-5 days to complete, so there can be quite an amount of progress made during this period. While Kaizen events are focused on improvement, they do not replace any continuous improvement programs.

Before the Kaizen event begins, it is worth spending a little time framing the issue and some key factors that may be examined during Kaizen event.

- What are the goals of the Kaizen event?
- Do the goals align with the strategic goals of the organisation?
- Is there a comprehensive understanding of the current performance of process, quality metrics etc.?
- Is there a high likelihood of success?
- Will the results will be highly visible?

As with any engineering project, the involvement of management and suitable team members is essential in ensuring success. The roles and responsibilities of participants should also be determined prior to the event. You will need (1) a team sponsor, someone who can support the work of the team, typically a manager or director. The team sponsor can ensure any financing is provided along with the right resources. (2) A team leader, a point person needs to co-ordinate the event and lead the team to its goals and objectives. (3) A facilitator may be used if the team is new to what constitutes a Kaizen event. The facilitator should be experienced in Kaizen and lean and can help guide and direct the team in the techniques required. (4) Team participants, depending on the scope and project charter, various skills may be required. Match the person to the task. (5) Create a team charter; this is a great way to get input from each team member and ensure everyone is one the right page. (6) Get the data prior to the event - save time and get off to a good start. It also makes a great impression with the wider team. You don't want to be waiting on information as the event is just about to start.

Kaizen Team Charter

Team charter clearly identifies rules of:

- Operation
- Objectives
- Scope
- Resources
- Authority of the team
- Deliverables
- Schedule
- Code of Conduct
- Created by leader and approved by teamKanban

Pull

Pull is a method used to control the resources by substituting only the material consumed. The customer initiates the process or pulls the product or service. During a "pull" manufacturing system, work is started in response to a signal from the customer (or an order being received). This prevents the occurrence of costly overproduction and the development of too much WIP.

Push

In a push system, the production process is started with the order. Production orders are based on the production plan. The production start does not depend on the current production.

Basic considerations

The introduction of Kanban systems can be difficult to implement and it is hard to know where to begin. The following points should be understood and answered prior to roll out.

- What is the right inventory size?
- What is the amount of inventory required? Where?
- What is the replenishment frequency?
- What is the batch size?
- Which information is essential for the Kanban signals?
- How can we manage number and location of Kanban?
- How can we make sure that the Kanbans are simple, visual and effective?
- How often should the Kanbans be evaluated?
- The system requires similar process steps and cycle times
- These considerations may help:
- Are there bottlenecks? Where?
- What is the minimum batch size?

- How can we optimise setup times?
- How often does the ConWIP system have to be evaluated?
- The appliance of the theory of constraints is the simplified version of a ConWIP system
- The introduction process is effected in 3 steps:
- Identify the bottleneck
- Develop a buffer of inventory in front of the drum to keep it turning
- Release orders as they are consumed by the drum signal – the rope
- In separate step the bottleneck has to be eliminated

Types of Kanban

Production Kanban: allows manufacturing to replenish removed material

Withdrawal Kanban: asks for transportation of materials/product

Signal Kanban: when a buffer stock is required between work stations, because the upstream process step produces batches bigger than one Kanban (such as long setup times but very short cycle times). These triangle Kanbans work as visual signals to production/ delivery

Kanban Space – a designated area that functions similar to a Kanban shelf. When the space is empty, the quantity in the shelf has reached a minimum and production is triggered. Kanbans are typically identified with the below information at a minimum:
Name and description of the product
Storage location for the final product
A Kanban number to identify it from other similar Kanbans

In a Pull system, the information for the production (after order) is only delivered to one process – the "pacemaker". The "pacemaker" starts the production and determines the rate. It is the starting point that enables a continuous flow to the customer / finished goods inventory. Inventories are essentially buffers which facilitate varying customer demand and help to absorb any demand variations to the supply chain. Inventory size depends on:
- Demand requirements
- Variation in supply
- Takt time – cycle time ratio
- Number of different products (if applicable)

Inventories are needed where replenishment time is bigger than lead time.
Calculating inventory size

Inventory size = Cycle stock X Buffer stock X Safety stock

Cycle stock: Average demand per period x Average lead   time per period.
Buffer stock: 2 standard deviations of the average demand per period.
Safety stock: Average scrap per period + 2 standard deviations from the average scrap.
Determining the number of Kanbans required for process or manufacturing line is done using a simple calculation. As a rule of thumb, the optimum number of Kanbans should be between 3-7 in number with a minimum of 2. The below equation can be used to determine the number of Kanbans required:

$$\text{No. of Kanban's} = \frac{\text{Inventory}}{\text{Kanban Size}}$$

$$\text{No. of Kanban's} = \frac{4000 \text{ Kg}}{1000 \text{Kg}} = 4 \text{ Kanban's}$$

6.4 Six Sigma

Introduction

Six Sigma is a business management strategy originally developed by Motorola USA in 1986. At its core, Six Sigma aims to improve the quality of process outputs by identifying and removing the causes of defects (errors) and minimising variability in manufacturing and other business processes. Product outputs include attributes such as diameter measurements and thicknesses of product features. Six Sigma is made up of a set of quality management methods and technical methodologies including statistical methods that create a pathway or roadmap to allow Six Sigma projects be implemented. Projects are typically divided into two categories (1) black belt projects and (2) green belt projects. A Six Sigma yellow belt curriculum provides an introduction to Six Sigma and the principals involved. Green belt is the next level of Six Sigma skills and competency. Green belt certification requires several days of classroom based training and a project on a specific area that needs improvement. Black belt level is even more in depth and requires substantial effort over several months. The black belt project will deliver far more cost savings than a green belt project. Every Six Sigma project will not only have a defined sequence of steps but specific and measurable targets such as cost reduction, profit increase and so on. The Six Sigma methodology provides the techniques and tools to improve the capability and reduce defects in any process (to the level of 99.9997% yield). At a yield of 99.9997% as little as 3.4 defects per million are produced by the process.

Six Sigma aka 6σ stands for six standard deviations from mean. It was started in Motorola, in its manufacturing division, where millions of parts are made using the same process repeatedly. The "symbol" sigma (σ) is the Greek letter used to represent standard deviation in statistics. Another reoccurring term used when discussing Six Sigma is "defect(s)". A defect is simply defined as anything that does NOT meet a pre-defined specification.

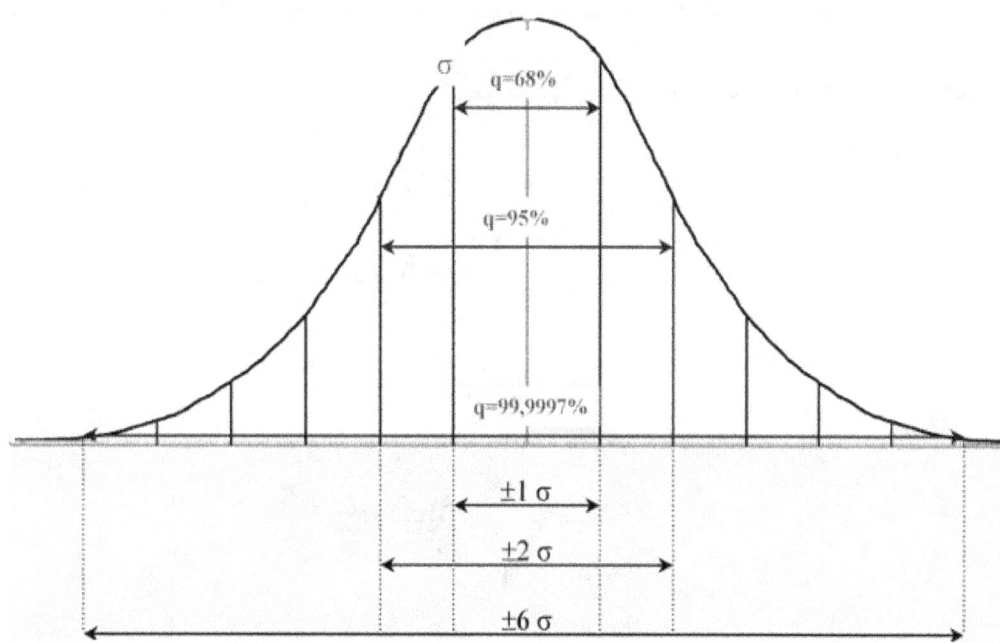

Figure: Normal distribution showing Six Sigma

Six Sigma (6σ)

At this point it should be clear that Six Sigma refers to a process having six standard deviations between the average of the process centre and the nearest specification. The goal of most Six Sigma projects is to eliminate costs. Costs can be incurred by wastage which programmes such as lean manufacturing aim to address. However, with 6σ, the aim is to reduce defects which in turn can provide financial savings and maximise profits. Before defects in a process can be eliminated, it is first important to be able to accurately measure the number of defects. Once the defect number or defect level is known, the aim is to get to "zero defects". Remember, a 6σ process has a defect rate of just 3.4 defects per million.

Process Sigma	Defects per Million	Yield
1	691.462	30.85%
2	308.538	69.15%
3	66.807	93.32%
4	6.210	99.38%
5	233	99.977%
6	3,4	99.99966%

Figure: Table of Sigma levels showing the corresponding defects/million and yield.

Key Elements of Six Sigma

Customer: The customer requirements or the Voice of Customer (VOC) is a central theme of Six Sigma. The customer is best positioned to attach the right value to the product. The value placed on a product by the manufacturer does not reflect the market; therefore, the customer must be in the driving seat. The customer also defines the quality of a product. As they are the user, they can critique the pros and cons of a product's features, functionality and look. Having a quality product at the right price best serves the customers' needs and increases the likelihood of customer satisfaction. Outside-in thinking helps a company understand the customers' perspectives.

Process: The manufacturing process, equipment, facilities and materials must be fit for purpose in order to deliver quality product in a consistent manner. A large part of satisfying customers is the ability to deliver the products consistently on time when they need them.

Employee: a company is only as good as its employees. This may be a platitude but there is a lot of wisdom and truth in it. The employee commitment to quality and to ensuring the customer gets a quality product will determine the success of any Six Sigma project more so than fancy charts or diagrams.

When the application of any system or methodology is new to a company or team, getting off on the right foot is key. Therefore, it is useful to understand what Six Sigma is not! First of all, it should not be viewed as "more work". This fails to appreciate the benefits available following the application of Six Sigma. It is also not an immediate fix for every problem.

Why Six Sigma?

Six Sigma is not only a theoretical programme or simply a statistical approach to measure variance. It is a continual process and culture of excellence which translates to real results for the customer. Yes, reducing variation is a key part of this as consistency will lead to fewer defects, but delivering quality to the customer is paramount. Fewer defects provide a measure of confidence in the manufacturing process. If defects are high this will impact upon the customer's experience and will drive up the costs of manufacturing. With this in mind there is a cost to quality. Quality needs systems that work, inspection that is effective and so on. However, investing the right resources will avoid more costly approaches to fixing issues after they occur.

The Cost of Quality (COQ) can be subdivided into different categories. They include

- Internal costs
- External costs
- Prevention

Internal costs refer to any rework, machine downtime, material losses, overtime and so on. External costs are associated with returns, liability claims, complaints, reputation loss etc. Prevention of quality defects can be mitigated against by firstly understanding the customer requirements and reviewing any market research on the product. More practically, process validation provides confidence in the consistency of a process and produces documented evidence of the same.

6.5 Polymer Processing

Introduction

Polymer can be divided into two categories (1)Thermoplastics and (2)Thermosetting plastics. These two types of polymers have different properties. Therefore, the type of plastic selected needs to be done based on the use and the type of product.

Both thermoplastics and thermosetting materials are man-made organic chemical compounds that soften or liquefy when heated. Thermoplastics can be reheated and re-shaped over and over again, while thermosetting plastics can only be heated and shaped once.

Injection moulding is a mass production process. It is generally not applicable unless 10,000 or more identical parts are to be produced. The reason for this limitation is the necessity of constructing a unique mould for each part. Production must be large enough so that the mould cost can be absorbed over the quantity manufactured. Even for smaller parts, moulds can be costly, in the region of several thousands of pounds. For larger intricate parts, they can cost tens of thousands or even hundreds of thousands of pounds.

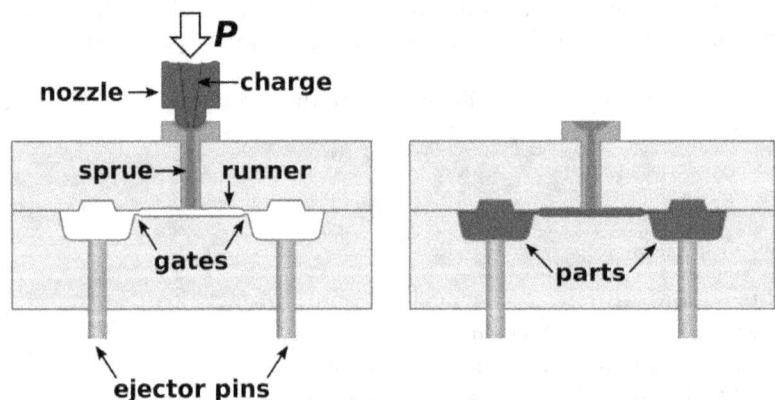

Non-Newtonian Fluids

In a fluid, when external stress is exerted, deformation occurs and continues indefinitely until the stress is removed. The Newtonian fluid is the simplest example, where the rate of deformation is directly proportional to the stress applied to the fluid. However, many fluids exhibit a non-linear response to stress and are called non-Newtonian fluids. Such fluids fall halfway between being a solid (where the stress depends on the instantaneous deformation) and Newtonian fluids (where the stress depends on the instantaneous rate of change in time of the deformation). For such 'soft solids' or 'elastic liquids', the stress depends nonlinearly on the history of the deformation. It is helpful to consider the definition of Newtonian behaviour since a liquid showing any deviation from the definition is considered non-Newtonian. Newtonian behaviour in experiments conducted at constant temperature and pressure has the following characteristics:

- The only stress generated in simple shear flow is the shear stress.
- The shear velocity does not vary with shear rate.

- The viscosity is constant with respect to the time of shearing and the stress in the liquid falls to zero immediately after the shearing has stopped. In any subsequent shearing, however long the period of resting between measurements, the viscosity is as previously measured.

- The viscosities measured in different types of deformation are always in simple proportion to one another, so, for example, the viscosity measured in a uniaxial extensional flow is always three times the value measured in simple shear flow.

Fluids containing long-chain polymer molecules of rigid particles have very different flow properties from ordinary (Newtonian) fluids. The stress in these fluids depends upon the configuration of the suspended particles, which in turn depends upon the flow history experienced by the particles. Polymer molecules behave like springs and become 'stretched' by the flow, giving rise to the elastic behaviour of polymeric fluids (for example the 'stringiness' of cheese on a pizza or 'silly putty'). These elastic stresses give rise to strange flow behaviour not seen in Newtonian fluids.

Non-Newtonian fluids have a wide variety of industrial applications. A huge variety of consumer goods are now made from injection-moulded plastics that often contain high concentrations of glass or carbon fibres. Many modern paints and lubricants contain polymer additives that are added to enhance their flow properties or the quality of the finished product.

The Injection Moulding Process

These materials are formed to specific shapes by injecting them when heated into a mould from which they take their final shape as they cool and solidify. The plastics are normally received by the moulder in granular form. They are placed in a hopper of an injection moulding machine from where they pass into a heated cylinder. As they heat in the cylinder they melt or plasticise.

A typical melting temperature is 180°C, although this varies with different materials and different moulding conditions. The mould, usually of steel, is clamped in the machine and is water-cooled. A plunger forces plasticised material from the cylinder into the mould. There it cools and solidifies. The mould is opened and the moulded part with its attached runners is removed. The process, with the usual exception of part removal, is automatic. It requires about 45 s/cycle, more or less, with most of that time being devoted to the cooling of the material in the mould. Very high pressures, on the order of 70,000 kPa or more are required during injection.

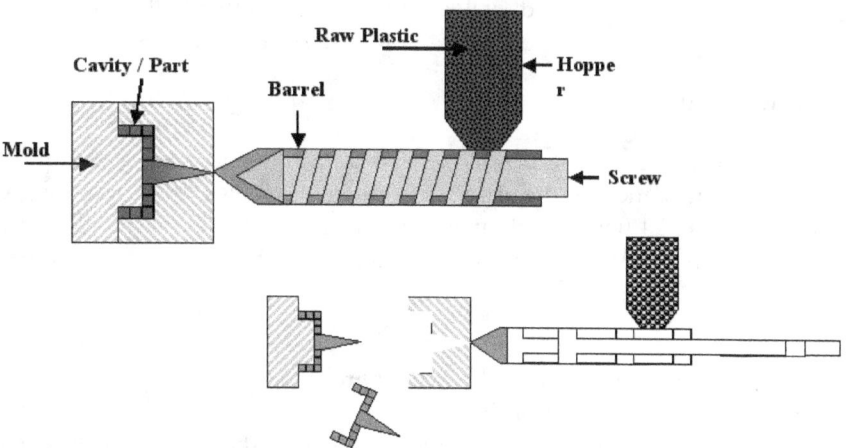

Characteristics of Injection-Moulded Parts

Injection moulding is particularly advantageous when intricate parts must be produced in large quantities. Although there are limitations, as discussed below, generally the more irregular and intricate the part, the more likely injection moulding will be economical. Injection-moulded parts are generally thin-walled. However, heavy sections and variable wall thickness are possible. Because thermoplastics are generally less strong than metals, they are more apt to be found in less highly stressed applications. Housings and covers are common uses rather than frames and connecting rods. However, thermoplastic materials are gradually being developed with better and better strength characteristics and are increasingly finding themselves used for moving parts and in more structural applications. The 'engineering plastics' nylon, polycarbonate, acetal, phenylene, oxide, polysulfone, thermoplastic polyesters and others, particularly when reinforced with glass or other fibres, are functionally competitive with zinc, aluminium and even steel. All thermoplastics exhibit shrinkage upon cooling and solidification. In addition to effecting a reduction in most dimensions, shrinkage of plastic material causes various irregularities and warpage in the moulded part. The most common such defect is the sink mark or surface depression, opposite heavy sections.

Suitable Materials

A large number of suitable thermoplastics are available to the injection moulder. Some of the more commonly used are polyethylene, polypropylene, polystyrene, polyvinyl, chloride (PVC), nylon, acrylonitrile butadiene styrene (ABS) and acrylic. Because of the importance of injection moulding to the commercial sale of thermoplastic materials, producers of these materials engineer them to be processable by injection moulding. Physical properties and cost rather than processability are normally the determining factors in the selection of materials for injection-moulded parts. Generally, the high-property engineering plastics are not as easy to mould as the commodity plastics such as polyethylene, polypropylene and polystyrene. In addition, polyvinyl chloride, though low in cost and having very good physical properties, is more difficult to injection-mould than many other materials. Its prime drawback is a narrow temperature range between its melting and degradation points.

Design Recommendations

Ejector Pin Locations

The designer should consider the location of ejector pins since they can impair surface finish. Ejector pins can usually be located on the underside of a part if it has an outside and an underside.

Gate Locations

Gates can be located in a number of locations. Centre gating of round and cylindrical parts and near-centre gating of other large-area parts are desirable for trouble-free mould filling. Fan gates allow easier filling but are harder to disguise after removal and require more material for the gate.

Suggested Wall Thickness

Generally, thinner walls are more feasible with small parts rather than with large ones. The limiting factor in wall thinness is the tendency for the plastic material in thin walls to cool and solidify before the mould is filled. The shorter the material flow, the thinner the wall can be. Walls should also be as uniform in thickness as possible. When changes in wall thickness are unavoidable, the transition should be gradual, not abrupt.

Holes

Holes are feasible in injection-moulded parts but are a complicating factor in mould construction. They also tend to cause flashing at the edge of the hole and to cause 'knit' or 'weld' lines adjacent to it. The minimum spacing between two holes or between a hole and sidewall should be one diameter. Holes should be located three diameters or more from the edge of the part to avoid excessive stresses. A through hole is preferred to a blind hole because the core pin which produces the hole can then be supported at both ends, resulting in a better dimensional location of the hole and avoiding a bent or broken pin. Holes in the bottom of the part are preferable to those in the side since the latter require retractable core pins. Blind holes should not be more than two diameters deep. If the diameter is 1.5 mm or less, one diameter is the minimum practical depth. To increase the depth of a deep blind hole, use steps. This enables a stronger core pin to be employed. Similarly, for through holes, cut-out sections in the part can shorten the length of a small diameter pin. Use overlapping and offset mould cavity projections instead of core pins to produce holes parallel to the die-parting line (perpendicular to the mould movement direction).

Ribs

Reinforcing ribs should be thinner than the wall they are reinforcing to prevent sink marks in the wall. A good rule of thumb is to keep rib width to one half or less of wall thickness. Ribs should be no more than wall thickness high, again to provide the extra reinforcement that would otherwise be provided by a high rib. The ribs should be two or more wall thicknesses apart. Ribs should be perpendicular to the parting line to permit removal of the part from the mould. Sink marks caused by ribs can be disguised or hidden by grooves or surface texture opposite the rib. Ribs should have a generous draft.

Bosses

Bosses are protruding pads that are used to provide mounting surfaces or reinforcement around holes. They should have generous radii and fillets. The rules indicated for ribs apply as well for bosses. Locate bosses in corners, if possible, to aid material flow in filling the mould. If a detached boss is necessary, a connecting rib will aid material flow. Bosses in the upper portion of a die can trap gases and should be avoided if possible. Use a 5° taper for bosses, the same as with ribs. If large bosses are needed, they should be hollow for uniformity of wall thickness.

Undercuts

Undercuts are possible with injection-moulded thermoplastic parts but they may require sliding cores or split moulds. External undercuts can be placed at the parting line or extended to the line to obviate the need for core pulls.

Screw Threads

It is feasible, though a complicating factor, to mould screw threads in thermoplastic parts. Several basic methods can be used. Use a core that is rotated after the moulding cycle has been completed. This unscrews the part and enables it to be removed from the mould. Put the axis of the screw at the parting line of the mould. This avoids the need for a rotating core but necessitates a very good fit between mould halves to avoid flash across the threads. Continuation of threads to the ends of the threaded sections should be avoided because they create featheredges and promote flashing. If threads with strong holding power are needed, use metal inserts. Internal threads can be tapped in almost all thermoplastics and if the thread diameter is small (5mm or less), tapping is usually more economical than moulding. Tapped or moulded threads finer than a 1mm pitch are not practical with thermoplastics.

Inserts

Inserts are useful and practical to provide reinforcement where stresses exceed the strength of the plastic material. Although they are economical, they are not without cost and should be used only when necessary for reinforcement, anchoring or support.

Drafts

It is highly desirable to incorporate some draft or taper in the sidewalls of injection-moulded parts to ease removal of the part from the mould. Draft may not be necessary if ejector pins can be properly placed but it is still wise to make an easily removable part with draft and generous radii.

The following are recommended minimum drafts for some common materials:

Polyethylene   1/4°

Polystyrene   1/2 °

Nylon   0-1/8 °

Acetal   0-1/4°

Corner Radii and Fillets

Sharp corners should be avoided except at the parting line. They interface with the smooth flow of material and create possibilities for turbulence resulting in surface defects. Sharp corners also cause stress concentrations in the part, which are undesirable from a functional standpoint. A desirable minimum under any circumstances is Ø 0.5mm.

Surface Finish

One significant advantage of the injection moulding process is the fact that surface polish or textures can be moulded into the part. No secondary surface-finishing operations (except of course, plating or painting) are necessary. High-gloss finishes are feasible if the mould is highly polished and if moulding conditions are correct. However, dull, matt or textured finishes are preferred to glossy finishes, which tend to accentuate sink marks and other surface imperfections. Painting of most thermoplastics is feasible but is not recommended of the colour can be moulded into the part.

Mould Parting Line

Every injection-moulded part shows the effect of the mould parting line at the junction of the two halves of the mould. The part and the mould should be designed so that the parting occurs in an area where it does not adversely affect the appearance or function of the part. One easy way to do this is to put the parting line at the edge of the part where there is already a sharp corner. However, removal of parting line flashing may destroy the sharpness of the corner.

Tolerances & Dimensional Recommendations

Though surprisingly tight tolerances can be held when moulding thermoplastics parts, dimensions cannot be held with the precision obtainable with close-tolerance machined metal parts. There are several reasons for this. There is shrinkage of materials including variation and unpredictability in the shrinkage itself. Plastics exhibit a high thermal coefficient of expansion. As a result, if the tolerances are extreme, designers should specify the temperature at which the measurements should be taken. Despite automatic control apparatus for pressure, temperature and time settings, there is some variation in these factors from cycle to cycle. These variations result in slight dimensional variations in moulded parts. Plastic parts are usually more flexible than metals.

The Polymer Extrusion Process

Polymer extrusion is a process for moulding polymer materials into sheets, tubes or shapes that have a constant and often irregular cross section. Dry plastic material, normally in the form of pellets or powder, is placed in a hopper, which feeds into a long carefully heated chamber. In the chamber, a rotating screw mixes the plastic to produce a uniform melt and forces it through a die orifice. As the extrudate leaves the die, it is passed through a cooling medium (air or water) by a conveyor or other takeoff mechanism. It solidifies to the cross-sectional shape of the die opening. The extrudate is pulled away from the die faster than it is extruded, thus causing it to extrudate straight as it cools and solidifies. It also permits slight adjustments in size to be made (by varying the drawdown rate). The ration of the die size to the final size of the product is defined as the 'drawdown ratio'. As the molten plastic leaves the die, it must be supported as it passes through the cooling medium.

Tubing or other cross sections with a hollow are made from dies that have a core (or mandrel) to form the hollow. Air is introduced into the core and injected into the hollow of the extrusion as it leaves the die. The air supports the inside of the extrusion and prevents it from collapsing during cooling. Generally, tubing is also run through a sizing die or a cooling mandrel to maintain concentricity.

Secondary operations down the line are often just as important in manufacturing an extruded product as the extrusion itself. The finished product is rarely the same as the original extrusion. Among typical in-line operations are cutting to length, application of films (such as simulated wood grain, foam rubber and protective tape etc.), punching (this allows for special holes, notches or cuts which are not possible to extrude), embossing, forming and assembly.

Characteristics and Applications

Extruded sections can be as small as a thread filament and as large as a 300mm pipe. Almost any shape with a constant cross section may be suitable for extrusion. The length can be as great as desired since material can be fed continually through the extruder. The finished part can be as short as a fraction of a millimetre if the extrusion is sliced into short pieces. Both rigid and flexible thermoplastics can be extruded and these can be either solid or cellular (foam).

Extrusion is often employed for door, window and floor mouldings. It is also used in wear strips for low-friction bearing components, tubing, seals, edge guards, wire harnesses and many other components.

Dual extrusion (two materials being extruded side by side and joined during extrusion) is similar to simple extrusion except that two extruding machines feed the die. This produces a single extrusion composed of two permanently bonded plastics. One of the most common uses of dual extrusion is to bond rigid to non-rigid vinyl. The rigid portion of the extrusion is used to preserve the overall shape of the profile. (It can also provide a strong base for attachment purposes.) The less rigid portion is suitable for sealing or cushioning. Almost any number of rigid and non-rigid members may be combined in a single extrusion. The extrusion of various colours simultaneously can eliminate the need for post-extrusion decorating.

Metal embedment is another common extrusion technique. It allows continuous lengths of solid wire or strips of metal to be embedded (either partially or totally) in the extruded plastic profile. The advantage of metal embedment is that it provides the ultimate in structural integrity while the extrusion has the warmth, colour and chemical resistance of plastics.

Economics

Plastic extrusions are most extrusions are most economically produced when production quantities are fairly large. Although tooling costs are very modest, a certain amount of trial-and-error development is required to achieve final dimensions with some accuracy. This process adds significantly to the cost of the extruded part if quantities are small.

Extrusion production rates are rapid. The size of the extruder equipment typically ranges from about 5 to 200kg/h. Higher rates are feasible in mass production applications. Standard shapes of many materials are available off the shelf from local plastics distributors. Special cross sections from manufacturers require a minimum production run.

Design Recommendations

Although virtually any profile may be extruded with thermoplastics, several design factors must be considered to produce high quality extrusions. The most important consideration is the wall thickness of the profile. A profile with a uniform wall thickness is easiest to produce. Non-uniform wall thicknesses may cause uneven plastic flow through the die and cause different parts of the extrusion to cool at different rates. This can cause warpage toward the heavier portion of the profile. If an uneven wall thickness is unavoidable, it may be necessary to provide additional cooling for the heavier sections. This increases tooling complexity and adds to production costs. In addition, non-uniform wall thicknesses usually require twice the tolerance limits of a similar uniform product.

Another disadvantage is that sink marks almost always occur in extrusions on a flat surface opposite a heavy area such as a leg or a rib. Often an unbalanced profile can be slightly redesigned to eliminate a heavy area or to undercut a leg.

Hollows are not as difficult to extrude in thermoplastics as in metals. However, they increase the cost of dies and also increase operating costs by introducing compressed air injection into the hollow. If possible, it is advisable to eliminate or minimise the hollow. Tolerances for hollow profiles cannot be held as closely as for profiles without them. Legs or projections inside a hollow should be eliminated or minimised. They greatly increase dimensional variations since there is no way to hold their shape while they cool. If an interior leg or rib is unavoidable, its projection into the hollow should never exceed the thickness of the wall around the hollow.

A hollow within a hollow should be avoided at all costs because this increases all the problems inherent in hollows. Tolerance control for nested hollows is very poor and production is especially costly. Sharp corners are very difficult to extrude since most thermoplastics bridge across sharp corners of the die and form radii. Sharp inside corners should also be avoided since they can form a notch that can be an easy breaking point for the more rigid plastics. Dies for sharp outside corners exhibit the same problem and in addition, form a concentration point for stresses caused by heating and cooling. Complex hollows increase tooling costs and dimensional variation. Round, square, half-round or rectangular shapes are preferred, since marks tend to occur on surfaces opposite ribs or walls. One way to avoid an appearance problem is to add serrations to the surface that is apt to be affected.

Material Selection

The choice of thermoplastic is very important in determining the extrudability of a particular profile. High-impact styrene is the easiest plastic to extrude. Cellulosics (cellulose acetate butyrate and ethyl cellulose) and acrylics are the next easiest. The most difficult plastic to extrude is nylon. Flexible plastics are not extrudable to as tight tolerances as rigid plastics.

As a rule, rigid materials (such as polystyrene, methyl methacrylate, rigid vinyl and cellulose acetate) are not quenched in cold water during the drawdown and cooling process. Rapid cooling in these plastics causes undesirable stresses and leaves a poor surface appearance. Crystalline plastics (such as polyethylene, nylon, vinylidene chloride and polypropylene) are generally cooled in cold water.

Vinyl, acrylonitrile butadiene styrene and polystyrene are easier to extrude through an unbalanced die than polyethylene and polypropylene. The latter have a low melt strength and leave the die in a more fluid condition (which is harder to control) than the former.

If the two thermoplastics used for dual extrusion are not the same (e.g. not both vinyl), the bond between them will probably not be complete. Radically different materials require undercuts, dovetails or mechanical joints to stay together. Care should also be taken when dual-extruding various plastics, as some are chemically incompatible. Plasticised (flexible) vinyl, for example, is not compatible with polystyrene because of plasticiser migration to the polystyrene.

ABS: Acrylonitrile butadiene styrene is a good rigid plastic that can be extruded very easily. A full range of colours is possible. ABS has an above-average tolerance control and complex profiles are possible. It can be used for slides, housings and handles.

EVA: Ethylene vinyl acetate is flexible, supports a full range of colours and has average tolerances for simple profiles. It is used for low-performance hinges, seals, gaskets and weather-stripping.

HDPE: Polyethylene, low density. It is somewhat flexible, non-toxic and difficult to extrude. It has poor tolerance control with good electrical properties; no known solvent at room temperatures. It is used for tubing, handles, straps, bumpers and edgings.

PVC/uPVC: Polyvinyl chloride (PVC) is flexible and available in a variety of hardnesses. It supports only average complexity of profiles with an average tolerance control. A full range of colours is possible. It is used for window frames, gaskets, seals and trims.

6.6 Quality and Engineering Tools

Plan-Do-Check-Act

PDCA (plan–do–check–act) is a four-step management tool often used in GLP and GMP impacting environments. It introduces a repeatable and structured process-approach to solving problems and helps to drive consistent practices.

Plan

The plan step is used to establish the objectives and desired goals of the proposed changes or modifications. Documenting these goals is important as it will drive all aspects of the next steps in the PDCA process. Plan may also involve defining the problem or issue at hand and the specific tasks or resolutions required to rectify the issue.

Do

Step two-implement the plan and the changes identified in the initial step. The "do" step may require data collection and/or analysis prior to the implementation of changes. Training may also be required. Responsibilities should be clearly defined.

Check

Review results and analysis against the planned and expected results or goals. This "check" may simply not be confined to reviewing initial results, it may require monitoring over an extended period of time.

Act

The act step ensures if any further corrective actions or modifications that are noticed in the check step, the process will require the person to "act" on the findings. However, any proposed changes are better captured by returning to the first step and starting the process, either way, the application of PDCA will drive continuous improvement and issue the problem is fully addressed.

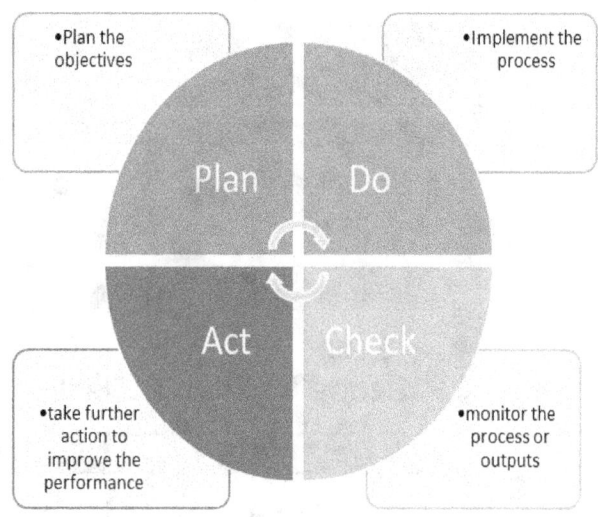

DMAIC Define-Measure-Analyse-Improve-Control

Define, measure, analyse, improve and control (aka DMAIC) is a Six Sigma problem solving methodology that is applied widely throughout the engineering industry. As with many methodologies, it provides a structure and a clear step-by-step approach to problem solving. It can also be used for continuous improvement activities within a factory or organisation. Often shortened to DMAIC, each step serves a purpose and each step produces an output or multiple outputs. The collective output of each step can subsequently be used as the input to the next, which creates the right focus and continuity within a project or improvement program.

Define

The first phase in the DMAIC methodology is focused on project definition. This phase is critical as it provides the basis of any problem solving activity – what is the problem or issue? What is the desired outcome or result? Many six sigma projects form a project charter as part of the definition phase. Other tools used to ensure the correct focus areas are identified from the beginning include capturing the Voice of the Customer (VOC) and creating an SIPOC table. The acronym SIPOC stands for suppliers, inputs, process, outputs, and customers.

Problem Statement: A problem statement is a concise description of the issue or problem that needs to be understood before the steps towards problem solving or identifying solutions begin. A problem statement should (1) state what is the problem (2) who experiences the problem? A problem statement helps focus attention of the wider problem solving team.

The 5 'Ws - Who, What, Where, When and Why is a simple tool that helps identify key information.

Who - Who does the problem affect? - Customers, suppliers, internal customers etc.

What - What is the issue? What impact is the issue causing?

When - When does the problem happen?

Where - Where is the issue or problem happening? In certain locations, at certain process steps, with certain products?

Why - Why does the issue need to be fixed? - What is the impact on the user?

> The **Problem** Statement describes what the issue is

> A **Goal** Statement defines the improvement objective

Problem statement example: "Over the past 10 months (when) 10% of mouldings (where) had cosmetic defects (what). This costs the company 30K per quarter on rework and inspection.

Goal statement example: "Reduce the scrap rate at moulding by 6% by the end of the year"

A common "checklist" for goal statements is to apply the following:

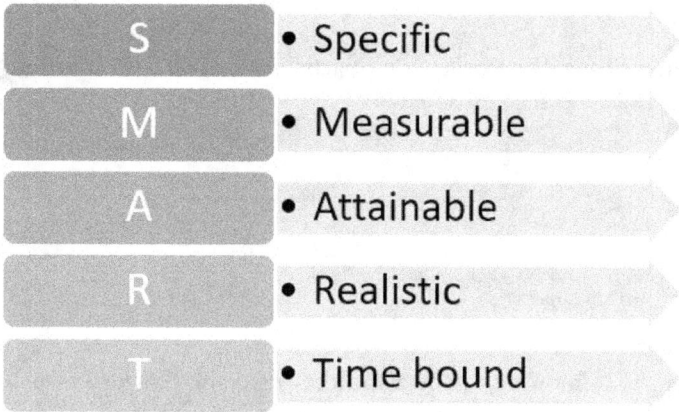

- S • Specific
- M • Measurable
- A • Attainable
- R • Realistic
- T • Time bound

Business Case: A business case is a justification or evidence-based argument that sets out a proposed course of action. The business case should summarise all the information that will influence the decision to initiate a project or a plan of action.

A business case also serves a purpose in linking the project with the higher level strategic priorities of the business by addressing the following:

- How does the project align with the company strategy?
- How will the project drive business goals?
- How will this project impact the customer?
- Why is it a priority (or why is it needed)?
- Why is it important to do now?
- What are the consequences of not doing it now?
- What are the potential financial benefits?
- What are the expected financial benefits?
- What are the limiting factors?

Scope: Scope of a process improvement project or problem solving project is typically described in concise statements in terms of what is in scope and what is out of scope. For example, only certain products or processes may be in scope. This narrows the focus and remit of a project.

Project Plan: A project is any planned activity that has a clear start and a clear end goal.

Metrics: Having some level of initial data is powerful in quantifying the scale of the problem or the potential for improvement.

Data Collection: The scope of data collection should be identified upfront before a project commences. The type of data and the source of data may impact the decisions and impact on any improvements.

Roles and Responsibilities: DMAIC teams can be made up of a range of expertise from engineering, management, finance and so on. The roles of each team member should be clearly defined and documented. For example, who has an approval responsibility and who is the decision maker?

Resources: Depending on the project, resources can include equipment, materials, funding, consulting and so on.

What Is IPO?

A process is a series of steps or a collection of activities that take one or more inputs and transform them into outputs that are of value to the customer.

I-INPUTS

P-PROCESS

O-OUTPUTS

An IPO diagram, also known as general process diagram provides visual representation of a process by defining a process and demonstrating the relationship between input and output elements together with the whole chain from supply to customer. The input and output variables are known as "factors (x)" and "responses (y)". An IPO diagram is a simple tool to define a process and focus on its key variables. With regard to DMAIC, an IPO also creates a problem specification or problem overview.

Input	Processing	Output
Temperature	Moulding	External dimensions
Material source	Hand finishing	(a), (b), (c)
Operator experience	Inspection	

SIPOC: SIPOC is a high-level process diagram of the whole supply chain, taking customer requirements into consideration and meeting those requirements. SIPOC is often presented at the outset of process improvement efforts of problem solving during the "define" phase of the DMAIC process. SIPOC maps or tables are simple tools that facilitate the documentation of any business process in a visual and concise way. They also help:

- Identify inputs and outputs of the process.
- Provide a high-level overview of the process.
- In defining a new process.

Supplier: The person or company that provides the input to the process, e.g. raw materials, components, labour, machinery, information etc.

Input: The materials, labour, equipment, machinery and information required for the process.

Process: The internal steps necessary to transform the inputs to outputs.

Output: The product or service being delivered to the customer(s).

Customer: The recipient of the product/output.

Voice of Customer (VOC)

The voice of the customer captures the features that the customer values and the intended use of the product or service. However, before the voice of the customer is accounted for, the customer base or segment must be crystal clear. Identifying customers and their needs will get the project moving in the right direction. It should be noted that customers are not always external parties; customers can be internal within a company or organisation.

The voice of the customer can be determined using a number of tools including interviews, customer surveys, user specifications, contracts and so on. Stemming from the VOC analysis, a clear picture of the goal and requirements can be understood. Often these specifics are referred to as critical to quality attributes.

Customers provide lots of information about products and services they use on an ongoing basis through the following:

- Complaints
- Product returns
- Product/service sales preferences
- Service contract cancellations
- Market share changes
- Customer-to-customer referrals
- Technical issues

Interviews: Interviews are a powerful way to obtain specific customer requirements and point of view with regard to a product or service and what they expect from them. Interviews are most powerful in the define stage of a project; however, they can also be used later on in the process.

Focus groups:

Purpose - Organise information from the collective point of view of a group of customers that represent a segment.

Some uses of focus groups include:

- They identify and define customer needs.
- To gain insights into the prioritisation of needs.
- To test concepts and get feedback.

▪ Sometimes as a next step after customer interviews or a preliminary step in a survey process.

Measure

After the define phase, the next phase is measure. It is important to measure the current state of a process, or the current issue or "state of error". Measurements will inform project teams with the facts while also creating a baseline of the process or problem. This baseline may prove critical after any changes have been implemented. Comparing the state of the problem before and after can be very helpful when illustrating data.

The measure phase cannot be skipped or delayed. As the mantra goes, if you cannot measure it, you cannot change it." Or more accurately, if you cannot measure it you will not be able to demonstrate that the changes or controls have benefited the problem or goal statement.

Most measurement will involve the collection of data. Therefore, some common tools will be used in order to capture and analyse data. Some simple tools include:

Check sheets
Scatter diagram
Cause and effect diagram
Pareto diagram
FMEA
Gauge R & R
Control Charts
Process Capability Analysis
Many measurement tools are applied through IT enterprise software packages, some of which include:

SPC
MS Excel
Statistica
Minitab

Data Collection

As part of the measure phase, it is typically necessary to develop a data collect plan. Again, the 5Ws methodology can answer a lot of the questions when creating a data collection plan.

Who: are the people or groups of people? What departments or expertise?

What: are the processes, machines and equipment that need to be assessed? What products, raw materials or services?

Where: is the data collected, is it a physical location at which the defect is manufactured?

When: does data need to be collected, at the start of a shift, at the end of a shift, weekends, when defects occur?

Which: information, just specifics or certain settings or parameters?

How much: how many data are enough?

Considerations for Sampling Plans

The amount of data, type of data and format of data collected should be driven by a sampling plan. This ties the data to a statistical rationale that provides a level of confidence in the calculations.

The sampling plan should address:

- The type of data (continuous or attribute)
- The type of statistical test (Cpk, Ppk, t-test etc.)
- The variability of the data (σ)
- A confidence level

Sampling

What: Specify the raw material data, component data or product data being collected that will be used or evaluated for the purpose of collecting data.

Where: Specify the sampling exact location (e.g. end of line, at station 2, after inspection and so on).

When: Specify the time or frequency that the raw material, component or product will be sampled at.

Sample Conditions: Samples may need to be packed or handled in a particular manner.

Recording: A template or data collection form should be available for recording the sample details. It is important to review any forms in advance of the sampling to ensure they are fit for purpose. Completing a simple dry run may highlight errors in the form or identify formatting improvements e.g. does the form allow all the information to be recorded?

Process Capability

Process capability and process performance are statistical tools used in engineering as a way of measuring the stability and consistency of manufacturing processes.

Cpk and Ppk are used during routine production, during verification and validation builds and as a product acceptance method. This short publication introduces the topic of process capability and provides clear definitions of Cpk and Ppk and how they differ. It also gives an overview of how to derive a suitable sampling plan for single inspection attribute data and double Sided specification for variable data.

What Are the Different Types of Data?

Understanding the different categories of data is fundamental to developing a suitable sampling plan. Data can be classified as variable or attribute, so what's the difference? Variable data is data that is measured e.g. measured with a ruler, a temperature probe, a conductivity meter and so on. Another way of putting it is that variable data represents a series of values, 1.2mm, 2.2mm, 3.1mm etc. Variable data is also known as continuous data. Attribute data is data which is pass or fail, yes/no or go/no-go type data. Attribute data is created from visual and cosmetic inspections. (e.g. are there marks or blemishes on a surface, is there mould flash?).

Variable data is more telling than attribute data. For example, take a diameter of a component that should be 30mm; a Vernier calipers is used to measure each component, the raw data can be trended and monitored. However, if the measurement was classified as pass or fail (to within a tolerance), all we would have is the quantity of "passing" components and the quantity of "failing" components. It does not tell us about the degree of variation. Generally speaking, more attribute data points are required than variable data points in order to be confident of making a decision. If the data collected is variable data but it is treated as attribute data (e.g. classifying each result as a pass or fail) any sample plan used must be a plan based on attribute data, not variable data.

Attribute data can be further categorised into the following.

Binary data: where there are only two outcomes or result types. For example,

    (1) Go – No go
    (2) Pass – Fail
    (3) Yes – No

Nominal data: can be categorised into groups (two or more)

    (1) Gender

    (2) Occupation

Ordinal data: sorting information where the distance between the data is unknown.

    (1) 1st, 2nd, and 3rd

    (2) A, B, C, D

What Is Process Variation?

Variation in a process output may be due to random causes inherent in the equipment. Even the most advanced equipment with a lot of automation will have limits to its accuracy and consistency.

Materials are also a source of variation; materials may come from different batches, different suppliers and may be made on different days on a range of machines. All of these factors can introduce variability in the material. Even if the materials are within specification and compliant to a certificate of analysis, one batch may be at the lower end of the specification and the next batch may be at the higher end of the specification.

Variation to some degree is unavoidable in most manufacturing processes and the level or variation should be consistent. Therefore, this is often referred to as variation within normal operating conditions or anticipated variation. However, if variation is excessive it may be a result of incorrect tooling, incorrect setup, tool wear, operator error or material deficiencies. Any excessive variation is unacceptable and should be eliminated. The two types of variation are defined and described below. Software packages are typically used to identify common cause versus a special cause variation, such as run charts and control charts.

Type	Definitions	Typical Characteristics
Common Cause	Cannot be removed Influenced by several sources e.g. material, man, method etc. (6M)	Always present Expected Normal Random
Special Cause	Can be removed Influenced by several sources e.g. material, man, method etc. (6M)	Not always present Unexpected Not normal Not random

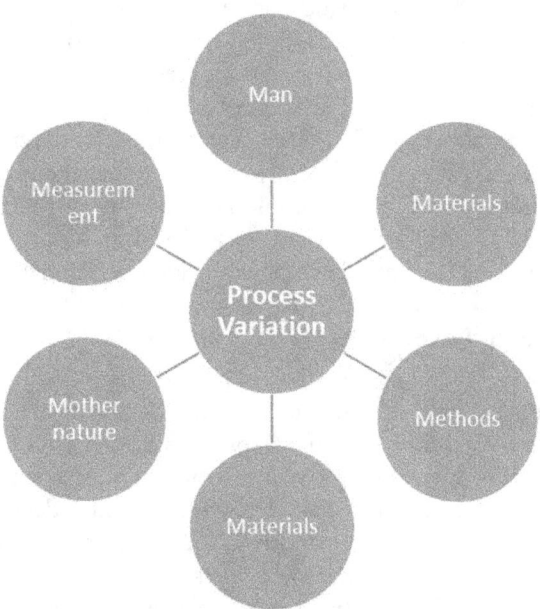

Figure: Often referred to as 6M, the above illustration shows the various sources of process variation.

Acceptable Quality Levels Explained

AQL is defined as the process performance level that the sampling plan will accept 95% of the time. This means processes with a process performance level at or better than the AQL are accepted at least 95% of the time and rejected at most 5%. It describes the risk (5%) associated with rejecting a good process.

Rejectable Quality Levels Explained

The RQL of a sampling plan is the process performance level routinely rejected by the sampling plan. RQL0.10 is defined as the process performance level that the sampling plan will reject 90% of the time. This means processes with a process performance level at or worse than the RQL are rejected at least 90% of the time and accepted at most 10% of the time. An RQL0.05 is defined as the process performance level that the sampling plan will reject 95% of the time. It describes the risk associated with releasing/accepting a bad process. The consumer would like the sampling plan to have a high probability of rejecting a validation with a process performance level greater than, or equal to, the RQL.

Note: Lot tolerance percent defective and rejectable quality level (RQL) can be used interchangeably.

Process Capability and Performance Indices Explained

Firstly, process capability and performance indices ONLY apply to variable data - not attribute data. The terms process capability and process performance refer to the ability of a process to meet specification limits and how consistently measurements fall within specification limits. Pp (Process Performance) and Cp (Process Capability) assess the stability of a process - the amount of variation in the output. Adding the letter "k" to Pp (Process Performance) and Cp (Process Capability) means that the value is now an index - **Ppk** and **Cpk**. With this addition, these terms now represent both the degree of variability and the degree that the output is centered between lower and upper specification limits.

What's the Difference between Capability (Cp/Cpk) and Performance (Pp/Ppk)

Cp and Cpk are both calculated using a sample standard deviation and represent a potential that could be achieved if normal sources of variation are eliminated. Pp and Ppk are both calculated using the standard deviation of the entire population and represent a long term performance. Pp and Ppk are typically measured over a number of batches and represent both normal and special causes of variation. Cp and Cpk are useful when looking at batches in isolation. Pp and Ppk are more beneficial when examining multiple batches. Pp and Ppk values are used to describe the Process performance for process performance qualifications as these values represent the process performance expected over the long term. Cp and Cpk are more commonly used during process optimisation studies as these represent the potential capability that could be achieved if the process was made stable by reducing special causes of variation. They may be applied during operational qualification.

Process Performance Level

The process performance level is a measure of the effectiveness of the process to produce conforming product on a consistent basis. It may be expressed in a number of different ways e.g. percent non-conforming, process capability index, process performance index and nonconformities per quantity (parts per million). Verification/validation studies demonstrate, with a degree of confidence that the level of non-conforming product delivered by a process is at, or below, a specified process performance level. If the study passes, a confidence statement can be made such as: "The data demonstrates, with 95% confidence that the defect rate is below the specified process performance level of 1.18 Ppk"

Run Charts

Run chars are useful in showing a graphical timeline of measurements over a period of time. Run charts are used to develop an understanding of the variation in a process and whether it is stabilising or changing.

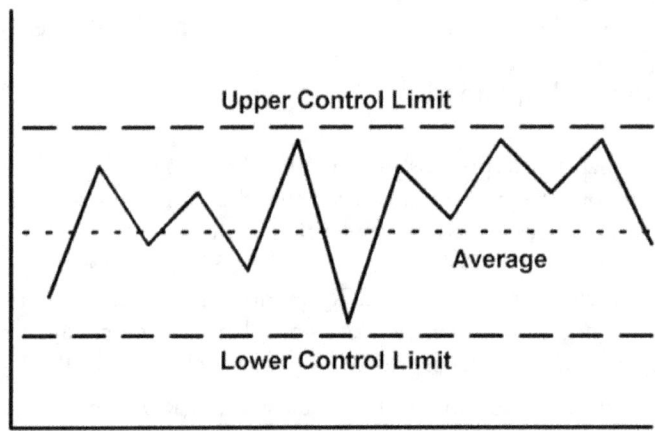

Figure: Layout of a run chart

Although run charts are useful, they have limitations and must be used with caution and experience. Viewing each variation as significant may put you on the wrong track or divert the focus from the real issues or special causes. The use of control charts will provide more statistically relevant results.

Analyse

During the analyse step the team should be focused on identifying the key root causes. There may be more than one cause and each one may have a varying impact on the process. Thus the analyse stage should be given adequate time to ensure all factors are considered.

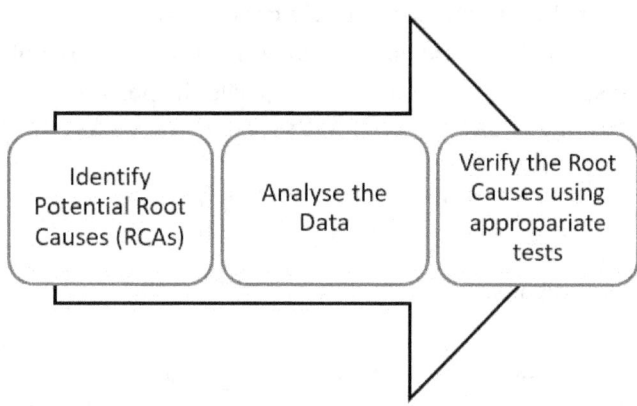

The Process Approach

The following methods are useful in analysing a process or system to identify sources of error. They are all based on a process approach.

- Process map
- Cause and effect diagram
- 5 Ws
- FMEA
- Cause and effect matrix

The Data Approach

The data approach provides an alternative to the above process approach. Examining data can help realise trends, issues and subsequent causes. Some data approaches include:

- Pareto
- Boxplot
- Histogram
- Run chart
- Scatter plot
- Control chart

Root Cause Analysis

Root Cause Analysis (RCA) is a systematic approach for effectively identifying the causes of a process failure or defect. The absence of an RCA can lead to:

- the current situation becoming worse
- a waste of resources
- the problem may be re-occurring

Variation Review

Variation is a result of (1) a special cause or (2) a common cause.

- **Special Cause**: something different happening at a certain time or place.
- **Common Cause**: always present to some degree in the process.

Analysis Tools

- Process map
- FMEA
- Cause and effect diagram
- 5 Ws
- Cause and effect matrix
- Control/impact matrix
- Nature of work

General Points for FMEA

- Use brainstorming and/or data analysis methods to identify key failure modes. It is useful to have data collected up front before you start to sit down and draft the FMEA.
- Document all the critical processes or steps in order to provide a reference and snapshot of the system.
- Identify any preventive steps to reduce likelihood of a failure.
- Create a recovery plan or response in case of failure.
- Use the FMEA to help prioritise future improvement projects or opportunities.

Cause and Effect

Cause and effect is a powerful visual tool used during improvement projects to brainstorm and organise possible causes for a specific problem issue or effect.

Points to note:

- Summarise all potential high level causes
- Provide a visual display of potential causes

Where "X" represents potential causes and "Y" represents the effect(s)

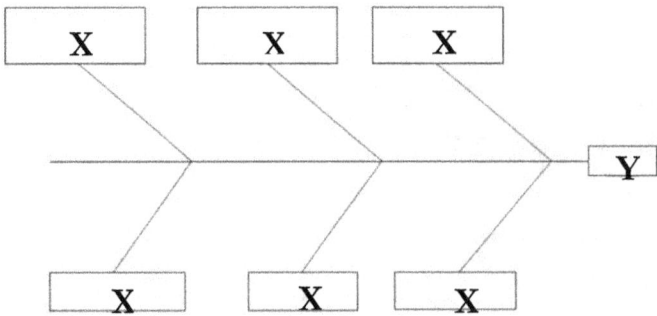

The Cause Chain

- Start with the event
- Work back to the direct cause
- Seek out the contributing cause(s)
- Continue the search down the chain to the root cause

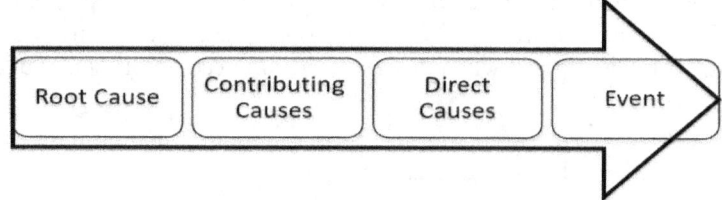

Improve

With the weight of the previous steps (define, measure, analyse), the project team must distil a practical, workable and lasting improvement solution.

- Develop potential solutions
- Evaluate, select, and optimise best solutions
- Develop the next state of value stream map(s)
- Implement proof of principle of the pilot study

Tools of Improvement
- Replenishment pull/Kanban
- Process flow improvement
- Process balancing
- Batch Sizing
- Design of experiments (DOE)
- Solution selection matrix

Control

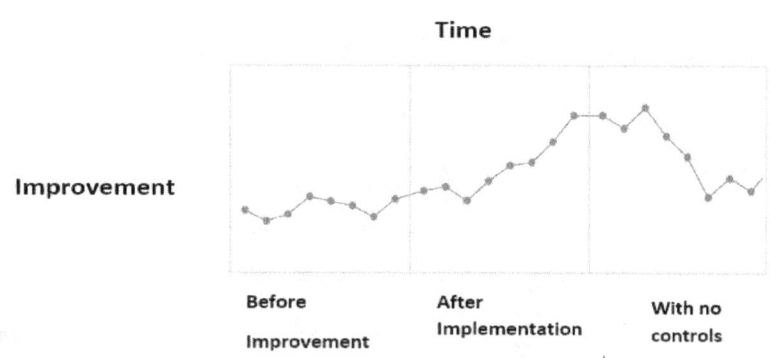

Why Control?

After improvements have been identified and implanted, effective controls must be applied and sustained into the future. As with many systems in engineering, if they are not maintained they can degrade over time. So too with controls, if they are not implemented and maintained, they can simply fall away or become less effective. Above all, the purpose of any control is to ensure quality and safety of the product, and if the process is in control, the customer is satisfied.

Process Management Chart

A process management chart is a flowchart and matrix, which helps manage a process with regard to (1) documentation, (2) monitoring and (3) the response plan. These 3 areas essentially allow controls to work effectively and provide support when process variation occurs during manufacturing.

Documentation
- Who completes the tasks or steps?
- Do standard operating procedures provide enough information to complete the job?

Monitoring
- Where is the data taken from?
- What CQAs or other metrics are captured with regard to process performance?
- How are measurements taken and recorded?
- When is the data collected?
- Is the data reviewed independently?

Response Plan
- When does a response plan get initiated?
- Who takes action based on the data?
- What courses of action can be taken?
- Is there sufficient information in the response plan or other documents to troubleshoot the problem?

What Does a Process Management Chart Look Like?

Documentation	(1) Deployment Flowchart	A diagram showing the process steps for each function and the transfers between functions
	(2) Key Process Steps	The process steps and how each step is completed should be documented here
Monitoring	(1) Process outputs	The CQA's for each process step. Or any key requirements e.g. visually clean, defect free
	(2) Monitoring Requirements	Is there corporate, regulatory or customer requirements
	(3) Data Collection Plan	For each process step, how is data captured and presented, e.g. scatter diagram, run chart, pareto etc.
Response Plan	(1) Containment	Describe the control and containment procedures that handle defective products
	(2) Procedure for Troubleshooting	If problems occur what troubleshooting steps should be completed, what adjustments are permitted.
	(3) Procedure for System improvement	If changes are needed outside current limits, what data is required to propose a change or modification

Documentation and SOPs

At the control stage of DMAIC, the introduction of new documents or updates to existing documents form part of the control strategy and aid its implementation. Typically, an SOP is required or at least updates to an existing procedure.

What are the typical elements of an SOP?

Purpose
Scope

Roles and responsibilities
References (internal /external)
Definitions
Overview
Materials and equipment
Procedure
Revision history
Approval signatures
Headers and footers

Purpose: The purpose details the aim of the SOP. What is the SOP to be used for? Is it a machine or is it required to describe a documentation activity?

Scope: The scope of the SOP identifies who is required to follow the procedure and where the procedure is to be used. Some SOPs will state what is in scope of the procedure and what is out of scope. For example, a HR department may not need to follow an SOP on the requirements of engineer reports.

Roles and Responsibilities: This should give clear and concise information on the responsibilities of the people using or consulting the SOP. For example, if the SOP is related to quality inspections, it may be the responsibility of all staff to follow the procedure properly.

References: Internal documents such as other SOPs or external documents such as industry standards or guidance documents should be listed in this section.

Definitions: This section should list any company terms or words that may not be familiar to a new employee. Remember an SOP should be clear enough so an inexperienced person can understand.

Materials and Equipment: Provide a list of all of the materials and equipment needed to complete the procedure.

Procedure: The procedure section should include all the information required in order to perform the task or process. The information should be presented in clear steps. The use of action words at the start of a sentence can be beneficial to the user of the SOPs (e.g. push, place, turn on, check). Use number formatting to help identify each individual step.

Revision History: The version number (aka revision) of the SOP, the effective date and description of the changes, along with the author/date of changes should be recorded within each SOP. Below are some examples of revision history. It is recommended that the format of the revision history should be controlled via a company template.

Writing Styles

Remember SOPs are documents that are used by personnel to execute specific tasks or actions. Therefore, they should be written in a concise easy-to-understand and step-by-step format. The author of the SOP should avoid convoluted explanations or complex words unless they are essential to completion of the task. Short and simple is the most effective way of writing an SOP. Many companies will have a procedure that sets out the requirements of SOPs. This should provide the author with guidance on formatting, font type, size and other information. The best way to achieve consistency is by having an approved template. This means that maintaining the same styles and formatting will be easier.

Standard Work

Standard work or standardised work is a particular type of work instruction. It is also a "lean" (see definitions and acronyms) tool as it not only creates a baseline but aims to create a balanced work flow with optimum product output.

Elements of Standard Work

- Each major step should be subdivided into key points using short concise sentences and action verbs (e.g. push the button, turn the lever, record the temperature).
- The time required to complete each task is documented.
- Reasons why key points need to be completed are highlighted to the operator/user.
- Quality actions are highlighted.
- Pictures may be used to help describe each step.
- Steps to Implementing a Standard Work Process
- Understand what constitutes best practice must be consistent and repeatable, while meeting quality requirements.
- Document the activities by identifying major steps and describing each major step using key points. Use pictures to help identify buttons, screens and options.
- Approval of standard work procedures is necessary, followed by training for each procedure. Controlled copies of standard work should be printed and made available at each relevant work station.

Review of Variation

- All repetitive activities of a process have a certain amount of fluctuation
- Input, process and output measures will fluctuate
- This fluctuation is called variation
- Variation is the voice of the process

Run Chart

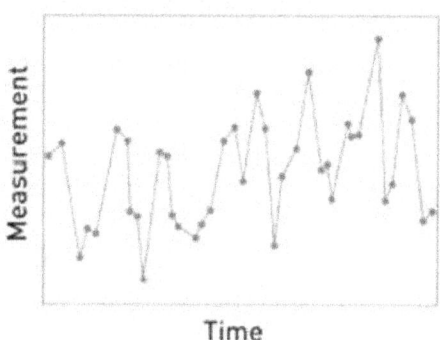

Histogram

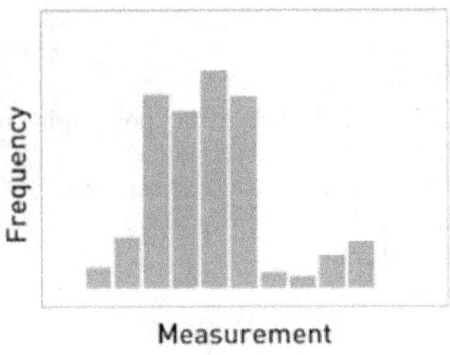

The 6 Ms

When variation occurs, it may not be evident as to what is the source or sources of the variation. It is best to approach such situations in a methodical manner.

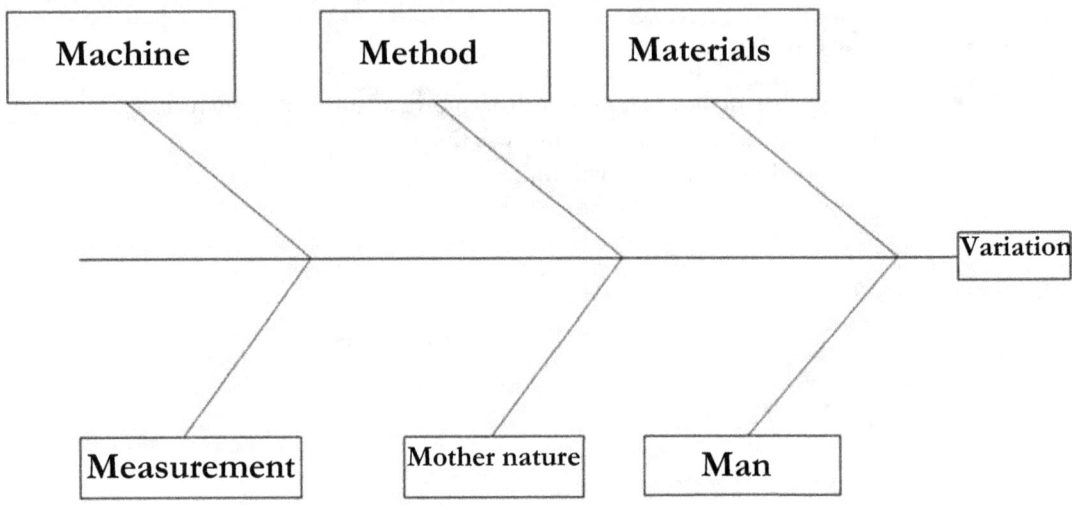

The 5Ps

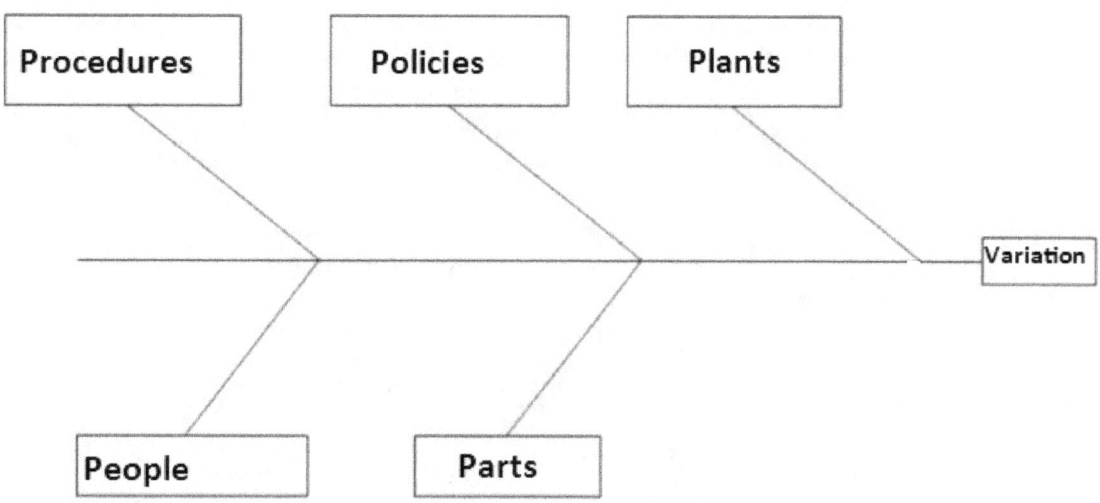

Whether using the 6Ms or the 5Ps to identify sources of variation, the type of variation must be determined. There are two types of variation sources, (1) common cause and (2) special cause. Common cause variation can be defined as a cause where there is no way to remove or eliminate the variation. This type of variation is expected to some degree and is normal. Special cause variation is when the source of variation can be removed. This type of variation is not always present and can be a result of changes or instability in any of the 6Ms (method changes, machine issues etc.) Depending on whether we are dealing with special/common cause variation, different tools will be applied for the improvement strategy.

Run Charts

Trends: A trend can be defined as a scenario where 7 or more consecutive data points continuously rise or fall.

Same Values: Where the values are static. When a sequence of data points are the same value. Again the pattern should have a minimum of 7 points.

Shifts: A shift occurs when 8 points are on one side of the median line, indicating a shift in the "centeredness" of a process.

Control Charts

Control charts are simply a more developed and useful way of presenting data that appears in a run chart. The control chart illustrates the average trend of the data being captured along with upper control limits and lower control limits of a given process.

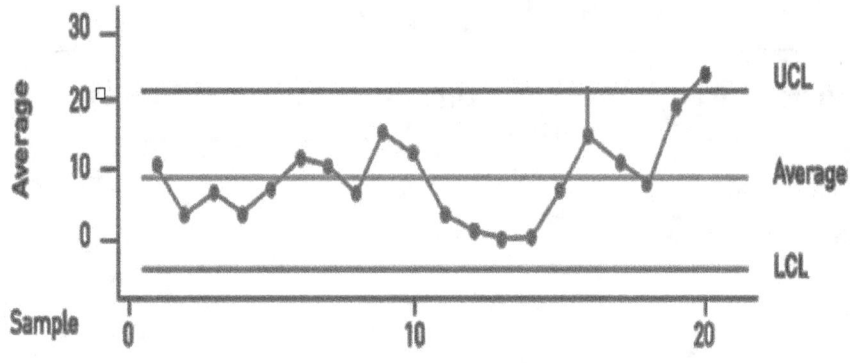

Figure: Control Chart

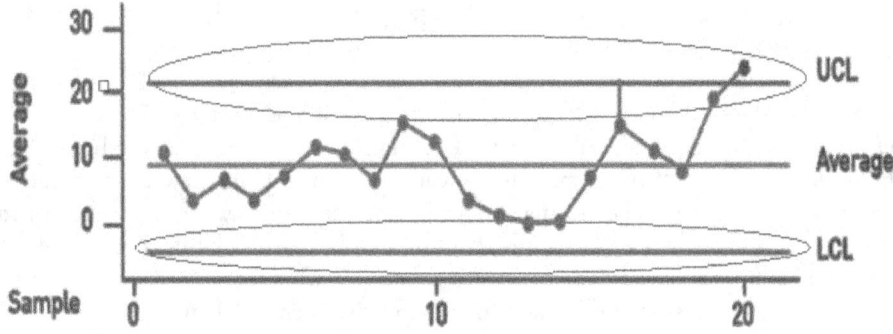

Figure: Shown in red are the UCL and LCL, the position of these limit lines are based on the upper and lower specifications of a particular process output e.g. surface roughness, a diameter, a dimension.

Control Limits and Customer Specification Limits

Control limits help determine if the process is in control and producing product within the desired specification. If there is a change made to the process, any adverse impact can be identified on the control charts. Likewise, with any desired change the improvement or shift in control chart information can be used to substantiate the impact of changes. Customer specification limits may also be applied when implementing process control systems and control charts. Customer specification limits may be used when the customer has additional requirements to a standard product. The limits themselves can be based on historical data or feedback from them. They are also more easily changed when the customer wishes to do so, where in contrast, a control limit may not be as easily switched or modified. While control charts are a powerful tool to monitor the process, they need to be current and up to date if they are to be effective. The application of the correct limits is also an important factor. Recalculation may be required and is permitted if perhaps an error was initially made. If other changes such as changes to the process or to the method of data collection are made, control limits may need to be adjusted.

Guidance on Types of Control Charts

For Continuous Data

- If sample size is 1 – use X and moving charts
- If sample size is 2-9 – use X and range charts
- If sample size is >9 – use X and S-charts
- For continuous data, control limits are calculated using control chart factors and the range bar

For Discrete Data

- If defects and equal sample size – use NP-charts
- If defects and unequal sample size – use P-charts
- If defects and equal sample size – use C-charts
- If defects and unequal sample size – use U-charts
- For discrete data, control limits are calculated using a formula that estimates sigma without the necessity to transform data

Response Plan

A response plan is a documented method for responding to any out-of-control conditions that may occur in a process. A good response plan will help ensure a timely, appropriate response to processing problems on occurrence – decreasing the risk of defects getting to the customer. A response plan for each monitored CQA/CTQ should provide the following information:

- Specific action to be taken
- Timing of action
- Owner of action / person responsible

PROCESS RESPONSE PLAN			
MEASUREMENT	ACTION	TIMING	OWNER
Diameter X1			
Diameter X2			
Diameter X3			
Surface Roughness X4			

After the activation of any response plan, there may be a requirement to revisit risk assessments and update accordingly. Perhaps the initial risk assessment missed the problem? Does it contain an error? Is there a need for extra controls or re-design? Apart from any formal intervention and activation of a response plan, it is wise to complete regular measurement reviews which get shared with key people and management. This ensures that the data being captured is accurate and represents the "real state" of the system.

Process Handover

After successful implementation of project findings, the process is typically "returned" to the process owner. A project may be executed by an R&D team; however, the manufacturing process may be the responsibility of the operations department. Prior to any implementation, there ought to be involvement or representation of the process owner in the actions and analysis of the project team. The below diagram illustrates that the involvement becomes more critical as the project progresses, in particular with the implementation of any solutions.

The process owner is responsible for the performance and monitoring of the new process OR implemented solutions.

Root Cause Analysis (RCA)

A standardised and effective approach to Root Cause Analysis (RCA) during investigations is required in order to deliver robust solutions to problems in a timely manner.

Corrective and Preventive Actions (CAPA) are essential to address quality and compliance issues and risks. A systematic and standardised approach to CAPA ensures:

- Deviations or other undesirable situations and events do not recur and are permanently corrected
- Identified potential failures are prevented from occurring in the first instance
- Reduction in recalls caused by known problems
- Regulatory corrective action commitments are met
- Improvement in right first time performance
- More robust and consistent production processes and increased customer satisfaction
- Reduced risk

Determining the need for RCA

The need to apply a formal RCA process can be influenced by the impact or severity of the defect or issue. If an event or issue is re-occurring, then an RCA should be completed. The more complex a process is, conducting an RCA may provide the structure and approach needed. If many people are involved in the process, they should be utilised in an efficient and organised manner.

RCA Requirements

Some typical minimum requirements for Root Cause Investigations are suggested below:

- Consult with relevant staff and gather documentation to collate facts and data.
- Use problem solving methodologies for best results as they bring a process and structure to the problem. (e.g. DMAIC, 5Ps, 6M)

Ensure the below information is available:

- full details of the investigation performed
- timelines of occurrence
- timelines of the investigation
- accurate flow of events
- remedial actions taken
- impact on product
- impact on patient
- rationale for instances where no CAPA plan is require

A 6-Step RCA Process

STEP 1: Clearly identify the need for an RCA Investigation. This may be mandated by an internal SOP or guidance document. If the source of the defect or failure is not apparent, then an RCA may be required. In addition, if there are conflicting theories on the source, the RCA methodology is useful in attributing the true cause or causes.

STEP 2:Gather all relevant information

This step should initially focus on mapping out the process – a clear map of what should occur. This is done by using established procedures and the knowledge of engineers and relevant staff. After the mapping of the "correct state" is completed, the next step is to map out what actually happed.

Verify the actual events that happened using data, interviews and any other valid sources. Map what actually happened using data available and "re-create" what actually happened.

STEP 3:Complete a Gap Analysis

Identify any differences between what should have happened and what actually happened and record.

STEP 4: Identification of Possible Causes

Each gap identified in step 3 must be treated as a potential cause of the problem Other possible causes must also be considered, these can be derived from a brainstorming process e.g. fishbone. Review and revise the problem statement based on the other identified potential causes. Each gap or identified other possible cause must be assessed for its potential impact on the problem under investigation, and using the available expertise within the investigation team, ranked for importance or impact.

STEP 5: Root Cause Determination

Each of the ranked likely causes should be reviewed to identify the root cause of the problem.

Where it is not possible to identify root cause each probable cause must be treated as if it were a root cause

The identified root causes should be evaluated against the problem statement to confirm that they completely explain the problem. This may require a series of trials or experiments to confirm decision to discount a likely cause. Methodologies such as 6M or Fishbone Diagrams are also effective in identifying root causes.

STEP 6: Determine the Solution -Corrective and Preventative Actions (CAPA)

Define the scope of the CAPA – could the issue occur in other departments, value streams, sites etc. ? If so, define actions for all applicable areas. Define the objective of the CAPA

- Problem statement
- Problem root cause.
- Be SMART (Specific, Measurable, Achievable, Realistic and Time-based)

Define the Corrective and Preventive Action Plan

Develop the CAPA plan and that enables a proper investigation to be completed. The following points should be addressed

- Processes, equipment and facilities
- Procedures and documentation
- Training requirements
- Communication of the issue and CAPA. This is to help prevent the same issue occurring elsewhere.
- interim controls if introduced and checks to ensure controls are removed or updated when required.

Definitions

Adverse Event: Any untoward medical occurrence in a patient or clinical investigation subject, linked in time with the use of a medicinal product or medical device

Contributory Cause: Other causes identified during the investigation that are linked to the root cause. A contributory cause on its own would not lead to the symptom.

Corrective Action: to eliminate the cause of a non-conformance, deviation, defect or other undesirable situation and prevent reoccurrence, typically short term (i.e. issue has already occurred)

Preventative action: Action to eliminate the cause of a potential non-conformity or other potential undesirable situation and prevent occurrence (i.e. can only apply where issue has not yet occurred)

Root Cause: Fundamental origin of an event which if corrected will prevent re- occurrence i.e. the underlying reason for the actual or potential occurrence of non-conformity or other undesirable situation. During the course of the investigation there may be several potential root causes.

GLOSSARY

A

Accelerated Ageing

When the deterioration of a device or product component from natural ageing is accelerated and simulated in the laboratory.

Accuracy

Accuracy or trueness. An expression of the closeness of agreement between the value that is accepted, either as a conventional true value or an accepted reference value and the value obtained. A system with low bias implies good accuracy and vice versa.

Adverse Event

A situation or condition that occurs when a data point, result, or process etc. is outside the expected or predetermined limits or ranges.

Air Exchange Rate per Hour (ACPH)

The rate of air exchange expressed as the number of air changes per hour and calculated by dividing the volume of air delivered in the unit of time by the volume of space.

Active Pharmaceutical Ingredient

Any substance or mixture of substances intended to be used in manufacturing a drug (medicinal) product and that, when used in the production of a drug, becomes an active ingredient of the drug product. Such substances are intended to furnish pharmacological activity or other direct effect in the diagnosis, cure, mitigation, treatment, or prevention of disease, or to affect the structure and function of the body. (ICH Q7A, Annex 18, Part II.)

ANSI

American National Standards Institute

Antimicrobial Resistance

Antimicrobial resistance corresponds to the emergence and spread of microbes that are resistant to cheap and effective first-choice, or "first-line" antimicrobial drugs.

Application

A term most often used in relation to software validation and computerised systems. It is any software installed on a defined platform providing specific functionality.

Approve

"Approve" means green-lighting the device after reviewing a premarket approval (PMA) application that has been submitted to FDA.

AVL (Approved Vendor List)

A list of all the vendors or suppliers approved by a company as sources from which to purchase materials.

Artwork

Electronic files or printouts containing the representation of a packaging item, graphical elements, and regulatory text. Approved artworks are used by suppliers for printing.

Aseptic (Conditions)

Conditions in the working environment under which the potential for microbial and/or viral contamination is minimised.

ASTM

Acronym for The American Society for Testing and Materials.

ATEX

An acronym derived from the French-titled 'Atmosphères Explosibles' 94/9/EC directive outlining what equipment and work environment is allowed in an environment with an explosive atmosphere. This European directive amends and adds safety requirements for hazardous areas in the relevant national legislation in the member states of the European Union, bringing in a common standard. Where equipment is to be used in potentially explosive atmospheres containing gas or combustible dust, it must comply with the ATEX directive.

Audit Trail

The audit trail is a control mechanism of a system that allows all data entered or modified to be traced back to the original data. A reliable and secure audit trail is particularly important in conjunction with the creation, change or deletion of GMP-relevant electronic records.

Acceptable Quality Level (AQL)

The AQL of a sampling plan is the Process Performance Level routinely accepted by the sampling plan.

B

Basis of Design

A design document that demonstrates a thorough understanding of the project and its intended output. Typically contains preliminary drawings and system descriptions etc. Together with the URS and the detailed design, it provides overall evidence that the design addresses the requirements of the equipment, system or facility.

Biocompatibility

A measure of how a biomaterial interacts in the body with the surrounding cells, tissues and other factors.

Bioburden

The level and type of micro-organisms that can be present in raw materials, API starting materials, intermediates or APIs. Bioburden should not be considered contamination unless the levels have been exceeded or defined objectionable organisms have been detected.

Biological Indicators

A test system containing viable microorganisms providing a defined resistance to a specified sterilisation process, e.g. vaporised hydrogen peroxide.

Biomaterial

Any matter, surface, or construct that interacts with biological systems. Biomaterials can be derived from nature or synthetic (manufactured). The active substance of a biosimilar medicine is comparable to a biological reference medicine. Biosimilar and biological reference medicines are used at the same dose to treat the same disease. The name, appearance and packaging of a biosimilar medicine differs to that of a biological reference medicine.

Bracketing

A bracketing (aka family or matrix) approach can be used where similar products are produced using the same equipment and processes. A particular product size or product configuration may be selected to represent the worst-case product. Therefore, by qualifying the worst case, all of the other products within the family are considered validated.

Body Orifice

Any natural opening in the body, as well as the external surface of the eyeball, or any permanent artificial opening, such as a stoma or permanent tracheotomy.

Borderline Classifications

In certain circumstances, it may not be clear if a product falls under the medical device legislation or whether to classify a device as a medicine, cosmetic, biocide and so on. The decision will largely depend on the particular intended use of the product, as assigned by the manufacturer, and on the demonstrated mode of action. The manufacturer's claims must be substantiated by relevant data.

Bulk Product

Any pharmaceutical form (liquid, powder, suspension) that is to be filled into either another container or its final container at the next process step or is already filled into its final container to be labelled and packaged at the next process step.

BOM

Bill of Materials.

BSI

British Standards Institute.

C

CAD, Computer Aided Drawing

A system used to create physical designs, usually three-dimensional. Some examples of CAD software are SolidWorks, Pro/ENGINEER and AutoCAD.

Calibration

A requirement that demonstrates a particular instrument or device produces results within specified limits by comparison with those produced by a reference or traceable standard over an appropriate range of measurements.

Campaign (Process)

A production strategy where consecutive batches of an API, a finished product, or intermediates are processed before the production line/system is cleaned.

Capability (Process Capability)

Process capability is a measure of how capable the process is of producing product meeting specified requirements. It is a measure of the actual variation in that product characteristic compared to the product specifications. Indices are used to represent the process capability such as Pp, Cp and Ppk, Cpk, depending on how the data is collected, e.g. multiple batches over time.

CAPA

A corrective and preventive action. A systematic approach that includes actions needed to correct, prevent recurrence and eliminate the cause of potential nonconforming product and other quality problems (preventive action) (21CFR 820.100).

Change Control

A formal system by which qualified representatives of appropriate disciplines review proposed or actual changes that may impact the validated status.

Change Notification (Agreement)

A signed declaration that states that the supplier agrees to notify the customer of changes in its product or process in order to allow the customer determine whether the changes can affect the quality of finished goods or quality system.

Change Management

An overarching approach to change control that is used during the preliminary planning and design stage of a project.

Cleaning

The process of removing potential contaminants from process equipment and maintaining the condition of equipment so that the equipment can be safely used for subsequent product manufacture.

Cleaning Validation

Documented evidence that provides a high degree of assurance that a specific cleaning process will consistently produce a result meeting predetermined requirements for cleanliness.

Cleaning Verification

Confirmation by examination and provision of objective evidence that specific requirements have been fulfilled.

Cocurrent (Flow)

This is when the fluids are applied in the same direction. Cocurrent flow is less effective as less heat can be transferred, therefore it is less commonly used.

Code of Federal Regulations (CFR)

Regulations issued by U.S. government agencies. The individual titles making up the regulations are numbered the same way as the federal laws on the same topic.

Competent Authority

A competent authority is the legally designated authority mandated to monitor compliance with directives and legal requirements within the industry. The competent authority has the power to grant and revoke licenses.

Compendial Organisations

Organisations certifying material standards that meet compendial requirements and acceptance criteria, e.g. the United States Pharmacopeia.

Commissioning

An engineering activity that includes all aspects of introducing a system, piece of equipment or process is installed and ready for use. Commissioning involves both requirements of installation qualification (IQ) and operational qualification (OQ).

Computer System

A group of hardware components and associated software, designed and assembled to perform a specific function or group of functions. [EU GMP Guide, Part II, ICH Q7.]

Computerised System

A system including the input of data, electronic processing and the output of information to be used either for reporting or automatic control. [EU GMP Guide, Glossary.]

Computer System Validation

A process that confirms by examination and provision of objective evidence that the computer system conforms to user needs and intended uses. System validation is a process for achieving and maintaining compliance with GxP regulations and fitness for intended use by adoption of life cycle activities, deliverables, and controls.

Concurrent Validation

Concurrent validation occurs when activities are executed at the same time as one another or concurrent to a product launch.

Confidence Level

Confidence level is expressed as a percentage and represents the probability that the conclusion of the test is correct. A 95% confidence level means you can be 95% certain that the conclusion is correct.

Conflict of Interest

A conflict of interest is a situation in which a public official's decisions are influenced by the official's personal interests.

Continual Improvement, CI

Ongoing activities to evaluate and positively change products, processes and the quality system to increase effectiveness

Consent Decree

A consent decree is a binding order issued by a judge that stipulates the voluntary agreement by the participants in a case of litigation. Decrees are sometimes issued after one party voluntarily agrees to cease a particular action without admitting to any illegality of the action to date.

Colony Forming Unit

One or more microorganisms that produce a visible, discrete growth on an agar-based microbiological medium.

Controlled Substances

Products that are categorised due to their potential for abuse, medical use and requirement for medical supervision.

Controlled Classified Areas

An environment supplied with HEPA-filtered air where materials, equipment, and personnel are regulated to control viable and non-viable particulates to an acceptably low level. Such areas are classified according to the maximum level of airborne particulate allowed.

CNC (Controlled Not Classified)

While these are not ISO-recognised room classes, they are generally used to describe non-GMP areas with a level of control in effect.

Clear (FDA)

The attainment of FDA 'clearance' for the device after reviewing a premarket notification, otherwise known as a 510(k) (named after a section in the Food, Drug, and Cosmetic Act) that has been filed with FDA.

Clean Room

An area (or room or zone) with defined environmental control of particulate and microbial contamination, constructed and used in such a way as to reduce the introduction, generation and retention of contaminants within the area.

Containment

A process or device to contain product, dust or contaminants in one zone, preventing it from escaping to another zone.

Contamination

The undesired introduction of impurities of a chemical or microbial nature, or of foreign matter, into or onto a starting material or intermediate, during production, sampling, packaging or repackaging, storage or transport.

Continued Process Verification

Once the initial validation is completed it is important that the system or process remains within the validated state. This is done by monitoring the performance and output of the system or equipment. Furthermore, any changes to this system or equipment must be assessed and documented in order to ensure the product is safe and meets acceptance criteria.

Critical Aspects

Critical aspects of manufacturing systems include the functions, features, abilities, and performance or characteristics required for the manufacturing process and systems to ensure consistent product quality and patient safety. They should be identified and documented based on scientific product and process understanding.

Critical Quality Attribute, CQA (Critical-to-Quality)

A property or characteristic with specific nominal value and appropriate limit and range providing a particular quality attribute. A CQA is typically classed as a high-risk requirement, where the safety or efficacy of the product depends on the CQA being within the specified limits.

CCC (Mark)

The China Compulsory Certificate mark, commonly known as a CCC Mark, is a safety mark for many products sold on the Chinese market. As of 2013, medical devices do not require this certification.

CDC

Centre for Disease Control & Prevention (USA).

CDRH

Centre for Devices and Radiological Health (USA).

CE Marking

CE Marking is a mandatory conformance mark on many products (including medical devices) placed on the single market in the European Economic Area. The CE marking certifies that a product has met EU consumer safety, health or environmental requirements. By affixing the CE marking to a product, the manufacturer declares that it meets EU safety, health and environmental requirements.

CEN

Communité Européenne des Normes (European Committee for Standardisation).

Clinical Trial

Clinical trials are conducted to allow safety and efficacy data to be collected for health interventions (e.g. drugs, diagnostics, devices, therapy protocols). These trials can only take place after satisfactory information has been gathered on the quality of the non-clinical safety, and health authority/ethics committee approval is granted in the country where the trial is taking place.

Clinical Trial Sponsor

The clinical trial sponsor is responsible for the safety of subjects in a clinical trial and informs local site investigators of the true historical safety record of the drug, device or other medical treatment to be tested, and of any potential interactions of the study treatment(s) with already approved medical treatments.

Cleaning

Removal of contamination or soils from an item or surface to the extent necessary for its further processing and its intended subsequent use.

CMDCAS

Canadian Medical Devices Conformity Assessment System.

CMDR

Canadian Medical Device Regulation.

Conformity

Fulfilment of a requirement or meeting a requirement.

Conformity Assessment Body (CAB)

A body, other than a regulatory (competent) authority, engaged in determining whether the relevant requirements in technical regulations or standards are fulfilled.

CRO

A "contract research organisation", also commonly known as a "clinical research organisation", is a service organisation that provides support to the pharmaceutical and biotechnology industries. CROs offer clients a wide range of "outsourced" pharmaceutical research services to aid in the drug and medical device research and development process.

D

Data Integrity

Refers to the degree to which data is reliable and without error. Data must be accurate, attributable, contemporaneous, original, legible and available. A breach of data integrity occurs when any person manipulates or distorts data and submits the results of that data as valid.

Dead Leg

A dead leg in the world of piping terminology refers to an area of piping where there is insufficient flow or a tendency for water build-up or stagnation. The formal definition of a dead-leg states that pipelines for the transmission of purified water for manufacturing or final rinse should not have an unused portion greater in length than six diameters (6D rule) of the unused portion of pipe measured from the axis of the pipe in use.

Debugging

The process of locating, analysing, and correcting suspected faults or machine issues.

Design Controls

Design controls are a collection of practices and procedures that are incorporated into the design and development process for a product such as a medical device. They provide a structure and clear path from the user needs assessment to product delivery through a step-by-step process. Design controls ensure proper assessment of the design is completed during the design and development phase. Design controls are a requirement of quality systems such as 21 CFR Part 820 (medical devices), and for certain classes of devices and per ISO 13485 - Quality Management Systems.

Decommissioning

When a system is taken out of production service and stored in an adequate environment for potential future use.

Depyrogenation

A thermal process used to destroy or remove pyrogens (endotoxins). Typically, primary packaging components such as glass vials are subject to depyrogenation.

Detection Limit

The lowest amount of analyte in a sample that can be detected but not necessarily quantitated as an exact value for an individual analytical procedure. (Ref: ICH Q2.)

Design History File

The DHF is a repository for all of the documentation generated as a result of the design control process. The DHF serves as a complete record of the design.

Design Validation

Establishing by objective evidence that device or product specifications conform to user needs and intended use(s) defined in design documentation.

Debarment

The FDA has the authority to disqualify or remove researchers from conducting clinical testing of new drugs and devices when the agency determines that the researcher has repeatedly or deliberately not followed the rules intended to protect study subjects and ensure data integrity. Further, the FDA can disqualify a clinical investigator who has repeatedly or deliberately submitted false information to the agency or study sponsor in a required report.

Under its statutory debarment authority, the agency may also ban or "debar" from the drug industry individuals and companies convicted of certain felonies or misdemeanours related to drug products. Once individuals have been subjected to debarment, they may no longer work for anyone with an approved or pending drug product application at FDA. Debarred companies may no longer submit abbreviated drug applications.

Design Qualification (DQ)

The documented verification that the proposed design of the equipment is suitable for the intended purpose. DQs are typical deliverables for facilities, systems and equipment and/or processes.

Design Space

The multidimensional combination and interaction of input variables, e.g. material attributes and process parameters that have been demonstrated to provide assurance of quality. Working within the design space is not considered as a change.

Directives

Directives are legal requirements. These must be met by manufacturers. Standard such as ISO 13485 help companies meet the requirements of directives, such as "Guidelines Relating to the Application of the Council Directive 93/42/EEC on Medical Devices."

Direct Impact (System)

A system that is expected to have a direct impact on product quality. These systems are designed and commissioned in line with good engineering practice (gep) and, in addition, are subject to qualification and validation. Such systems include HVACs and clean utilities such as WFI (Water-for-Injection)

Diffusion Blending

A process in which particles are reoriented in relation to one another when they are placed in random motion and interparticular friction is reduced as a result of bed expansion (usually within a rotating container). Also referred to as tumble blending.

Deviations

A deviation can be simply described as an unintended event which causes a test or verification to fail to meet expected acceptance criteria.

Degree of Invasiveness

A device, which in whole or in part, penetrates inside the body either through a body orifice or through the skin surface, is invasive. Invasiveness is generally categorised as invasive of a body orifice (including the surface of the eye), surgically invasive devices and implantable devices.

Device Master Record (DMR)

A compilation of records containing the procedures and specification for a device. The contents of a DMR can contain local procedures such as SOPs and work instructions along with global or divisional specifications used to detail manufacturing processes, intermediate product or final product.

Drug Product

The dosage form in the final immediate packaging intended for marketing. The finished dosage form that contains a drug substance, generally, but not necessarily in association with other active or inactive ingredients. (FDA)

Duration of Contact

In determining the classification of a device, the duration that the device is in continuous contact with the patient is defined as transient, short term or long term. The longer the device is in contact with the patient or user, the greater the risk and therefore this has to be taken into account when determining classification. Continuous use is defined in MEDDEV 2.4/1 as the uninterrupted actual use for the intended purpose. Where use of a device is discontinued in order that the device is immediately replaced with an identical device (e.g. replacement of a urethral catheter) this shall be considered as continuous use of the device.

E

Electronic Signatures

Electronic signatures are computer-generated character strings that count as the legal equivalent of a handwritten signature. The regulations for the use of electronic signatures are set out in 21 CFR Part 11 of the FDA. Each electronic signature must be assigned uniquely to one person and must not be used by any other person. It must be possible to confirm to the authorities that an electronic signature represents the legal equivalent of a handwritten signature. Electronic signatures can be biometrically based or the system can be set up without biometric features.

Encapsulation

The division of material into a hard gelatine capsule. Encapsulators should all have the following operating principles in common: rectification (orientation of the hard gelatine capsules), separation of capsule caps from bodies, dosing of fill material/formulation, re-joining of caps and bodies, and ejection of filled capsules.

Endotoxin

A pyrogenic product (e.g., lipopolysaccharide) present in the bacterial cell wall. Endotoxin can lead to reactions in patients receiving injections ranging from severe fever to death.

Equipment Qualification

Qualification means the process to demonstrate the ability to fulfil specified requirements. EQ consists of proving and documenting that equipment or ancillary systems are properly installed (installation qualification, iq), work correctly (operations qualification oq), and the different sub-systems work together as a system (performance qualification pq) and actually lead to the expected results. Qualification is part of validation, but the individual qualification steps alone do not constitute a validated process.

Excipient

Substances other than the API which have been appropriately evaluated for safety and are intentionally included in a drug delivery system to provide a specific role in manufacturing, shelf-life or physical property.

Equipment Range

The full range that equipment is capable of performing, as per the manufacturer specification and tolerances. (a process may not utilise the full equipment range, operating over a narrower range).

F

Factory Acceptance Testing (FAT)

An FAT or Factory Acceptance Test is an engineering activity that inspects and verifies that the equipment or system meets the requirements of the URS.

Failure Mode and Effects Analysis (FMEA)

A risk assessment tool that provides for an evaluation of potential failure modes and their likely effect on outcomes and/or product or process performance in order to prioritise risks and monitor the effectiveness of risk control activities. It is often used to identify areas within a given process, product, or system that render it vulnerable.

FDA 483s

An FDA 483 letter typically includes a summary of findings and observations in relation to an audit or inspection where the FDA representatives have reason to believe GMP or other regulations have been violated or are not being met. In response to an FDA 483 letter, the company should address each item and provide a timeline for correction or request clarification of what changes are required.

Functional Design Specification (FDS)

A functional design specification is a document that specifies how particular requirements are met – this can be a combination of how the equipment/process operates mechanically/automatically etc. An FDS is typically written in response to a URS

Fluid

A fluid is a substance that undergoes continuous deformation when subjected to a shearing force.

G

GAMP

Good Automated Manufacturing Practice (GAMP) is a set of guidelines for manufacturers and users of automated systems in regulated industries, specifically the medical device, pharmaceutical and biopharmaceutical industries. The application of GAMP and validation of automated systems in manufacturing helps ensure that regulated medical devices and medicinal products have the required quality and are manufactured according to good practices, meet regulatory and legal requirements and ensure patient safety.

Good Documentation Practices, GDP

The handling of written or pictorial information describing, defining, specifying and/or reporting of certifying activities, requirements, procedures or results in such a way as to ensure data integrity.

Granulation

A process of creating granules. The powder morphology is modified through the use of either a liquid that causes particles to bind through capillary forces or dry compaction forces.

Grade A Areas

Aseptic processing areas, critical in nature where sterile products are exposed to the environment receiving no further sterilisation. High-risk operations (for example aseptic stopperage, filling, loading of the lyophiliser) occur in Grade A areas. They are considered ISO 5 under both dynamic and static conditions.

Grade B Areas

Aseptic processing areas where the sterile product is protected from the environment. Grade B processing areas are the background environments for Grade A areas and are considered ISO 7 environments in the dynamic state and ISO 5 environments under static conditions.

Grade C Areas

Non-critical areas where bulk product or materials are exposed to the environment, yet final sterilisation has not yet been performed. Grade C areas are support areas for non-sterile production activities; purification, formulation, and preparation of components, equipment, etc. for sterilisation. They are considered ISO 8 (Class 100,000) environments in the dynamic state and ISO 7 (Class 10,000) environments under static conditions.

Grade D Areas

Non-critical production areas, support areas, airlocks, or corridors. They are support areas for non-sterile production activities in closed systems; cell culture, or buffer and media preparation areas. Grade D airlocks are used for the movement of product, materials and personnel into classified areas.

GHTF

Global Harmonisation Task Force.

GxP

GxP is a general term for good practice with regard to quality guidelines and regulations. These guidelines are used in many fields, including the pharmaceutical, medical device and food industries. X is used as an umbrella letter representing different subjects or disciplines in industry. Some prime examples include GLP (Good

Laboratory Practice), GDP (Good Documentation Practice), GEP (Good Engineering Practice) and GMP (Good Manufacturing Practices). Furthermore, the use of a lower case "c" as a prefix indicates "current" or "up-to-date".

H

Harm

Damage to health, including the damage that can occur from loss of product quality or availability.

High Level Risk Assessment (HLRA)

A high-level risk assessment that can be used at the beginning of a project to estimate the risk, such as the risks involved with bringing in new computerised/automated equipment.

HVAC

Heating, ventilation and air-conditioning (HVAC) systems are used to control the environmental conditions within an area or manufacturing facility. HVAC systems also provide comfortable conditions for operators based in the manufacturing environment. Temperature, relative humidity (RH) and ventilation should not adversely affect the quality of products during their manufacture and storage, or the proper functioning of equipment.

Hydrogel

A biomaterial made up of a network of polymer chains that are highly absorbent and as flexible as natural tissue.

I

ICH

International Conference on Harmonisation of Technical Requirements for Registration of Pharmaceuticals for Human Use.

Intended Purpose

Intended purpose means the use for which the device is intended according to the data supplied by the manufacturer on the labelling, in the instructions and/or in promotional materials. (Chapter I section 1 of Annex IX of Directive 93/42/EEC.)

Impurity

Any component of the new active pharmaceutical ingredient which is not the chemical entity defined as the new active pharmaceutical ingredient _or_ any component present in the active pharmaceutical ingredient or final product which is not the desired product, a product-related substance, or excipient including buffer components.

Invasive Device

A device, which, in whole or in part, penetrates inside the body, either through a body orifice or through the surface of the body.

IQ/OQ

Equipment IQ/OQ is defined as establishing documented evidence that all key aspects of the process equipment installation adhere to the manufacturer's approved specifications and any recommendations of the supplier of the equipment are suitably considered. The process/equipment must also operate as intended and all user requirements must be adequately fulfilled.

IFU

Instructions for Use.

Injunction (Plant)

An injunction is a judicial process initiated to stop or prevent violation of the law, such as to halt the flow of violative products in interstate commerce and to correct the conditions that caused the violation to occur. (FDA 21 U.S.C. 332; Rule 65, Rules of Civil Procedure.)

If a firm has a history of violations and has promised correction in the past but has not made the corrections, the injunction is more likely to succeed. However, the freshness of the evidence is critical.

For an injunction action to be credible in the eyes of the Department of Justice (DOJ), the U.S. Attorney and the court, the evidence must be current. Timeliness is an important factor when considering an injunction action, with or without a Motion for Preliminary Injunction or a temporary restraining order (TRO). However, case quality and credibility must not be sacrificed to meet guideline time frames. The purpose of the guideline time frames is to limit, as much as can reasonably be expected, the need to update evidence. Updating entails extra work at all levels of the case development and review process and more importantly, delays obtaining an injunction which is intended to stop violations that adversely affect the safety or quality of products in commerce.

ISO

International Organisation for Standardisation. Agency responsible for developing international standards, e.g. ISO 13485 Medical Devices.

Isolator

A sealed enclosure, which provides full physical separation between the critical processing zone and the other surrounding processing zones. The internal surfaces of the isolator and its contents are decontaminated in accordance with defined objectives, by highly effective cycles, e.g. vaporised hydrogen peroxide. The enclosure must be capable of preventing ingress of contaminants by means of physical interior/exterior separation, and be capable of being subject to reproducible interior bio-decontamination.

Isoelectric Precipitation

Isoelectric precipitation works by reducing the electrostatic forces to near zero, allowing the proteins to precipitate out.

ISO 13485

ISO 13485 is an ISO standard, published in 2003, that represents the requirements for a comprehensive management system for the design and manufacture of medical devices.

ISO 14971

An ISO standard, published in 2007, that provides a framework and requirements for a risk management system for medical devices. This standard establishes the requirements for risk management to determine the safety of a medical device by the manufacturer during the product life cycle.

ISO 9001

ISO 9001 is an ISO standard that represents the requirements for quality management systems. It is used across industries and is not specific to medical devices like ISO 13485.

Item Master

The item master is a record of all components that a manufacturer buys, builds or assembles into its products. The item master includes information like the size, shape, material, manufacturer, manufacturer part number and vendor for each component.

IVD

In vitro diagnostic tests are medical devices intended to perform diagnoses from assays in a test tube, or more generally in a controlled environment outside a living organism.

IVDD

The in vitro diagnostic device directive delineates requirements that in vitro diagnostic devices must meet before they can be sold in the EU market.

Intermediate

A material produced during steps of the processing of an API that undergoes further molecular change(s) or purification before it becomes an API.

J

JIT (Just in Time)

A strategy used to monitor inventory levels with the goal of reducing inventory and associated carrying costs.

K

Kanban

A scheduling system that advises manufacturers what to produce, when to produce and how much to produce. Pioneered by Toyota, the approach is based on demand. Inventory is replenished only when visual cues like an empty bin, trolley or cart show that it's needed.

L

Laminar Flow

Laminar flow is when fluid particles move in parallel layers at a constant velocity.

Life Cycle (Validation)

The validation life cycle refers to the requirement to control and document all validation activities from conception and URS stage to the retirement of equipment or a process. The life cycle approach ensures compliance throughout the life of the process/equipment while maintaining a validated state throughout the application of change control.

Linearity

The ability of an analytical procedure (within a given range) to obtain test results that are directly proportional to the concentration (amount) of analyte in the sample.

Line Clearance

The act of performing and documenting the removal of materials from a production or packaging line and cleaning prior to the introduction of a new batch or lot.

Lyophilisation (Freeze Drying)

Lyophilisation is the removal of ice or other frozen solvents from a material through the process of sublimation and the removal of bound water molecules through the process of desorption.

M

Maximum Allowable Carry Over (MACO)

The amount of allowed product residue (carry-over) from lot-to-lot, batch-to-batch. This limit is based on the most conservative or lowest level of three MACO calculation methods: (1) limited based on toxicity, (2) limit based on smallest therapeutic dose, and (3) worst-case dose.

Measurement Capability Index (MCI)

The Measurement Capability Index (MCI) represents the capability of the measurement system. It is used to evaluate the capability of the gauge to classify product against predetermined specifications.

Measurement System Analysis (MSA)

A study to determine the degree of error involved in measuring the given parameter. The measurement system involves the combination of operations, procedures, gauges, instruments, environmental conditions, people and software.

Medical Device

A medical device is "an instrument, apparatus, implement, machine, contrivance, implant, in vitro reagent, or other similar or related article, including a component part, or accessory which is:

•   recognised in the official National Formulary, or the United States Pharmacopeia, or any supplement to them,

•   intended for use in the diagnosis of disease or other conditions, or in the cure, mitigation, treatment, or prevention of disease, in man or other animals, or

•   intended to affect the structure or any function of the body of man or other animals, and which does not achieve any of its primary intended purposes through chemical action within or on the body of man or other animals and which is not dependent upon being metabolised for the achievement of any of its primary intended purposes."

Medicinal Drug Products (Finished Products)

Finished dosage forms (e.g. tablet, capsule, or solution) that contain the active pharmaceutical ingredient usually combined with inactive ingredients. Medicinal products are intended to furnish pharmacological activity or other direct effect in the diagnosis, cure, mitigation, treatment, or prevention of disease or to affect the structure and function of the body.

MDD

The Medical Device Directive is intended to harmonise the laws relating to medical devices within the European Union. Medical Device Directive 93/42/EEC was most recently reviewed and amended by 2007/47/EC.

MHRA

The Medicines and Healthcare Products Regulatory Agency (MHRA) is the UK government agency which is responsible for ensuring that medicines and medical devices work and are acceptably safe.

MSDS

Material Safety Data Sheet.

N

NCR

Non-Conformance Report.

NIH

National Institutes of Health (U.S.)

NOEL

No Observed Effect Level. In relation to cleaning validation.

Non-Conformity

A deficiency in a characteristic, product specification, CQA, process parameter, record, or procedure that renders the quality of a product unacceptable, indeterminate, or not according to specified requirements.

Non Parametric Data

Where the type of data is non-variable. Also referred to as attribute data, e.g. visual inspection resulting in a PASS/FAIL result.

Notified Bodies

A notified body is a certification organisation which the national authority (the competent authority) of a member state designates to carry out one or more of the conformity assessment procedures or audits described in the annexes of the medical devices directives or GMP legislation.

NPI (New Product Introduction)

The market launch or commercialisation of a new product. NPI takes place at the end of a successful product development project.

O

Open System

An environment in which system access is not controlled by persons who are responsible for the content of electronic records on the system (21 CFR, Part 11).

Outlier

A test result that is statistically different compared to a set of other test results obtained from the same sample or samples from the same lot of material.

Out-of-Specification

A recorded result that falls outside the established specification(s) or acceptance criteria.

Out-of-Trend

Analytical result, which is within specification or acceptance criteria, but different from those usually obtained or expected. Out-of-trend results should be investigated by the same general principles as out-of-specification results.

Quantitation Limit

The lowest amount of analyte in a sample which can be quantitatively determined with suitable precision and accuracy for an analytical procedure. The quantitation limit is a parameter of quantitative assays for low levels of compounds in sample matrices and is used particularly for the determination of impurities and degradation in products.

Overall Equipment Effectiveness (OEE)

A calculation for measuring the efficiency and effectiveness of a process by equipment breaking it down into three constituent components. (The OEE factors: Availability x Performance x Quality.)

Overkill

A sterilisation process that is demonstrated as delivering at least a 12 Spore Log Reduction (SLR) to a biological indicator having a resistance equal to or greater than the bioburden level.

P

Pan Coating

The uniform deposition of coating material onto the surface of a solid dosage form while being translated via a rotating vessel.

Particle Count Test

This test covers verification of cleanliness. Dust particle counts are measured. The number of readings and positions of tests should be defined in accordance with ISO 14644-1 Annex B5.

Performance Indicators

Measurable values used to quantify quality objectives to reflect the performance of an organisation, process or system, also known as performance metrics in some regions. (ICH Q10.)

Performance Qualification (PQ)

Establishing by documented evidence that the process, under anticipated (controlled) conditions, consistently produces a product which meets predetermined requirements.

Precision

The degree of agreement (scatter) between a series of measurements when a method is applied repeatedly to multiple samplings of a homogeneous sample or artificially prepared sample under the prescribed conditions. There are three types of precision; repeatability, intermediate precision and reproducibility.

Pressure Cascade

A process whereby air flows from one area, which is maintained at a higher pressure, to another area at a lower pressure.

Piping and Instrument Diagrams (P&IDs)

Engineering technical drawings that provide details of the connections and integration of equipment, services, material flows, plant controls and alarms. The P&IDs also provide the reference for each tag or label used for identification.

PMA

Premarket approval by FDA is the required process of scientific review to guarantee safety and effectiveness for Class III devices.

PMDA

The Pharmaceutical and Medical Devices Agency in Japan reviews applications for marketing approval of pharmaceuticals and medical devices. It also monitors their post-marketing safety and provides relief compensation for people who have suffered from adverse drug reactions from pharmaceuticals or infections from biological products.

PMS

Post marketing surveillance is the practice of monitoring a pharmaceutical drug or device after it has been released on the market.

Process Design

Defining the commercial manufacturing process based on knowledge gained through development and scale-up activities.

Process Qualification

Confirming that the manufacturing process as designed is capable of reproducible commercial manufacturing.

Process Window

The selected operating range of machine settings/parameters that will produce product to meet all quality and product specifications.

Product Recovery

Product recovery is a critical and important step in the process. It is also referred to as "downstream processing". It is often the most expensive step in the process. For recombinant-DNA derived products, purification can often account for 90% of the total production costs.

Prospective Validation

Prospective validation is when validation is done in advance of commercial manufacturing.

Procedures

Also known as Standard Operating Procedures (or SOPs), procedures give directions for performing certain operations.

Protocols

Protocols give instructions for performing and recording certain discreet operations. (Examples include engineering protocols, validation protocols etc.)

Pure

A term typically used within pharmaceutical manufacturing, a product or substance is pure if it is free of contaminants, foreign matter, chemicals and harmful microbes.

Q

Quality Management System

A Quality Management System, often abbreviated to QMS, is any system based on a collection of business processes that are primarily focused on providing safe and quality products that consistently meet customer requirements.

Quality

The degree to which a set of inherent properties of a product, system, or process fulfils requirements. (ICH Q9.)

Quality by Design

This is a systematic approach that begins with predefined objectives and emphasises product and process understanding and process control, based on sound science and engineering principles.

Quarantine

The status of materials isolated physically or by other effective means pending a decision on their subsequent approval or rejection.

Quality Policy

A document in which a company or organisation outlines their commitment and approach to quality. It usually sets out how they plan to achieve a high and consistent standard of quality. It should in some way speak to the customer or end user.

Qualification Plan

A Qualification Plan (QP) describes all the qualification measures and at which stage of the qualification the verification will be completed. It typically contains detailed descriptions of the necessary test measures and a description of the interdependencies of the individual tests. In some instances, there may not be a need or a requirement for a qualification plan. A validation plan can also serve to detail the qualification strategy.

QP

Companies that intend to manufacture or import medicinal products or intermediate products for use in clinical trials or for the EU market must appoint a qualified person in order to comply with EU good manufacturing practice standards.

QPM

Quality Policy Manual.

QSP

Quality System Procedure.

QSR

Quality System Regulations.

R

Range

Range is defined as the interval between the upper and lower measurements required. The minimum specified range should be within the equipment range and validated to operate at all points within the range.

Recall

As defined at 21 CFR 7.3(g), "recall means a firm's removal or correction of a marketed product that the Food and Drug Administration considers to be in violation of the laws it administers and against which the agency would initiate legal action, 2 21 CFR 806.2(h). e.g., seizure. Recall does not include a market withdrawal or a stock recovery." Recall does not include routine servicing. Recall also does not include an enhancement, as defined by this guidance.

Relative Humidity

The ratio of the actual water vapour pressure of the air to the saturated water vapour pressure of the air at the same temperature expressed as a percentage. More simply put, it is the ratio of the mass of moisture in the air, relative to the mass at 100% moisture saturation, at a given temperature.

Reusable Medical Device

A device intended for repeated use either on the same or different patients, with appropriate decontamination and other reprocessing prior to re-use.

Reusable Surgical Instrument

An instrument intended for surgical use by cutting, drilling, sawing, scratching, scraping, clamping, retracting, clipping or similar surgical procedures, without connection to any active medical device and which is intended by the manufacturer to be reused after appropriate procedures for cleaning and/or sterilisation have been carried out.

Re-Qualification

Requalification is designed to verify and ensure that the equipment/instrument/system is maintained in a qualified state after modification or after a stipulated time period (downtime).

Residual Risk

The risk level remaining after applying the identified controls on a high risk of harms and hazards manifestation.

Resolution

The smallest change in quantity that can be detected or provided by an instrument.

Residual Solvent

Organic volatile chemicals used or produced during the manufacture of APIs or excipients, or in the preparation of medicinal products.

Retain Samples

Samples that are kept for potential investigations and retests. It should be noted that retained samples are not a regulatory requirement as per Annex 10 or 21 CFR part 11.

Retrospective Validation

Retrospective validation is used for facilities or processes that have not completed formal validation. Historical data or a retrospective review can provide the evidence that the process or facility is operated as intended.

Rinse Sampling

Using a solvent to contact all surfaces of the sampled item to quantitatively remove target residue. The solvent can be water, water with pH adjusted, or organic solvent.

Right First Time

Right first time strives to create a culture of excellence. People are challenged with performing their tasks always in the correct manner to achieve the correct results always — right the first time.

Risk

The combination of the probability of occurrence of harm and the severity of that harm.

Risk Management

Risk management involves the systematic application of management policies, practices and procedures that identify, analyse, control and monitor risk. It is important to recognise that risk management should begin at the outset of the design and development phase of a project. The first step is to identify the user needs and intended use and application of the device.

RoHS

"Restriction of Hazardous Substances in Electrical and Electronic Equipment 2002/95/EC". An initiative that was adopted by the European Union (EU) in February 2003 and put into effect July 1, 2006.

Ruggedness

An indication of how resistant a test method or process is to typical variations in operation, such as those to be expected when using different analysts, different instruments and different reagent batches.

S

Scaffold

A structure of artificial or natural materials on which tissue is grown to mimic a biological process outside the body.

SKU

SKU is an acronym standing for **s**tock **k**eeping **u**nit. It represents a unique sales stock identifier.

Specifications

An approved document detailing the requirements with which the products or materials used or obtained during manufacture have to conform to. They serve as a basis for quality evaluation.

Specificity

The ability to assess unequivocally the analyte in the presence of components which may be expected to be present.

Stability

Stability studies are used to demonstrate and justify assigned expiration or retest dates.

5S

5S is a Japanese methodology of organising and storing items in a work or lab environment. It has been adopted by many Western companies as a tool to help maintain standards and reduce errors and mix-ups. The "5s" represents each stage of the method:

Sort

Sorting out any items that are not in use and removing them to a more appropriate area such as a storage facilities or the bin.

Set-in-Order

The idea behind "set-in-order" is to be always organised. It requires "a place for everything and everything in its place". By setting things in order, we can help to make live processing and testing more efficient and reduce the risk of errors, omissions and accidents.

Shine

Regular cleaning is an important practice and it is always helpful to "clean as you go."

Standardise

Implement standard practices through SOPs and training. Standardisation can also be applied to workstation layout.

Sustain

Make it a habit! After implementing a 5s methodology, it is only effective if continuous efforts are made to "sustain" the changes.

Sterility Assurance (SAL)

SAL or "sterility assurance level" refers to the probability of a single viable microorganism occurring on an item after sterilisation. For a terminally sterilised medical device to be designated as "sterile", the minimum sterility assurance level must be SAL = 10-6 or better. When applying this quantitative value to assurance of sterility, an SAL of 10-7 has a lower value but provides a greater assurance of sterility than an SAL of 10-6 .

T

Tableting

The reconstitution of a powder blend in which compression force is applied to form a single unit dose (tablet).

Tableting Press

Tablet press subclasses primarily are distinguished from one another by the method that the powder blend is delivered to the die cavity. Tablet presses can deliver powders without mechanical assistance (gravity), with mechanical assistance (automation), by rotational forces (centrifugal), and in two different locations where a tablet core is formed and subsequently an outer layer of coating material is applied (compression coating).

Traceability Matrix

A traceability matrix is a document that links the user requirements and specifications to where the verification and testing have been documented within the validation activities. It also illustrates that all user requirements are traceable to the evidence-based test.

Turbulent Flow

Turbulent flow is when the movement of fluid particles are varying in velocity and direction.

U

Uniform

The product is manufactured consistently and will have the same quality between batches manufactured on different days.

UDI, Unique Device Identification

The UDI is a series of numeric or alphanumeric characters that is created through a globally accepted device identification and coding standard. It allows the unambiguous identification of a specific medical device on the market.

Uninterrupted Power Supply

An uninterruptible power supply (UPS) is a system for buffering the main power supply. If the power supply fails, the battery of the UPS supplies the required power. When the power supply returns, the UPS battery stops supplying power and is recharged.

Unit Operation

Unit operations are the individual steps in the process that modify materials and their properties at each step of the process. Each unit operation comes together to create a complete process.

User Requirement Specification, URS

The URS is a critical document that defines the requirements of a particular system, equipment or process. Requirements such as the functional and operational aspects of the system are typically documented here.

USP

United States Pharmacopeia.

V

Validation

Validation is confirmation via documented evidence that the particular requirements for a specific intended use can be consistently fulfilled under anticipated conditions.

Validation Master Plan

A document providing information on a company's validation work programme. It typically details timescales for the validation work to be performed along with the key deliverables.

Verification

Verification means confirmation by examination and provision of objective evidence (i.e. documentation) that the specified requirements have been fulfilled.

Vaporised Hydrogen Peroxide (VHP)

Vaporisation of liquid hydrogen peroxide which results in a mixture of VHP and water vapour. The VHP mixture is used to decontaminate isolators.

W

Warning Letter

A warning letter is a correspondence that notifies regulated industry about violations that FDA has documented during its inspections or investigations.

WEEE Directive

Waste Electrical and Electronic Equipment Directive. European Community directive 2002/96/EC where manufacturers are responsible for disposing of electrical/electronic waste.

WFI (Water-for-Injection)

WFI is sterile and pyrogen-free water containing no less than 10 CFU/100ml (Colony Forming Units) with a sample size of between 100 and 300 ml and an endotoxin level < 0.25 EU/ml.

WHO

World Health Organisation.

WI

Work Instructions.

Witnessed By

When signed or initialled is legal proof that the individual signing is physically present and observes the step, calculation, or operation being performed by someone else, and that all entries of data are true and accurate.

Worst Case

A set of conditions or parameters which, in combination with product specification or attributes at their limits, pose the greatest challenges to the process.

X

--

Y

--

Z

Zone Classification

Zone classification refers to GMP areas which include controlled (aka classified) and non-controlled manufacturing areas. Areas may be classified based on EU Grades A–D and/or ISO Class 5–8 (in the US - Class 100–Class 100,000 areas).